net
LOSS

NATHAN NEWMAN

net

LOSS

Internet Prophets, Private Profits, and the Costs to Community

the pennsylvania state university press I university park, pennsylvania

Library of Congress Cataloging-in-Publication Data

Newman, Nathan, 1966–
    Net loss: Internet prophets, private profits, and the costs to community /
    Nathan Newman.
        p.   cm.
    Includes bibliographical references and index.
    ISBN 0-271-02204-3 (cloth: alk. paper)
    ISBN 0-271-02205-1 (pbk.: alk. paper)
    1. Internet industry—Government policy—United States.
    2. Internet—Government policy—United States.   3. Industrial
    promotion—United States—Regional disparities—Case studies.
    4. Computer industry—California—Santa Clara Valley (Santa Clara
    County).   5. Computer industry—Developing countries.
    6. International division of labor.   7. Globalization—Economic
    aspects—United States.   8. United States—Economic
    conditions—1981—Regional disparities.   I. Title.

    HD9696.8.U62 N48 2002
    338.4'7004678'0973—dc21

                                                            2001053633

Contents

Analytical Table of Contents

You hold in your hands an ancient document—at least by Internet standards. It is a relatively minor rewrite of a 1998 manuscript published as my doctoral dissertation, whose research in turn was begun back in 1993. The vagaries of life and traditional publishing timelines means that parts of what you are reading date back almost a decade.

Which is a good thing.

Too much of what passes for analysis of technology takes six-month-old trends, projects them twenty years into the future, and discovers that the world will be remade completely. What is ignored is that not only does technology on its own terms not evolve in linear fashion, but technology exists in the context of broader economic structures and policy decisions that can and are modified in response to that technology. Hopefully, if this book's basic analysis has survived and is still compelling after the past four years of turmoil in the technology world, it may maintain its relevance into the future.

One message of this book is that technology eludes easy prediction based on short-term deterministic trends. Technology is not destiny, either positive or negative, but its fate and its effects are based on contingency and collective decisions by society. This book places the evolution of the Internet and Silicon Valley in a much broader history dating back decades and even back centuries. It also situates the technology in the broader economic forces of the "old economy" and of the government policies that birthed the Internet, pointing to realities of which many technologists are in almost pathological denial.

This is not done to claim some deeper predictive power based on a broader vision, but to challenge the easy assumption that any writer can second-guess the future, since the future can be changed. I have my proud two-bit Nostradamus predictions from that first version four years ago, including my general expectation of the coming technology bust and my specific prediction of utility blackouts following the Internet-driven deregulation of energy markets. (For those interested in such predictions or the general evolution of the work, check out http://www.nathannewman.org/diss/index.html for the original draft.) But even the conditions that made those events likely are being rapidly altered as new political and social decisions are made.

Just in the past six months, the attacks on the World Trade Center and the Enron scandal have inspired social reactions that will have far-reaching and unpredicted and unpredictable effects on everything from the movements organizing around global justice to the shape of Internet security protocols to the structure of America's regulatory regime. Predictions in such a rapidly changing world are a fool's game.

However, a healthy respect for contingency does not mean that social analysis is useless; quite the contrary. But the point of good social analysis is not to predict the future, but rather to give people the tools to participate in changing it. Soothsaying essentially brings readers to the world as the audience of a spectator sport; social criticism invites them into active participatory action to improve that world.

The goal of this work is to encourage the latter and it is dedicated to the community activists, labor organizers and social visionaries seeking to channel these technological changes into bringing about a more just world.

In writing this book, I owe a debt of gratitude to a wide range of people, who in conversation and in their writings inspired its creation. Many of them are acknowledged in the footnotes, whether through books and articles they wrote or from the interviews I conducted with a range of business executives, union organizers, government officials, and community leaders. There are no doubt numerous others who in conversation or writing influenced me over the years.

I do want to say a good word for journalists, who often get little respect from academic writers (at least until they are dead, whereupon they may be elevated to the higher plane of "primary source"). One side effect of the technology boom was a flourishing subculture of technology journalists who lived and breathed the environment of the emerging Internet world. They no doubt helped promote many of the public misunderstandings of that world—and fed the financial bubble of the 1990s—but within that group were some admirable social analysts whose deep firsthand knowledge was invaluable in any attempt to write a wide-ranging synthetic work such as this one.

A few individuals deserve special mention and appreciation for their assistance.

First, I want to thank my longtime friend and colleague Anders Schneiderman, whose technical knowledge and social insight was invaluable in shaping this work. For years, on almost a daily basis, our conversations and political work together helped develop the ideas that became this book.

Second, I want to thank my dear friend Aimie Gresham, whose social insight was always an inspiration and whose loving support was always critical.

Last, but certainly not least, I want to thank Peter Evans, my doctoral advisor, intellectual guide, de facto unpaid literary agent, and most of all a good friend throughout this process. He was everything an advisor should be: supportive of my ideas while challenging their flaws and attentive to what was needed to improve their development throughout the process. His help and friendship was priceless.

January 15, 2002

Acronyms

A book dealing with both technology and government inevitably is infested by the acronyms that dominate both fields of endeavor, so the following is a partial list of acronyms used in this book. In general, it contains those that are most important and are repeated in the text.

Acronyms without fuller names in their definition reflect the habit in the world of technology of creating such acronyms, even while the acronyms had never had full names or these names had been forgotten.

ABAG: Association of Bay Area Governments, Northern California regional planning agency

AEA: American Electronics Association, technology business association

AFL-CIO: American Federation of Labor–Congress of Industrial Organizations, national labor federation

AFSCME: American Federation of State, County, and Municipal Employees, public employees union

AMTEX: American Textile Partnership, government-business textile research consortium

ANS: Advanced Network and Services, first commercial Internet backbone

APC: Association of Progressive Communications, international nonprofit Internet service provider

ARC: Augmentation Research Center, Stanford center pioneering Internet and computer innovations

ARPA: Advanced Research Projects Agency, federal agency that developed Internet

ARPANET: original designation of what would become the Internet

ATM: automatic teller machine

B2B: business to business, jargon for Internet commerce between businesses

BAMTA: Bay Area (then Broad Area) Multimedia Technology Association, business consortium

BARRNET: early Bay Area provider of Internet services for nonprofits and universities

BBN: Bolt Beranek and Newman, consulting firm hired to design original ARPANET computer routers

BSD: Berkeley Software Distribution, openly licensed version of UNIX operating system

CAD: computer-aided design, software for graphics-based design

CERN: European Organization for Nuclear Research, research center where World Wide Web originated

CIX: Commercial Internet Exchange, first association of commercial Internet service providers

CPU: central processing unit, electronic brain of a computer

CPUC: California Public Utilities Commission, utility regulatory agency

CRA: Community Reinvestment Act, law to evaluate bank service to community

CSNET: Computer Science Research Network, online network of universities in 1980s

CTWO: Center for Third World Organizing, community-based organization

DARPA: renamed ARPA with *Defense* added to its name

DoD: Department of Defense

DOS: disk operating system, first IBM personal computer operating system

DVD: digital video disk, successor to compact disk with greater storage capacity

eCo: XML-based electronic commerce standard developed by CommerceNet organization

EDD: Employment Development Department, California state employment agency

EIT: Enterprise Integration Technologies, key Internet commerce firm in CommerceNet consortium

EPRI: Electric Power Research Institute, utility research consortium

FCC: Federal Communications Commission, national regulatory agency for many Internet issues

FERC: Federal Energy Regulatory Commission, national regulatory agency for energy industry

FTP: file transfer protocol, networking standard for exchanging files on Internet

GAO: U.S. General Accounting Office, congressional research arm

GIS: geographic information systems, software for linking information to geographic location

GNU: GNU is Not Unix, set of free software with freely readable code

HTML: HyperText Markup Language, language for Web pages for displaying information in browsers

HTTP: HyperText Tranfer Protocol, standard for transmitting information over the Web

IAB: Internet Activities Board/Internet Architecture Board, government network overseeing Internet development

ICANN: Internet Corporation for Assigned Names and Numbers, named nonprofit regulator of Internet domain names in 1998

ICCB: Internet Configuration Control Board, oversight network preceding IAB

ICFTU: International Confederation of Free Trade Unions, international labor federation

IETF: Internet Engineering Task Force, subcommittee of IAB that set Internet standards

IGC: Institute for Global Communications, organization providing online services to nonprofits in the United States

IMF: International Monetary Fund, global institutional loan provider

IMP: Interface Message Processors, first computers used to route first ARPANET connections

InterNIC: ARPANET organization at SRI that managed Internet addresses

IP: Internet Protocol, standard for exchanging information between computer networks

IPO: initial public offering, event where public shares are issued for a privately held company

IPTO: Information Processing Techniques Office, ARPA division that developed the Internet

ISP: Internet service provider, company or service providing access to the Internet

JTSIN: Joint Transmission Services Information Network, utility group managing online energy transactions

LAN: local area network, electronic networks between local computers

LINUX: free version of UNIX operating system based on GNU tools

MAI: Multilateral Agreement on Investments, failed proposal to protect corporate investment power

MCC: Microelectronics and Computer Technology Corporation, Austin-based electronics consortium

MSN: Microsoft Network, originally proprietary online system that became Internet Service Provider

MTA: Metropolitan Transit Authority, Los Angeles agency running public transit

NAPs: network access points, connection points between local ISPs and Internet backbones

NASA: National Aeronautics and Space Agency

NBI: National BankAmericard Incorporated, credit card network later renamed Visa

NCSA: National Center for Supercomputing Applications, government computer center at University of Illinois that created first major graphics-based Web browser

NII: National Information Infrastructure, Clinton-era designation for vision of networked nation

NSF: National Science Foundation, government agency funding scientific research and Internet expansion

NSI: Network Solutions, Incorporated, private company given control of Internet addresses in early 1990s

NWG: Network Working Group, first network of university researchers overseeing ARPANET

NYSERNET: early New York provider of Internet services for nonprofits and universities

OASIS: Open Access Same-Time Information System, utility online transactions system required by federal government

PARC: Palo Alto Research Center, Xerox center that created first personal computer

PBX: private branch exchange, non–phone company telephone systems

Project MAC: MIT's Project on Mathematics and Computation, ARPA-funded computer research center

PSINet: Performance Systems International Net, early Internet Service Provider and CIX founder

PURPA: Public Utility Regulatory Policies Act, which began nationwide energy deregulation in 1978

R&D: research and development, investments in technology

RAND: Defense-connected private nonprofit policy research institute

SAGE: Semi-Automatic Ground Environment, ground-based missile program in the 1950s

SAIL: Stanford Artificial Intelligence Lab, computer research center

SDC: Systems Development Corporation, spinoff from SAGE, trained many programmers in the 1950s

SEC: Securities and Exchange Commission, national regulator of financial sector

SET: Secure Electronic Transaction, protocol for online financial exchanges

SPARC: Scalable Processor Architecture, SUN-designed computer design for UNIX

SRI: Stanford Research Institute, Stanford center that eventually became private company

SURANET: early Southeast provider of Internet services for nonprofits and universities

TCP: Transmission Control Protocol, first computer network standard, later combined with IP protocol

TELENET: first privately run network company based on ARPANET-style protocols, later acquired by Sprint

TURN: The Utilities Reform Network, California consumer advocacy group

UNIX: operating system used primarily on larger computers

USENET: early shared electronic discussion bulletin boards

UUNET: spinoff company from Defense Department seismic research lab; became major backbone Interet provider, acquired by WorldCom

VAN: Value-Added Network, private proprietary electronic networks

vBNS: very high-speed Backbone Network Service, "Internet II" university-based research project

VLSI: Very Large Scale Integration, 1970s Japanese computer research consortium

W3C: World Wide Web Consortium, MIT-based Internet standards organization run by Tim Berners-Lee

WSPP: Western States Power Pool, West Coast organization of energy utilities

WTO: World Trade Organization, international institution governing rules of trade

XML: eXtensible Markup Language, successor Web language to HTML

Introduction

The new "information economy" seems to evoke a contradictory debate on regions and decentralization. On the one hand, technologists such as Nicholas Negroponte believe that local regions are disappearing as important entities in the face of the "spaceless" technology of information exchange. On the other hand, futurists like Alvin Toffler and his political disciples, such as Newt Gingrich, have argued that the microchip is the midwife of regional rebirth and the death knell for central political decision-making.

How do we explain this contradiction?

The Internet has emerged as the focus for much of the strongest hype and substance in debates on this new economy. It has become the defining economic event of the end of the twentieth century—a fact reflected by the obsessive media attention and by the raw economic explosion of companies associated with it.

The Internet is seen as the metaphor for, even the embodiment of, the new information age, of a postindustrial economy, and of a new para-

The post-information age will remove the limitations of geography. Digital living will include less and less dependence upon being in a specific place at a specific time, and the transmission of place itself will start to become possible.
—Nicholas Negroponte, director of MIT's Media Lab, in *Being Digital*[1]

National economies are swiftly breaking down into regional and sectoral parts—subnational economies with distinctive and differing problems of their own.
—Futurist Alvin Toffler on regional economies in *The Third Wave*[2]

digm in workplace and company organization. According to this view, information rather than raw materials has become the substance of commerce and the Internet is the highway of the new era.

Most strikingly, the Internet is seen as the herald of the globalization of the economy and the triumph of a deregulated marketplace. In this vision, the economics of place have given way to telecommuting, global production, and just-in-time delivery of goods and information from all points on the globe. In such a world, *economic regions* become an oxymoron as the economy becomes a matter of bits and e-mail in cyberspace, not transit and meetings in local space. The "Third Wave" in this scenario leaves economic regions as the archaic leftovers of the industrial age. Governments, those stalwart institutions tied to such geography, become impotent and unimportant in this new global information society.

Now, there are truths in each of these ideas, but the truths obscure the underlying reality of transformation rather than decline in both the vibrancy of local economic activity and the importance of government action. On the face of it, it is nonsensical to argue that new information technologies such as the Internet show the irrelevancy of national governments and economies. The Internet is one of the crowning achievements of central government in the past few decades—planned over decades, funded by a series of federal agencies, and overseen by a national network of experts. And its success is not merely an exemplar of technical achievement; it is also an example of the efficiency of government planning over purely private economic development. Almost all analysts admit that in the absence of the Internet's open standards, which were developed and promoted by the federal government, the private vision of toll-road information services promoted by industry would not have created the surge of explosive economic innovation that we are currently seeing in connection to the Internet. It is only with the success of the Internet (and the profits to be made) that industry is now decrying the interference of government in information markets.

The most striking counter to the vision of global placelessness is the very existence of Silicon Valley, the region most closely associated with the rise of the Internet. If any region were to collapse on the wave of cybercommunication, it would be Northern California's "hot-wired" Silicon Valley. Contrary to what some might expect, Silicon Valley not only survived but thrived and expanded its role as the geographic focus of a supposedly geography-free revolution in the 1990s. From network router companies such as Cisco to Web-tool makers among the dot-coms to the multimedia upstarts of San Francisco's "multimedia gulch," companies in Northern California seem to be refusing to let geography die its proper death. While the dot-com meltdown raises the issue of the long-term role of the region, it is clear that in the short term, the Internet and regional strength went hand in hand.

But at a deeper level, the vibrancy of Silicon Valley's regional economy does not defy the Internet's globalizing trends; rather, its regional strength was in many ways the precondition for the triumph of the Internet. Such a fundamental technological innovation as the Internet requires more than the introduction of new products; it requires transformations in an array of mutually supporting institutions, goods, services, and standards, which must all advance together. While this process can take place between people and companies in different places, the organic trust and interaction of those living in the same region has always been a key factor in such broad-based technological advancement, whether in the car industry in Detroit or in the financial district of Wall Street.

As economic theorists dating back to Alfred Marshall have noted, regional "industrial districts" have always been a breeding ground for specialized innovation where day-to-day activity supports the trust and human interaction needed for such codependent innovation. If anything, the intense technological details needed in high technology and the conditions of rapid technological change that we live under only accentuate the need for ongoing local interaction, and Silicon Valley has just emerged as the premier space for innovation in networking technology.

Silicon Valley itself is largely the creature of federal spending and effort; its pioneering firms, among them Hewlett-Packard and Varian, grew as a result of defense contracts entered into during World War II and its aftermath, contracts that pumped billions of dollars into the Bay Area economy. This occurred just as federal research dollars were pouring into the region via the University of California at Berkeley, Stanford University, and government laboratories such as the Ames Research Center, run by the National Aeronautics and Space Administration (NASA). The Internet itself was a project directed for a quarter of a century by federal government agencies largely in association with regionally based university computer science departments.

Yet despite what might be seen as the continuity from the past in the role of both regions and government in advancing technology and its associated economic benefits, there is a justified sense that something has radically changed in the economy. While Silicon Valley designers may cluster together at Palo Alto bars, production of the computer components powering their tools has scattered to factories throughout the world. Industry is using the new technology to extend itself globally as production becomes a global process.

Business-to-business interactions are in turn reshaped as the cost of communication at large distances drops to virtually zero. The Internet promises a global marketing venue reaching consumers around the world. For industries such as software or banking in which the transfer of ideas and commitments (rather than physical goods) is the key, the Internet promises an even more radical reshaping of where and how core services are distributed.

Community in regions increasingly takes the form of regional business associations emerging like kudzu across the economic landscape. It is through these business-based associations, tied to local, state, and federal government, that the innovations of specific regions are harvested to leverage corporate profits and global economic changes such as engendered by the Internet. This horizontal approach of business-to-business community alliances has largely supplanted the vestiges of the vertical cross-class collaborations that had once somewhat tied the economic fates of rich and poor together within regions. It is these local horizontal business linkages, supported by the federal government, that were essential to the emergence of the mutually reinforcing technologies and institutional changes that sped the dominance of the Internet in economic life.

Inequality within economic regions has increased, just as inequality has increased across the country and the globe. What is disappearing is not the importance of geography but the singularity of a "region," of the shared economic fate of those inhabiting the same physical space. Instead, information technology is being used to link the professional elites of regions within a space of shared innovation in order to market that space to a global marketplace, even as the less skilled workers of regions find themselves locked in geography that whipsaws wages downward through that same global competition.

The institutions that once linked investments and broad-based economic development within regions—local banks, power utilities, and the local telephone company—are being rapidly supplanted by global corporations competing in and fracturing local markets in favor of global niches serving different economic strata within regions. This in turn has undermined the shared regional economic-development strategies tied to such institutions that had once linked labor unions, community groups, and elite businesses in some degree of cross-class collaboration around regional goals.

In this transformation, government is not merely the victim of a deterministic technological trend but has been the trend's enabler by means of specific political decisions. Beyond creating the Internet, the federal government promoted a program largely mislabeled "deregulation" that deliberately fractured regional banking, utility, and telephone institutions for the benefit of national and global competitors. But government did not disappear in this change of policy: in fact, federal regulation of telecommunications activity, crucial to the new information age, has accelerated as a range of subsidies, interconnection rules, and antitrust interventions have reshaped the economic map at the behest of government regulators and judges.

What has changed is the relative power of global corporations in dictating local government policy and the wage levels of lower-skilled workers within specific regions. Technological innovation may happen overwhelmingly within local venues, but because of the new technology, corporations have the ability to quickly pick and choose new venues outside the control of local government and beyond the influence of grassroots activists who desperately attempt to negotiate with these global partners. The lack of traditional regional economic anchors such as community banks and local utilities that once mediated some degree of alliance in regional growth has left local actors with few allies for broader integrated economic development.

With this, we see the present reality of local governments teetering on the edge of insolvency and austerity while abandoning any serious commitment to equality. We end up with a form of local government that increasingly markets services to global corporations over the needs of local lower-income citizens while using tax breaks to lure and keep businesses in their regions. The Internet and related information technologies promote an increasingly national and global retail market, thereby further undercutting local government revenues, which are dependent on sales taxes levied on locally purchased goods.

For local government, the promise of the new technology to enhance democracy is increasingly giving way to a blurring of the lines between government functions and business interests as "public-private partnerships" and privatization undermine local political control. Desperate for revenue, local governments have begun marketing information about their own citizens to global corporations, even as those same corporations use the Internet to rapidly survey and play off local governments against one another in bidding for corporate location decisions. The fragmentation of utilities leads to increasing inequality in telecommunications between richer and poorer towns and between schools serving richer and poorer students.

There is a sad irony (and a political agenda) in calls for returning budgetary decision-making powers to local governments, prostrate before the power of global corporations to dictate local policy. That this "decentralization" agenda is occurring even as federal regulators increasingly displace local government control over banks, utilities, and telecommunications only emphasizes that the ambiguity about the globalizing and decentralizing effects of the new information economy are not merely technological contradictions but also political and ideological ones that are shaping the economic landscape.

The Focus of This Book

This book is intended to be a case study in the interactions of government, technology, and the changing role of regions in our economy, using the emergence of the Internet in Silicon Valley as the focus. The more modest goal is to tell that history in the context of the issues raised in the previous section and to throw new light on the dynamics of a region and a technology too often discussed in purely economistic or technological terms.

The more ambitious goal is to use this case study to build a broader case for how technology, government, and regions are interacting with one another in this new economic era. Obviously, Silicon Valley as an early consumer as well as producer of networking technology is a key region in understanding these dynamics, even as its uniqueness makes it problematic for complete generalization to other areas. The Internet is a radically unique innovation whose lessons are only partly applicable to lesser breakthroughs. Still, Silicon Valley's very precociousness as a high-tech region makes its evolution a credible model for insights into the fate of other regions where technological innovation is increasingly supplanting raw-commodity production. The dynamics in this region of economic inequality and the corporate undercutting of integrated regional economic development to be explored in this book will only highlight the even more pronounced effects seen in more peripheral regions, which are even more at the mercy of global production.

The study of the Internet and Silicon Valley illuminates the highly mediated nature of regions—mediated by the technology that shapes new industries, by the federal investments that fuel the growth of new population sectors and new innovations, by the shaping of new business relationships that grow around such new industries and by how global markets themselves interact heavily with core regions that produce the innovation fueling those global markets. The particularity of the story of the evolution of the Internet and its interaction with the Silicon Valley region, like the unique story of all technologies and regions, helps to undermine the simplistic models of universal economic development, models that favor abstract "market rules" while ignoring the specific history of government and of the social interaction that lies at the creation of each new market.

The focus on the federal government's role in the evolution of both Silicon Valley and the Internet inevitably raises more universal issues of how and why the federal government acts in the development of technology and of the economy. As specific controversies are detailed in this volume, the experience of other regions will be used to highlight similarities and contrasts to throw greater light on these universal dynamics. Although no region will be treated with the same integrated and comprehensive manner as will Northern California, comparisons will help to enrich the overall case study of the region.

Since this book highlights some of the bleaker implications of technology, it is worth emphasizing that my view is not antitechnology in any sense. In fact, one of my main purposes is to refute the technological determinism of both the optimists of the Right and the technological pessimists of the Left in favor of an analysis that recognizes the crucial interaction between political choices and the direction of technology with its specific social outcomes. It is through the application of technology and through the social structure created to implement the technology that the positives and negatives of technology manifest themselves.

In *The Visible Hand*, Alfred Chandler wrote that the wholesale transformation of capitalism as a system between the nineteenth and twentieth centuries was caused by the combined effects of communication and transportation technology along with radical changes in managerial systems. In the case of the Internet and related information technologies, relations of production are being as deeply reshaped in the transition from the twentieth to the twenty-first century.

At the heart of any changes in production are changes in power relations, and the Internet itself embodies changing social forces, which are shaping the political and economic forces that will hold power in the new era. Karl Polanyi emphasized the way that the underlying government-created infrastructure of rules of exchange under capitalism shaped all actors in the economy; those rules set the environment for how economic conflict and technology played out in the rise of industrialization. In the same way, the Internet is less the cables and wires tying homes and offices together than a system of contested rules for information exchange. It is these rules that are constraining the shape of power in the new information age. At the heart of this book is the story of how the battle over those contested rules are reshaping regional economies and politics in the modern era.

Why Does Inequality Grow in the "New Economy"?

A basic argument in this book will be that government, particularly the federal government, played the key role in the origins of the Internet, its emergence in Silicon Valley, and its impact on inequality in power and economic status within those regions. But to argue that point, it is important to first outline the major debates on why inequality and regions themselves persist as critical attributes of the new economy.

In approaching this research, starting back in 1993, I came to an understanding of the Internet not as a computer "techie" but as a former union organizer of low-income hotel workers who was struggling, as part of a network of community and labor groups, to define an alternative economic policy for California during the recession of the early 1990s. Northern California was home to the world's most cutting-edge technology, yet the state government was proposing to slash spending on education and other basic infrastructure that might help those who were not benefiting from Silicon Valley's growth.

Older generations of Californians, of Americans generally, had seen the growth of new industries lead to broad-based gains in income for the general population in regions where new industries grew and prospered. Yet in the past generation, something dramatic had happened as economic inequality soared, average worker wages stagnated, and the poorest workers dropped off the economic map altogether into joblessness and poverty.

For the 82 percent of American workers classified by the Bureau of Labor Statistics as "production and nonsupervisory" employees, real hourly wages had essentially peaked back in 1972, after nearly doubling from what they had been in 1948. Real wages have been dropping by fits and starts ever since.[3] By 1998, average hourly wages had fallen by 9 percent from their postwar high in the early 1970s.[4] Yet as wages fell for average workers, the richest 20 percent of families saw their incomes soar by 41 percent in the same period.[5] The wealthiest 10 percent of Americans now own 73 percent of all wealth compared with just half of all wealth in the 1970s.[6]

And the human costs of this inequality are all too apparent, from persistent poverty and hunger in inner-city and rural areas to expanding prison populations, as jails house those excluded from economic opportunity.[7] Yet despite the waste of human potential as people are undereducated and consigned to poverty or imprisonment, inequality persists and economic opportunity within regions remains fractured between the haves and have-nots. Despite the boom in the tech economy—or rather partly because of it, as I will argue in this book—Northern California is no exception to this pattern of increasing inequality.

How to explain this upturn in inequality, after generations during which the "American Dream" seemed to promise broad-based economic gains for all working families?

At least four broad explanations of this expanding inequality have been promoted by various analysts:

- Increasing corporate power used to undermine labor unions and general employee bargaining power.[8]
- A lurch in government policy because of conservative electoral and ideological mobilization.[9]
- Increasing relative returns in income and wealth to those with greater education or other skills, sometimes phrased as the growth of a "winner-take-all" economy.[10]
- Globalization of the economy.[11]

All these factors played a part and reinforced one another in many ways, but the fact that each made such a difference just in the last generation deserves more explanation, since all of them—from corporate power to even globalization—are not new phenomena. As economic historians Karl Polanyi and Immanuel Wallerstein document, global economic trade under capitalism dates back centuries.

In fact, as a result of recession and world wars, global trade fell in the mid-twentieth century at the same time that economic equality grew in many nations. This mid-century system of high wage, high growth state-centered capitalism had been dubbed Fordism—in honor of Henry Ford's famous pledge of high wages to his autoworkers[12]—and analysts have argued that both rising inequality and economic crises of the last few decades have been due to the unraveling of that state-based Fordist system in favor of new global multinational power.[13] Yet recent inequality in developed countries is not explained by some kind of "evening out" of incomes as their wages converge with the populations of developing nations. According to the United Nations Development Program (UNDP), in 1998 the poorest 20 percent of the world's people accounted for only 1.3 percent of private consumption, down from 2.3 percent three decades ago, even as the share consumed by the richest 20 percent of the world's people had grown to 86 percent.[14] So why global growth ended up being directed into the hands of the wealthiest global citizens itself still needs explanation.

Many analysts have targeted the role of technology specifically as the exogenous force playing a new role in the economy—speeding a global economy, increasing the power of corporations, and shifting income and power to those with the education to master it.[15] The arguments of this book will support the idea that new networking technology did in fact play this role in the past decades, but I will challenge the teleological odor of some of the debate, which has made these changes seem the inevitable result of the new technology, rather than a contingent result based on the policy decisions made in how to implement it. Far from new government policies being the passive result of technological changes, government policy has in fact been front and center in shaping the use of technology in the economy, so that different policies could have led to very different outcomes. Rather than losing importance, as some conservatives argue, government has become if anything more central to economic development—the creation of the Internet being the prime example detailed in this volume—along with the way government shaped the economic growth that was tied to its expansion.

Inequality and the Polarization of Regional Development

Along with the general debate on inequality, controversy exists about the shifting role of regions in the economy. There is a paradoxical sense of the disappearance of geography in the face of global commercial culture, even as regional cultural communities assert their own importance, often to the point of cultural and armed conflict. Using contrasting metaphors, such as those found in the recent titles *Jihad vs. McWorld*[16] and *The Lexus and the Olive Tree,*[17] a range of analysts have noted how those left out of the gains of cosmopolitan growth return to regional communal affirmation to maintain their dignity.

But beyond this local-versus-global contradiction is the polarization of economic communities within the same regions. In the United States, analysts have noted the pulling apart of communities into different fates within the same geography, especially in discussions of the inner-city poor, and they have blamed a number of factors:

- Deindustrialization of the cities, heightening both economic and racial stratification.[18]
- Government zoning and suburban politics[19]

• Elite professionals enjoying the benefits of the global economy, while segregating out the poor from shared participation.[20]

Yet any story of the retrograde nature of community contrasting with globalization misses a deeper contradiction, since economic regions themselves play an increasingly important role in today's technological economy. Silicon Valley and its mirrors around the world are only the most evident examples of this phenomenon. However, there is a real sense of change in the nature of regions and in who benefits from them as economic engines. As I will argue, poorer working families, far from existing in isolated economic backwaters, are often more at the mercy of global economic forces than are the elite, who often harvest the advantages of community dynamics for themselves.

Industrial Districts in the Global Economy

Neoclassical economists have long had little room in their theories for specialized regions such as Silicon Valley, older industrial centers such as Detroit, or even older craft regions in Europe: traditional economics assumed that firms would locate wherever wages were cheapest, workers had training, transportation costs were lowest, or a combination of these. However, a range of economic and social thinkers have built on the original insight of economist Alfred Marshall that firms could gain in both productivity and innovation from geographic concentration. Michael Piore and Charles Sabel are two key recent proponents of the idea that such "industrial districts" have historically been an important alternative to large-scale vertically integrated firm production. In fact, as the title of their major book, *The Second Industrial Divide*, indicates, they see the choice between such localized "flexible" production and corporate integration not as some natural economistic result but as a fundamental social choice. In their historical view, it was the combination of corporate power backed by government structures that forced vertical integration on vibrant industrial districts in the nineteenth century: from silk production in Lyons to ribbons and specialty steel in Saint-Étienne to cutlery in Sheffield to cotton goods in Philadelphia and Pawtucket. Ideologically, the logic of how these districts operated disappeared under the twin onslaught of classical capitalist and Marxist socialist views that stressed economic centralism at the expense of local craft production—alternatives promoted by cooperative movements such as the Knights of Labor in the United States and France's Proudhon in Europe.[21]

However, new technology has revived the competitive strength of regional production districts. Specialized "batch" production is now easier and less costly because of computerized machine tools that flexibly shift from one product line to another, in contrast to traditional mass production. Whole new regions have appeared with small companies working together to produce products ranging from specialty steel to chemicals to fabrics. The star region for theorists (aside from Silicon Valley itself) are new industrial districts in Italy known as the "Third Italy," whose heart is the province of Emilia-Romagna. The 3.9 million residents of Emilia-Romagna, once one of the poorest provinces in the nation, have recently emerged as the fastest-growing economic powerhouse of the country. The remarkable fact is that in 1990 its 1.7 million active workers were divided between 325,000 firms—an average of five workers per firm in a region where 90 percent of firms had fewer than one hundred workers and such small firms accounted for 58 percent of the workforce.[22] Such industrial districts have been dubbed "collective entrepreneurs" where firm relationships can be rearranged as needed to rapidly address changes in the global market. Writes analyst Michael Best, "Coordination in a dynamic industrial district is not planned in that the initiatives, responses, networks, and aggregated constellation cannot be specified in advance, but instead are developed in processes of mutual adjustment in unforeseeable ways."[23]

But this is not just traditional market coordination, since not only money and goods but also ideas without set market value are exchanged. "To the extent that product and process innovation is based upon new ideas," notes Best, "and that the creation of ideas is a social process involving discussion, then geographical proximity is important in innovation."[24] In the knowledge-intensive industries of computers and networked information, the collaborative nature of industrial districts seems an attractive model to refer to in understanding regions like Silicon Valley.

Berkeley scholar Annalee Saxenian has used this view of industrial districts to specifically analyze the success of Silicon Valley in the global high-tech industry. Since its early days, Silicon Valley has had a tradition of small-firm start-ups and close collaboration between firms. Saxenian noted that as many small semiconductor firms died out in the late 1970s and early 1980s, when basic computer chips became a commodity item produced by large firms such as Intel and Advanced Micro Devices (AMD), there was a burst of new specialized producers, among them Cypress Semiconductors and Chips and Technologies, that addressed niche markets. These newcomers produced small batches of complex, high-value-added components, often produced in collaboration with customers to enhance the performance of specialized items ranging from cars to machine tools to missiles to ultrasound machines. Local knowledge and relationships could be converted into innovative products and services. Because of these kinds of collaborative contracting relationships, Silicon Valley became home to a diverse array of specialized equipment and component makers, working in areas ranging from disk drives to networking products to computer-aided design. In fact, Saxenian argues that it was the self-sufficiency of firms in Massachusetts's Route 128 that led to that region's decline as a high-tech bastion in favor of the more flexible, collaborative Silicon Valley. Where Route 128 minicomputer and workstation firms tried to survive as stand-alone companies and failed to enjoy the innovation of working with one another, Silicon Valley firms blossomed by working in a whole environment supporting new products.[25]

Regions in the Web of the Multinational Corporate Enterprise

However, while other analysts grant the importance of these industrial districts, they reject the vision of these areas as self-contained economic engines in favor of a view that situates them within a global production system. As they point out, most of the component inputs for Silicon Valley hardware firms are from sources outside the Valley and most of the rest from divisions and departments within the firm doing the final assembly. What usually comes from local production linkages are in nontechnology items, such as cabinets, casings, power supplies, raw materials, and documentation. More technically sophisticated inputs come mostly from outside the region.[26]

One of the most far-reaching attacks on the idea that local regional networks are central to understanding the "new economy" comes from Bennett Harrison, who has argued that concentrated corporate power is growing, not diminishing, in relation to smaller firms. What is new is that corporate power is "changing its shape, as the big firms create all manner of networks, alliances, short- and long-term financial and technology deals—with one another, with governments at all levels and with legions of generally (although not invariably) smaller firms who act as their suppliers and subcontractors."[27] Harrison notes that even in the Third Italy, "lead firms" from both inside and outside the industrial districts are disrupting the collaborative nature of interfirm relationships. The temporary success of a regional specialization just feeds into a global system populated by big companies on the prowl for profit and opportunities.

Harrison emphasizes that the money trail in Silicon Valley and other industrial districts leads away from seeing the region as in any sense a concentration

of local firms operating for "regional advantage." The outsourcing of manufacturing content and other goods is one indication, but the heavy financial
investment in the region by outside firms, including by those outside the
United States, signals that a quite different dynamic is at work. "Growing
direct foreign investment in the Valley," writes Harrison, "seems aimed more
at tapping into the region's exceptionally rich knowledge base in order to
improve home-country performance . . . than at furthering the foreign investors' desire to join a local club of firms interested mainly in the economic
development of Silicon Valley, per se."[28]

Theorists have argued for a number of decades that the world is being reshaped by an international division of labor where regions are not so much
self-sufficient production areas unto themselves, but rather geographically dispersed nodes on a global hierarchy in the production process.[29] The rise of
such regions as Silicon Valley and the Third Italy are a modification of the
idea of a simple hierarchy of corporate headquarters spoking out from home
offices to branch production in a simple global division of labor. In this case,
there is the birth of more flexible strategic alliances that rearrange corporate
organization to fit collaborations around the world that maximize the exploitation of advantages in any particular locale.

A few analysts take a middle ground. While arguing that multinational corporations tie these regions together, they also argue that regions often develop
an "autonomous systems of goals," in the words of urban theorist Manuel
Castells. He sees these flexible regional relationships as a response to the uncertainty of technological change and an attempt by corporations to external-
ize the costs as much as possible outside the core global corporation.[30] Former
labor secretary Robert Reich has argued that these regions are often less about
divisions of labor within the firm than about harvesting labor from regions
with unique skill sets. The idea is to bring firm employees in contact with
other specialist problem-solvers where "each point on the 'enterprise web'
represents a unique combination of skills."[31]

However, the result has inevitably been regional loyalties evolving into multinational corporate loyalty. Around the third world, many countries have
been sorely disappointed that initial support in getting high tech firms off the
ground did not lead to national champions, but instead just facilitated the
growth of such international webs into new countries. Initial government support for high-tech firms in places such as India, Korea, and Brazil has led to
local firms, once established, quickly turning against investments that would
further strengthen those regions and instead looking to global investors and
alliances. "With the growth of local companies," writes development analyst

Peter Evans, "came more, not less, involvement with international markets, global technology, and transnational capital. . . . In the end, greenhouses [for local firm start-ups] turned out to be an indirect strategy for internationalization, not a means of escaping it."[32]

While new networked technology has allowed global corporations to more easily coordinate these far-flung outposts of economic control, this process has been further facilitated by changes in national and international law that allow corporations to move beyond the nation-state-based legal organization of the past toward what some have called a "postimperialist" stage of global corporate power.[33] These changes allow firms to transcend the state loyalties they once had, whether those of multinationals to core states such as the United States or those of "local champions" to the developing states that birthed them.

So while the vibrancy of local industrial districts may be very real, the politics of those regions are inevitably shaped by global economic players seeking their own advantages in harvesting their benefits.

The Elitist Nature of Modern Industrial Districts

If there is any consensus on what does fuel the persistence of regional industrial concentration, it is the sharing of workers and skills. A whole set of skills for particular industries congregate in particular cities, from music and film in Los Angeles; science and engineering in the San Francisco Bay and greater Boston areas; global finance in New York and Chicago; international affairs, government relations, and the worldwide marketing of weapons in Washington, D.C., and law, advertising, and publishing in New York. Beyond these are smaller pockets of even more superspecialized workers from medical-device research near Minneapolis to molecular biotechnology near Little Rock.[34]

Such a shared high-skill workforce facilitates the firm start-ups and contingent employment that accompanies the turmoil and innovation of a place like Silicon Valley. The concentration of companies meant that job-hopping was less of a risk for workers. "The region's social and professional networks were not simply conduits for the dissemination of technical and market information," writes Saxenian. "They also functioned as efficient job search networks. . . . People change jobs out here without changing car pools." This network of workers acted as a "kind of meta-organization" for rearranging skills and talents as needed.[35]

The definition of a region can be largely determined by identifying which people can actually work together on a daily basis. Corporations can change products lines, move products globally, and transfer subcontracts to new suppliers worldwide, but, as geographer David Harvey notes, "unlike other commodities, labor power has to go home every night and reproduce itself before coming back to work the next morning. . . . Daily labor markets are therefore confined within a given commuting range. The geographical boundaries are flexible; they depend on the length of the working day within the workplace, the time and cost of commuting. . . . The history of the urbanization of capital is at least in part a history of its evolving labor market geography."[36]

However, a shared geographic labor pool among workers does not necessarily mean that all workers in a region are integrated into the same production regime. While proponents of industrial districts emphasize the unity of regions, others notes that production regimes ultimately include workers far from the region upon whom elite knowledge workers ultimately depend. Writes Bennett Harrison:

> There are whole neighborhoods of Los Angeles—hundreds of miles away from Santa Clara County—where both documented and undocumented workers perform unskilled and semiskilled assembly tasks, often at home, for contractors to the high-tech firms of Silicon Valley. . . . These Urban ghettos are as much a part of the famed "Silicon Valley production system" as are the engineering laboratories at Stanford, or the military R&D facilities within Lockheed's Missiles and Space Division in Santa Clara County.[37]

The creation of elite industrial districts is often driven by corporations seeking to separate out the skilled workers, who need special treatment, from lower-skilled work that can be moved at will to undermine the demands of employees for better pay. Even the vaunted districts of the Third Italy were driven partly by the actions of corporations such as Fiat that decentralized production away from traditional centers, luring skilled workers by the prospect of high earnings in runaway shops, even as lower-skilled workers would see their pay drop or outsourced altogether.[38]

This spatial division of high-skill from low-skill jobs is intimately tied to the ongoing social policies of local governments. Zoning laws tend to reinforce real estate prices in these privileged areas, while other local governments extend tax benefits to lure particular levels of skilled jobs. Argues Castells: "In this way, the spatial division of labor is self-reproductive and self-expansionary. Increasingly valuable spaces first segregate and later expel function and people that are not worth the cost of keeping them in the gold mines of our technological age."[39] One reason we have urban segregation and polarization, rather than complete geographic separation, between high- and low-skilled communities is that information-based services, especially the high-end skilled versions, have traditionally needed both the support and infrastructure that urban centers provide.

The Technological Shift of Urban Politics and Social Infrastructure

If regional unity is being undermined and inequality increased, the question raised is, What is changing in the political and economic landscape from the past?

One key to exploring this question is to look at how the interplay of class forces have shaped urban politics in the past and how new technology may be reshaping them now and in the future. In parallel to the general view of a midcentury economy tied to state-based Fordist class compromise, urban politics in the immediate post–World War II period can be understood as promoting "growth machines"—alliances made up of various interests including strong unions and regional businesses—to create jobs and economic progress for all residents. Whether in attracting railroad depots, building ports to attract shipping, or, in more recent decades, seeking more general capital investments and tourism, cities act as common agents for bringing growth to their citizens. The strongest proponents of such growth were usually those with the most fixed capital invested in a region, whether utilities, local newspapers, or banks with a range of loans invested locally.[40]

Capitalism is always in the process of producing both labor and capital surpluses that need to be reabsorbed; urbanization has been one key method to assure rapid turnover of both into new production. However, the expansion of those flows of trade and commerce repeatedly threatens to dissolve the coherency of local communities. Government steps in to refocus that local coherency. Where individual capitalists systematically underinvest in public works, government can step in to pick up the slack to assure proper investment of excess capital that might otherwise be lost in wasteful speculation.

The changing nature of the city in the process of consumption as well as production is highlighted in David Harvey's analysis of the shift from mass production to newer models of flexible capitalism. Mass production created the need for mass markets along with fixed investments that were not easily redeployable outside their urban settings. This "Keynesian city" focused on the debt-financed consumption of both private and public goods as a key to regional and national economic health.

All these complications created a space for a relatively autonomous urban politics in which skillful politicians could forge coalitions to govern and unify the community. Harvey notes that such coalitions were not necessarily synonymous with local government, since local jurisdictions often divide real labor and production markets. Other means were often used to build higher tiers of cooperation, whether through a range of support from state and federal powers or from a base of local government reaching out to civil society in bringing together business and community. This politics created a cross-class alliance of businesses, labor unions, and government in urban politics that Harvey calls a "structured coherence defined around a dominant technology of both production and consumption and a dominant set of class relations."[41] These alliances were by their nature unstable as different partners in the alliance sought power over others through the threat of exit or by pulling new resources into the region that changed the balance of power. Yet for most of the postwar period, these alliances seemed to be the face of urban politics in the United and States and much of the developed world.

With new technology and the fracturing of traditional mass production has come the weakening of the power of local actors and governments to bargain effectively against the global reach of multinationals. Instead of concentrating on regional social consumption, urban politics increasingly gives way to competition between urban areas for jobs. This competition has led to rapid shifts in urban fortunes as urban areas have accommodated much more flexible labor processes in order to attract urban development.[42] The relative autonomy of most urban politics is more and more disciplined by this interurban competition as each region shapes itself to the needs of multinational capital. This does not mean spatial location is less important; in fact, the freedom of capitalists to move around makes minute differences even more important in daily decision making and in strengthening capitalist domination over labor. Each area's unique attributes are exploited for advantage by different capitalists, which in turn leads local areas to seek to make their areas more attractive to free-ranging companies.[43] However, even as cities seek unique images in this heightened interplace competition, the whole process ends up producing a "serial" monotony along established patterns such as in New York's South Street Seaport or Baltimore's Harbor Place. The end result is the ephemeral fashions of Benetton being sold in similar malls across the world along with food markets that are now broadly diversified, with the same goods from local markets around the world being delivered through global distribution.

The public goods needed by new technology firms are not primarily the road construction and other public works that once tied capital and unskilled labor together in an alliance. Instead, the public goods of the new era are basic research and information standards supplied by combinations of universities and government support for business consortia. Left out of this equation is any real need for agreement by non-elite members of the community or the creation of goods that serve their day-to-day economic needs.

Of course, industry has always tried to tilt public spending toward their needs, and the evolution of the technology in some ways just accentuates that tradition. But what has decisively fractured regional economic development and made it practically an oxymoron is the disappearance of what Harvey calls "fixed capital"—public utilities serving regional markets and the holders of public debt tied to specific regions. It was always fixed capital that acted as an anchor across the divide between globally oriented industry and the working-class public. Such companies found a cross-class interest with labor unions and other community organizations in building a growth coalition that expanded wealth across the region—an interest tied to the fact that companies whose markets were circumscribed by geography could grow most easily with the widest possible expansion of consumption.

The retreat from these "growth coalitions" largely reflects (and reinforces) the economic fault lines between less-skilled and more skilled workers in the new, flexible economy. Robert Reich has argued that, politically and economically, "America's symbolic analyst class have been seceding from the rest of the nation."[44] One form of secession has been the shift of the tax burden to lower-income Americans, through both physical separation of where rich and poor live and what kinds of income is taxed. This is linked to a decrease in funding for the public investments that would help the unskilled. In a nation that does not rise and fall together, the elite sees less interest in general social investment for their own well-being.[45]

Worldwide, the hope had been that alliances between government and local business in promoting regional economic development would build a cycle of profit and reinvestment in public goods that would eventually reach all citizens. Instead, often the most successful state regimes in supporting economic development have been the gravediggers for broad political support for continued public investment. The more successful local firms become around the world, the more transnational corporations see them as viable economic partners and the less those local firms see an interest in local government.[46] Similarly, local politics is unlikely to have much effect when firms' fates in the United States and Europe are voluntarily tied to decisions made far away in corporate headquarters. Saxenian herself worries in her writings that Silicon Valley's individualistic heritage could easily lead straight back to the autarchic economic strategies of production that almost killed it in the early 1980s. Cuts in social funding of education, research, and training; congestion; and soaring housing prices have undermined the local conditions that originally created its economic revitalization.[47]

In recent years, "reinventing government" has largely meant privatizing public services and undermining the very public authorities that are needed for autonomous economic local planning. Instead of local government acting as a referee or coordinator of local economic development, it increasingly becomes merely the servant of the corporations that are blackmailing it for tax breaks and services by threatening the company's departure if its demands are not met. Even as the elite may seek to conserve the benefits of "civicness" through purely business-based associations, its secession from the general public leaves inequality in its wake. The very technology and economic growth that have fueled these regional engines of innovation may end up being their undertakers as broad civic society is undermined.

The Core Role of Government in Technological Innovation

This would be the end of a somewhat depressing teleological economic story if we did not step back and ask, What was the role of the United States government in this technological change and in the political decisions in how it would be implemented?

"Technology" is not equivalent to scientific advances, but itself entails a whole range of social practices and decisions in implementing knowledge advances. And those knowledge advances themselves are not merely economic epiphenomena, but are shaped by the social organization of research and development (R&D).

While there are a host of factors that have driven new technology such as the Internet, it is government policy that fundamentally drove the Internet's creation and its implementation in our society. At the most obvious level, government spending has been crucial in providing the public goods necessary for technology innovation in which private-sector markets systematically underinvest. The computer industry is a notable example of civilian and defense spending leading to a wide range of innovations that the private sector could then commercialize. Private companies innovate rapidly around developments with immediate commercial payoff, but it is left to public-sector investment for the breakthrough research and public infrastructure that allows transformative innovations. The Internet is just the most recent (albeit dramatic) permeating technology that was the child of centralized government planning and economic support. From aerospace to biotechnology, the government has played a crucial and usually a leading role in technological advance and with corporate research labs cutting back on basic research, that role is unlikely to diminish in the future.

The Internet may be highly identified with the Silicon Valley region, but, much to the annoyance of cyberlibertarians, it was born in the bureaucratic halls of Washington, D.C. It came out of a whole set of institutions and a milieu of innovation directly and purposefully funded by federal agencies, primarily in the Advanced Research Projects Agency (ARPA) but also out of a host of other technology-oriented bureaucracies. The federal government created a national network of experts who could guide the Internet to economic viability, laid the wires and funded the computers where it was tested and developed its initial critical mass, and funded many of the Internet companies in today's headlines as contractors for federal government projects or agencies.

The reality is that no corporate research laboratory and no local government could operate on the decades-long time frame needed for the development of the Internet. Breakthrough innovation requires generations of improvements with little immediate commercial payback. Often, the need for public discussion and collaboration on the basic science makes it impossible for any one company to enjoy the fruits of proprietary discoveries. Analyst Kenneth Flamm has noted that, despite large gains in the horsepower of the processors and increases in memory capacity, the basic architecture of computer design is relatively unchanged from concepts developed under the period of most massive government support.[48] Federal funding created the advances of time-sharing minicomputers and most of the networking technology that is the heart of high technology to this day. Commercialization increased the speed and lowered the price of each of these innovations, but the driving engine for its creation was the federal government.

The Cyberlibertarian Myth and Contesting Government Policy Options

Although this history is contested by some cyberlibertarians, the argument against the government's role is largely the product of a self-interested social amnesia to justify private actors harvesting the profits of innovation while denying any responsibility to the society whose investments made those profits possible. Before going further with outlining the basic arguments on the role of the government in shaping these technology advances, it is worth noting that there is a another group of analysts, conservative and radical, who might accept the role of government described in this history, but treat such government decision-making as inevitable, as economically determined by the needs of business as others treat technological change itself.

Whether stated in the more deterministic strain of Marxist analysis or by way of a conservative "public choice" approach, the assumption is that government policy is a somewhat inevitable result of economic forces with little room for political agency or contested results. For all practical political purposes, the government's role disappears in such analyses almost as much as in the libertarian myth, since the government is always the object of economic forces and therefore plays no real independent role in social change. Government ends up merely as a secondary step in a deterministic pattern, an inevitable useful handmaiden or bureaucratic obstacle to the "real" story of economic and social forces shaping policy to reflect the power of those forces or the needs of technology itself.

One should have a healthy respect for the deterministic power of social forces; and government policy, including technological policy, does reflect an overwhelming deference to the needs of corporate power. However, such a story is not complete, since it ignores the inevitable shifting struggle between employers, workers, and other social actors that government policy reflects. This contested policy exists not only because corporations are not all-powerful politically, but also because a degree of consent is required in any advanced technological economic system.[49] This was a lesson that the Soviet Union and Eastern Bloc countries, with all possible power vested in their elite, learned as their earlier advances in economic industrialization collapsed into stagnation as they tried to assimilate new technology into their social systems.[50] With possible democratic challenges to distribution through the market in our system, the very legitimacy of the economic system requires some degree of independent mediation by the state.[51] In a sense, the very fact that corporate interests feel the need to promote the myth of government's limited role in the development of the Internet reflects the fact that corporate power is as much based on mobilizing ideological consent as on exerting raw economic power.

At a more fundamental level, some degree of independence by the state benefits corporations themselves, since such independence from direct control of corporate interests allows the state to rein in short-term economic opportunism in favor of longer-term investments and growth. The basic form of this public-private synergy is the provision of public goods to supplement private-market activity. But at a deeper level of private-public engagement, the government promotes a strong civic culture that itself becomes an engine of

economic growth by creating the trust needed for companies and their communities to engage in new kinds of economic relationships. The fundamental
dilemma for any firm, especially one engaged in technology innovation of any
kind, is the tension between learning—changing patterns of organization to
advance economically—versus the need to make sure in the short term that a
company is getting a good deal from its economic relationships. By nature,
learning and innovation upsets established patterns and undermines traditional assurances of a fair deal previously negotiated. In an industry or environment of heavy learning and innovation, it is hard to write private contracts
that cover all contingencies or to create stable hierarchies that create enforce-
able rules. So only some degree of trust developed among competitors can
encourage radical innovation across a region or society. As analyst Charles
Sabel argues, "The preponderance of historical evidence is that, regardless of
their level of development, economies seldom pull themselves out of long-
term, low-equilibrium traps by the bootstraps that market prices theoretically
provide individual firms."[52]

Helping firms escape outdated safe "equilibrium traps" can be accom-
plished by the complementary relations of the provisions of public goods, but
they are also transcended by what sociologist Peter Evans labels "embeddedness," in which the participation of public officials across the public-private divide helps prevent opportunistic behavior by either private or public
entities and encourages mutually supportive endeavors. Instead of finding a
strict separation between government actions and private economic activity,
Evans has noted the success of those governments and societies that create
civic networks that transcend the division between government and private
firms.[53]

Political scientist Robert Putnam, in his pioneering studies of the relation
of civic organizations to good government and economic development, has
noted that strongly "civic" regions have become increasingly wealthy in the
modern era compared with areas without strong civic networks. Using as his
test the contrasting success of different regions in Italy, Putnam has noted that
not only does a history of civicness correlate strongly with economic development, but also "[c]ivics is actually a much better predictor of socioeconomic
development than is [past] development itself." Rather than strong government undermining civic trust and cooperation, the right kind of government
action reinforces civic society and the "social capital" to build the trust necessary for both innovative government and economic development.[54]

All this points to the sterility of debates over whether industrial policy by
government is useful when the real question is what kind of government

involvement is most fruitful. Across the world, modern governments shape whole industrial sectors, from industrial-district policies in Italy to Japanese-style "administrative guidance" that promotes cooperation among companies.[55] As Evans notes, state involvement in the economy is a given and the only question is "what kind" of involvement. Where government has been most aloof from technology development, as in India, early on, local information industry was least able to thrive while Korea's focus on coordinating the basic research needed to get private firms involved in commercialization has been tremendously successful. Across successful societies, government has played a key role in building the social capital needed for the synergy around which innovation thrives.[56]

Setting Standards and Guiding Internet Growth

Part of this "social capital" role for government has historically been to encourage technical standards that encourage innovation based on improved efficiency growing from shared standards rather than opportunistic rents connected with incompatible and proprietary goods.

Economic historians point to examples of the early nineteenth-century role of the U.S. armory in Springfield, Massachusetts, noting that the standardization of gun part sizes and specifications encouraged U.S. firms to pioneer interchangeable parts and allowed collaboration between a host of small gun producers who could specialize in different parts using those standards. Notably, when the War Department withdrew its involvement in the late 1840s, large integrated firms forced out small companies that could no longer depend on open government-backed standards.[57] This emphasizes that the nature of corporate hierarchy in economic markets is not simply an issue of technology or economics, but rather of social choices tied up in political decisions on such standards.

In the case of the Internet, beyond its investment in the basic research and infrastructure to launch the network, the crucial element the federal government supplied was precisely the social capital required to promote the technology standards needed to encourage rapid decentralized innovation. At its heart, the Internet is just a system of protocols and information-exchange rules that all computers involved recognize. In the 1980s, private companies, each seeking opportunistic advantage, had promoted a stew of conflicting proprietary standards for electronic communication, but none was able to engender the broad-based trust in their approach to reach a critical mass of users. Instead, it was the work of the federal government, backed by decades of planning and social networking, that would create the Internet that would replace or subsume all those proprietary approaches.

As will be explored in Chapter 2, over the decades the Advanced Research Projects Agency and other federal agencies like the National Science Foundation (NSF) concentrated their funding and support on standards that facilitated the most open network connections possible. Through strategic support for those protocols in the popular UNIX operating system, they would then help spread them throughout the computing world. The federal government organized and funded an emerging professional network of computer experts to oversee and guide the emerging connections between government, universities, and the slowly building commercial sector involved in the Internet. And through the creation of public space and the harnessing of volunteer energies in its early stages, the federal government encouraged a stream of free, quickly shared software that promoted continual innovation on the network. Far beyond traditional conceptions of industrial policy in which a few dollars are invested in promising industries, the federal government fundamentally created the entire framework of a new electronic marketplace of common standards, thereby breaching the monopolistic divides of what had been rather stunted proprietary systems.

The initial commercialization of the Internet would be done largely by direct government spin-offs or companies relying on government contracts for their origin. The companies that took over the management of the backbone wires carrying most of the traffic would include UUNET, a direct spin-off from the Department of Defense; Bolt Beranek and Newman (BBN), with its longtime contract ties with the government; and MCI, which was involved in contracts throughout the 1980s in building the original NSF backbone. Government defense contracts were often the key point of entry of the first generation of computer start-ups into the industry, especially in Silicon Valley.

The Federal Role in the Regional Development of Silicon Valley

This last point highlights the inescapable link between federal policy and the success of regional industrial districts like Silicon Valley. In a sense, when looking at why a particular region like Silicon Valley became such a concentrated center for the industry, the most simple (but not simplistic) answer is to ask, Why Not? and chalk it up to dynamics that could as easily have favored some other region. Economist Brian Arthur, who has been a pioneer in applying the insights of mathematical complexity theory to economics, has argued that in looking at the success of firms in the information economy, too much

ex post facto analysis obscures the unpredictability and multiple likely outcomes of these new markets. In what Arthur calls an "Increasing Returns
World," the new economy is one where early innovation is rewarded with ever
increasing dominance of a market because of the up-front development costs
of entry and customer "lock-in" to standards established by the market leader.
In the "casino of technology," inferior products can dominate the market
purely because of unpredictable events that give them a legup at the beginning
of the process of technological lock-in. Once a lead is established by a technology, all sorts of other factors come into play to reinforce its economic dominance.[58]

From this viewpoint, many of the factors used to explain the rise of technological dominance by a region—supportive services, available capital, business
culture—can as easily be seen not as a cause but as the result of that dominance. In a global context of trade, trade theorists have noted that along with
lock-in, complexity theory encourages a viewpoint of spontaneous self-organi-
zation within geographic spaces that will tend to concentrate centers of production in specific locations, not based on present factors that can be easily
duplicated but based on initial conditions that predate present dominance.
Economist Paul Krugman argues that "in a more realistic model of the world
economy . . . there would be many possible outcomes, depending on initial
conditions; given a slightly different sequence of events, Silicon Valley might
have been in Los Angeles, Massachusetts, or even Oxfordshire."[59]

Overwhelmingly, the "initial conditions" that attracted high technology to
regions across the country were earlier rounds of government investment in
technology, largely through defense spending. Across the spectrum of high-
technology industries, the single overwhelming factor correlating with the rise
of technology firms in any region is the level of defense spending. In the most
comprehensive national studies of high-tech industries, scholars Ann Markusen and Peter Hall found that whereas various business "amenities" such as
airports or other services played a minor role, defense spending played a disproportionate role in the regional location of high-technology firms. Strikingly, in the absence of defense spending, the existence of a research university
had no effect on high-tech employment.[60] This highlights the fact that merely
concentrating engineers and scientists in the same location does not deliver
automatic benefits to a region, a fact that a number of local regional investments in "science parks" have discovered.[61]

The economic development of the Sunbelt was inextricably tied to the government spending that was fueled by World War II and Cold War imperatives.
For a range of reasons, from building a military presence on the Pacific Coast

to establishing military commands far from established pockets of population, as in Colorado Springs, to taking advantage of talent located in areas such as New England that had seen a wave of industries leave before World War II, military spending largely abandoned the traditional industrial heartland of the Midwest for the Pacific Coast, Southwest, and New England. This "gunbelt" policy (as Markusen call it) also acted as an "underground regional policy" that allowed backward economic regions such as the South and areas with vulnerable resource-dependent economies such as the West to stabilize their economies and increase jobs. While there was no national policy to respond to plant closures, defense-related communities qualified for impact-aid programs to smooth adjustment and help such areas gain new defense activities. This regional policy was highly selective and it left most inner-city areas in need of economic revitalization outside this defense-driven industrial policy and may even have drained resources from those traditional areas in duplicative community building. But the results are the high-tech enclaves of Silicon Valley and other areas.[62] Globally, similar studies have shown that the rise of technology centers around the world has followed similar specific government development policies.[63]

Similarly, national government policies in regard to standards and other trust-building policies have important effects on the strengths of regional industrial districts. Especially in an age of enterprise networks and the restructuring of firms to take advantage of regional opportunities, building regional trust becomes a decisive difference between innovation and stagnation. Firms often fear trying to decentralize decision making in order to take advantage of regional resources outside the firm; they worry that if they depend on regional cooperation rather than internal vertical integration, they will adopt the wrong approach to innovation through mistakes or bad faith by partners. By providing uniform standards, government authorities help to overcome managerial obstruction and fears of inequality in dealings between larger and smaller firms. Economist Charles Sabel cites as an example the promotion by the National Institute of Standards and Technology (NIST) of technology centers where small and medium-size firms collaborate on technical design and organizational problems, and in line with Putnam's argument, these new business networks inevitably push new functions on the public authority in an interactive process that Sabel labels "bootstrapping."[64]

Along with describing the initial defense funding of Silicon Valley, in Chapter 3 I will detail the kind of government action that has made such industrial district approaches more or less successful—with the Internet a prime example of that process. It has been government support and commitments to the diffusion of technology and broad open standards that has allowed Silicon Valley and other regions to thrive over those tied to more hierarchical firms that based their strength on proprietary approaches.

In fact, the origins of the region as a center for semiconductor production would have been impossible had the federal government not forced AT&T to license its transistor technology widely and barred the company from the industry for the following three decades. Without that act (and a continued commitment by the government to pushing open technology diffusion over the years), peripheral regions, among them Silicon Valley, would have had little chance of overcoming the strength of AT&T and other larger corporations. They would have used patents and their existing economic position to keep technology within traditional industrial enclaves—exactly the pattern in most of Europe, where old-line industrial companies were more favored by their national governments' policies.

Saxenian and other analysts have painted the success of such Silicon Valley firms as Sun Microsystems and their commitment to regional networked relations and nonproprietary standards as following from an inherent economic advantage of nonproprietary approaches in the new economics of technology. In her story the firms in Route 128 were inherently doomed to failure given their proprietary approaches. There is little question that Silicon Valley firms hitched their fates to nonproprietary standards at a rate far beyond that of any other region (although not without many firms attempting otherwise and a culture of lawsuits hardly in keeping with a pure image of cooperation), but the success of that strategy over the long term was as dependent on a supportive national government context as the region's high-tech origins had been dependent on government investments. The cooperative Silicon Valley industrial-district system would have been useless without the creation and support by the federal government of the standardized computer systems around which that cooperation was organized. One need only look at the alternative of the proprietary lock-in of the Windows-Intel duopoly that developed in the microcomputer market in the late 1980s and the 1990s—notably a technology market where the federal government had the least involvement in setting standards in the period.

The key to the success of Silicon Valley in the networking marketplace was the leveraging of the federal government's investments through a system of regional cooperation to commercialize the Internet. Silicon Valley firms that would be at the heart of that commercialization, such as Sun, Cisco, and Oracle had all got their start based largely on selling to government buyers and agencies. Or, as in the case of Netscape, such firms would raid the talent of the government centers that built the Internet to commercialize government-created software such as the Mosaic Web browser and servers.

The New Politics of Regions in the Age of the Internet

As the federal government withdrew funding and coordination of Internet standards in the 1990s, the question then became, What role would regions like Silicon Valley continue to play and how could local policy, both by industry itself and by local government, work to sustain the regional economic dynamism that federal support had created? The lock-in of technological advantages and the concentration of talent gives such regions advantages when federal support is withdrawn, but that federal withdrawal changes power relations within regions. The global context of multinational interests then shapes the options of regional actors, largely at the expense of working-class and less-skilled interests within the region.

Education and training, a key attribute of local and state government in the United States, is seen as a prime tool for local regions in sustaining regional success. Given corporations' lack of loyalty to any particular region in the global economy, the only factor really under the control of a region is the skills of its residents.[65] However, as important as the skills themselves is the environment of ideas and innovations promoted by local educational institutions. Universities both provided the initial crop of engineers and technology for many high-tech regions and continue to serve as a place for diffusion of ideas and a "corner café" at which to meet and to set community standards. As the industries grow, new companies depend on local and state government for lower-level vocational training and other employment services[66] along with the "milieu of innovation" that helps attract corporate R&D centers and other research networks.[67] Much of the focus of such local policy is to build up the "social capital" and civic networks that analysts such as Evans and Putnam see as so crucial to regional economic success.

All this said, the issue becomes whether such policies are either effective for many areas or actually benefit the broad range of taxpayers and workers supporting such public incentives. Analysts have noted that without the initial support of federal money in attracting firms to a region, there is an outright failure of most local investments in science parks to attract and keep a critical mass of tenant firms.[68] Taxpayers have seen little for their money, and worse, the investments end up contributing to polarizations between older industrial workers, usually working class, and the more suburban regions where the high-tech, usually middle-class workers are subsidized.[69]

For established technology regions like Silicon Valley, such local investments do seem to work but it is unclear who benefits from such policies. Regional concentrations end up less as an indigenous relationship of local firms than as a new kind of venue for multinational organization to make strategic alliances in a global production system.

As will be explored in Chapter 4, instead of the Internet becoming a tool used by semiconductor manufacturers for improving supply networks in the region, it became an explosive industry all by itself. And counter to the predictions of Saxenian and other celebrators of industrial districts, hierarchical mergers, global alliances, and corporate acquisitions, rather than flexible small-firm alliances, became even more the norm in the region. Although the more limited view of regional advantage—shared workforces and business amenities—continued to play a strong role, in the region multinationals and leading firms seemed to be imposing increasing economic concentration. In fact, through new tools such as electronic catalogs and other global connections facilitated by the Internet, the need for close local supply relationships have been diminished in serious ways because of the new technology.

Yet this later process was also accompanied by an intense and accelerating process of regional organizing by businesses in an alphabet soup of manufacturing and software alliances based in the Valley. The key to this new explosion of regional business consortia is the need to substitute regionally agreed-to public goods such as standards for earlier clear federal intervention. Especially in a networked economy where software and networking hardware need to be compatible, such business consortia become increasingly common despite the globalization of production.

Privatization of standards makes economic regions an efficient way for combinations of private corporations to establish the trust needed to commercialize open standards. This is ultimately an economically beneficial result for all the cooperating companies as a whole, since open standards tend to create a far larger market than do proprietary standards, which dampen innovation and new entries to the marketplace. In Silicon Valley, this process was aided by a federal government still involved enough to support a range of industry networking efforts tied to the Internet. The efforts of the consortia and other regional alliances were critical in promoting the standards around new applications that would sell Silicon Valley products tied to the Internet. A network of allied companies being based within the same region meant that the spillover of growth in the Internet industry was most likely to redound to the economic benefit of those supporting the "public good" of open standards.

What is radically new in this arrangement and outside the traditional industrial-district view of the world is the way Bay Area companies have used standards-setting cooperation to leverage a host of local companies, including those in the entertainment and banking industries, into a more globally dominant position. As multinationals increasingly made Silicon Valley a global center of technology production, that dominance of technology could be leveraged into a much broader and more diverse corporate position in the global economy.

As corporate coordination of innovation, rather than production links, becomes the key to understanding the business politics of regions, this elite version of cooperation leaves little need for serious concessions to the needs of non-elite workers in a region. Implicit (and often explicit) in the older views of industrial-district advocates was the idea that shared production links require active involvement of all members of the production stream in decision making, creating a vital role for workers and communities as a counterbalance to the power of companies. To the extent that companies are small and tied to indigenous growth factors, they are dependent on and will support broad community economic development.

However, if those small start-up companies are tied to global corporate needs, as is apparent in the dynamics of innovation detailed in this book, then the much darker vision of economic polarization is likely to be the norm. Even as some wages rose in the Silicon Valley areas, housing and other costs rose faster for the average workers as poverty expanded rather than fell within the overall prosperity of the region. In the end, for most workers, the Silicon Valley boom gave little sense of security, but rather, with the rise of temporary agencies and contingent employment for as much as 40 percent of workers, a sense of the ephemerality of growth. Even as elite engineers invested the dividends of initial public offerings (IPOs) for their long-term security, other workers watched as continual outsourcing of lower-end jobs eroded any sense of stability. And as the recent downturn in the Internet economy shows, even the security of higher-end engineers was far less guaranteed than many had assumed.

The new business leaders of the region, instead of investing in transportation or other public goods that would encourage regional consumption and therefore benefit all economic players, used their global leverage to demand subsidies to stay within their affluent suburban enclaves. With local municipal budgets often teetering on the end of bankruptcy, such demands for elite investments fed a competition between urban areas that bid away resources from lower-income citizens. Urbanization took on new patterns and new divisions as cities gentrified, even as conspicuous consumption and poverty both deepened. Technology firms promoted limited mass transit serving their high-technology corridor, even while spending millions to defeat proposals that would support bus service or other needs in poorer communities. All this reflected the evolution of Silicon Valley into what the *Economist* labeled a grander version of a gated community—the ultimate expression of private public goods created at the expense of the broader public.

Federal Regulatory Policy and the Fracturing of Regional Growth Politics

This fracturing of traditional growth coalitions was not merely the passive result of federal government withdrawal. Instead, it was facilitated by active government regulatory policy that deliberately undermined the traditional anchors of regional development coalitions, namely the regional banks and utilities that once lay at the core of old urban growth machines. Local banks have given way to interstate and global banking, local power utilities are moving to national and even global competition, and most decisively for the shape of the Internet, telecommunications is fracturing into a whole range of global competitors with little regard to the needs of regional development or equal access. Whatever the economic effects, the political results have been the disappearance of the "fixed capital" anchors of urban politics.

While such free market advocates as George Gilder have hailed new market competition in financial services, power utilities, and telecommunications as the natural order of things, Steven Vogel and other analysts have stressed both the government imperatives and the interventionist character of so-called deregulation. Even while markets have grown in these sectors, national government power and control has not diminished. While globalization is a real process, the fact that governments enacted quite different reforms reflects their room for negotiation. Referring to the process as "reregulation" and "liberalization," Vogel notes that advanced industrial countries have reorganized their control by private-sector behavior, but "not substantially reduced the level of regulation."

This process served various government-based imperatives, from fattening government budgets in the case of privatization in Europe and Japan to feeding internal bureaucratic power politics as separate regulatory agencies fought over control of converging markets in financial services, computers, and tele-communications. The creation of markets actually increased the power of many regulatory agencies by decreasing the power of regulated utilities and state monopolies, as in the breakup of AT&T or privatizations in Britain and Japan. Each new marketplace needed a welter of new rules to promote compe-tition against the previous monopoly incumbents and to structure access to customers over now shared infrastructure.[70] What is clear is that markets in the networked economy are even more the product of social and political relationships and rules. To merely laud or blame technology or "economic globalization" for the rise of these new markets is to ignore the crucial and deliberate acts by the central government that made them possible and that continue to sustain them.

As global companies have assumed greater economic and political control in these sectors, they have forced through regulatory changes that have under-mined those protections that had safeguarded the needs of lower-income resi-dents within regions. As global market competition envelopes retail banking, power utilities, and phone service, it is the power of global actors that defines local politics rather than any indigenous cross-class collaborations. Where funds from business and wealthier customers had once supported universal services, the newer political regimes promote the exact opposite: low banking and service rates for the wealthy against escalating fees for lower-income con-sumers. And as once local "fixed capital" becomes controlled by footloose global players, it becomes unclear where the political will and economic re-sources for public goods will come from for local economic development that is not purely aimed at the needs of the business elite.

As Chapter 5 will detail, the federal government not only repealed banking laws that stood in the way of this globalization of local money, it actively supported this globalization through tax laws that encouraged savings and loan associations to resell their loans on the global bond market and supported new global banking expansion. Increasingly, banks and other financial firms could pick and choose their customers not by region but by income and occu-pation across the nation and the globe, thereby dissolving any fixed pots of capital tied to geography.

At the same time, banks in the Silicon Valley region increasingly forged ties with the computer industry to increase their power within that global banking market. Much as regional collaboration on standards strengthened the role of Bay Area computer companies within the general Internet industry, banking collaboration within the region became a vehicle for the economic elite to strengthen their power in the global online banking system. So even as the ties between the poor and the wealthy within the region have unraveled, the economic and political ties of the regional elite have only strengthened.

In the case of the electric power industry, the Internet is becoming the vehicle for a breathtakingly rapid collapse of the regional power market that once served as a crucial engine of regional industrial planning and growth. Spurred by Federal Energy Regulatory Commission regulations mandating a real-time Internet wholesale market for power sales, the power industry is beginning to forge national and global links that interpenetrate regional markets. The results of creating national markets for energy sales is an expansion of regional polarization as large business users grab the fruits of technology innovation in the form of hundreds of billions of dollars in energy price discounts, even as small users face the costs of bailing out local utilities who invested too heavily in obsolete nuclear facilities and other white elephants. Local regulators are increasingly stripped of the ability to govern local energy markets as, against the rhetoric of "decentralization," the federal government increasingly invests national regulators with micromanagement powers over all aspects of energy production, distribution, and marketing.

As critics, including myself, argued when new deregulation laws were passed in California and other states, these new national microregulations favor short-term production and profits at the expense of long-term investments in the electricity grid. Grid blackouts in the core of the network were easily prophesied as the result, as short-term market competition replaces the cooperation between utilities that has maintained the public energy grid over the years.[71] The rolling blackouts in California by 2001 were a predictable result.

Nothing highlights the internal polarization of regional economics more than the trajectory of telephone "deregulation," which itself is tied up in the birth of the Internet. Despite the rhetoric of the Internet as a new free market, its emergence and the profits of Internet Service Providers (ISPs) have been dependent on regulations that allowed the Internet to cannibalize local telephone infrastructure at the expense of general phone users. Worse, the fragmentation caused by the breakup of AT&T of different telecom services, including the Internet, has ended up preventing a general digital upgrade of that local infrastructure. High-end business and professionals have profited from an array of new companies and technologies given free access to the shared network, even as underinvestment in that infrastructure has led to higher costs for average users of the phone service.

It is this mandate for interconnection to public networks by those looking for marginal profit that lies at the heart of economic polarization and the abandonment by the elite of the need for any serious involvement in regional agreements that include the needs of working families. Each step in the breakup of AT&T, from the rise of MCI and Sprint to private branch exchange (PBX) services, saw the erosion of investments focused on the general network serving all customers. In their place arose ad hoc private solutions for elite customers that used the existing public network with little responsibility to fund its maintenance. Worse, as new electronic networks attached themselves to the public network, they each developed their own internal proprietary standards, thereby creating the fragmentation of electronic networks in the 1980s. And this interconnection to the public network by proprietary services was mandated by federal regulators over the objections of not only AT&T and later the Baby Bells (AT&T's local phone company offspring), but also over the opposition of most state regulators, who bemoaned the erosion of their ability to promote the upgrading of integrated regional networks.

In a sense, the Internet became the government-backed solution to a fragmented mess created by the federal government–backed interconnection mandates of the 1970s. But while a technically innovative solution, it was an economically regressive implementation that further eroded regional coherency around public goods and the economic planning that would assure equality of access. To leverage the success of the Internet, the federal government turned over the national backbones it had created to competing private companies while subsidizing local Internet providers through nearly free access to local phone infrastructure. Hemorrhaging costs in their core business, local phone companies turned to their own high-end services for profit and in the 1990s went on a merger and acquisition spree that essentially obliterated any real role as regional economic anchors. The 1996 Telecommunications Act would just ratify the competitive pressures that were driving costs up for lower-end users of phone and cable services while delivering discounts for global services to business and upper-income customers.

The irony is that even as federal regulatory mandates have helped liquidate regions as functioning economic units, we still see the same conservatives who have pushed those mandates arguing disingenuously for the "decentralization" of government economic responsibilities to local government. The problem is that the same global economic forces that have accelerated economic commerce over the Internet and through other venues have undermined the tax revenues and political capabilities of local governments that are dependent on local commerce.

Prostrate Local Governments in the Age of Internet Commerce

Local economic development once created a virtuous cycle of local planning and investment leading to local jobs where residents would shop at local stores and generate revenue for further economic development, both privately and publicly. As will be explored in Chapter 6, the technologies of cyberspace are making it increasingly easy for consumers to shop for goods far from their home, thereby depriving local governments of the tax revenue they could once count on. As out-of-state Internet-based commerce continues to expand, local and state governments lose billions of dollars in revenue. The federal government has done nothing either to assist local governments in collecting lost sales tax revenue or to help replace the revenue with other forms of taxation. If anything, they have made the problem worse through heaping new spending responsibilities on local government and, through measures such as the Wyden-Cox bill, further restricting their revenue options.

As commerce has been globalized by the new technology, within regions the economic and political ties between rich and poor communities have been frayed even further. In a sense, this process started with the globalization of financial networks, which, as we have noted, helped dissolve regional pools of capital. With global speculation on the rise, housing prices and thereby property taxes accelerated out of control in the 1970s; richer property owners responded by pushing through Proposition 13 in California and similar measures throughout the country. Combined with escalating divisions between rich suburbs and urban centers, the ability of regions to coordinate economic development has been tightly constricted. In this context, regional corporate strategies allied purely to suburban and upper-income customers make sense in the context of local governments without the revenue to mount broad-based economic development initiatives.

Dependence on sales tax revenue has biased many local governments against promoting industry in favor of retail outlets while encouraging governments to offer tax breaks to companies to relocate—a further drain on local revenues in a destructive zero-sum competition for revenue. Internet commerce undermines that sales tax–dependent strategy, leaving local communities scrambling for revenue against one another in ways that often further undermine intelligent regional planning.

However, in the context of the elite regional business politics more broadly detailed in Chapters 3 and 4 of this book, the scramble for tax dollars may only be a sideshow in the transformation of local politics in the age of the Internet. Integrated public utility networks and cross-class growth coalitions had defined the social space in which Progressive reformers in the early part of this century had built modern civil services that served regional economic management goals. However, as industry has built its "gated community," and elite economic networks selectively connect rich suburbs and professional urban enclaves across the globe with the most advanced technology, poorer communities and urban sections have been left with little more than virtual dirt roads. We are seeing new ideologies of privatization and corporate servicing by local governments that end up doing little or nothing for the general population. Instead, cities and towns are pitted against one another in an endless competition to spend what little resources they have serving those with the most capital, while eroding democracy to make government services one more set of amenities that corporations choose from in conducting branch site selections.

At the most basic level, the invisible regional geography of communication serves to polarize already existing economic and racial divides as cities rush to support business with public networking goods. Technology investments in schools end up overwhelmingly in the hands of more privileged communities as business finds concentrated support for schools in their suburban enclaves a more cost-efficient approach than general revenues for all schools.

And just as networking has eroded firm barriers separating firm from firm, the Internet is helping to blur the lines between government and business. Global firms scoop government contract bids off the Net as local services become merely part of the business plan of multinational corporations. Conversely, government services respond ever more precisely to the demands of those businesses operating in the region, whether in expediting construction permits electronically or aiding in the wholesale marketing of government data for the benefit of firms doing business in the area. And in those areas such as job searches where electronic information could serve the public, the lack of access to information on the private sector (partly the result of cutbacks in government funding) means that the Net fails for lack of political will. At best we see local governments seeking to extract small economic concessions for the wholesale benefits they deliver in their desperate recruitment of business.

Grassroots Organizing

As blackouts have rumbled through Northern California, and multinationals look to outsource jobs throughout the world, it is ultimately unclear whether industrial districts like Silicon Valley can sustain themselves over the long term. Where such districts have lasted historically, as in the Third Italy or other regions, it has required a much more broad-based politics to restrain the excesses of opportunism and speculative markets. The regimes that have managed to escape this divergence of public and private interests have focused on the mobilization of non-elite sectors to keep the state balanced between capital and civil society. Citing both Austria and the Indian state of Kerala, Peter Evans has noted that such regimes have strong mass-based political parties with a vibrant civil society to keep the focus of state action on the broad interests of all citizens in society. Without that trust engendered by broad community participation and sanction against exploitation, the advantages of industrial districts dissolve in the face of opportunism to "overgraze" the commons—a collapse of innovation and trust as multinational capital captures the best resources of the region for internal global strategies.

However, such a result is not inevitable. While corporations use resources from outside a region to increase their bargaining power and undermine previous local alliances for general growth and prosperity, grassroots organizations such as unions and community groups can also build their capacity to bring in outside resources and alliances in support of their interests. Strike funds supporting local labor actions and generalized boycotts promoted beyond the region are historical examples of the ability to counteract the power of multinationals. General political mobilization at central government and even global institutional levels is another option to redress the balance of power at local levels, as well. While this ability of grassroots organizations has been undermined in recent decades by the power and new flexibility of the multinationals in using new technology, it is an open question whether this is a permanent condition or merely a failure by the grassroots to readjust to a more global strategy as quickly as corporate capital has been able to do.

Already, as will be shown in the concluding chapter, there is a countertrend in which grassroots organizations use the new networked technology, particularly the Internet, to strengthen their power at national and international levels. Even as the Bay Area has been a center for promoting networked technology for global corporations, its parallel history as a source of radical civic energy has manifested itself in the region's support for global electronic networking. Small urban community groups are finding ways to focus national attention on local problems, while unions nationally are increasingly using the resources of the Internet to build solidarity and strengthen their organizing drives.

Most dramatically, with the experience of "deregulation" undercutting local economies and political power, grassroots activists, ranging from union locals to environmental groups, have become more focused on the dangers posed by international trade deals in lessening grassroots power in favor of global corporate power. With local power undercut by global corporations, more and more organizations are using the new technology to build the global alliances that, perversely, are seen as the only way to preserve local sovereignty.

The reality of this new grassroots power exploded on the global consciousness in the 1999 "Battle in Seattle" that helped derail a new round of World Trade Organization trade talks, although this was in fact the culmination of years of smaller mobilizations that the Internet had helped facilitate. Traditional defenders of corporate-led globalization bemoaned this result and, after years of triumphalism on behalf of neoliberalism, began to admit that alternatives were possible as new democratic forces mastered the politics of the new technology. As the *Economist,* one of the bellwethers of international coverage, expressed its new worries, "Globalisation is not just the inevitable result of technological change; it is also driven by trade liberalisation, which could be reversed."[72]

The growth of the Internet embodies a broad change in the urban space where control of time and communication, especially to resources and people outside one's region, is becoming the ongoing and critical issue in local power, making semipermanent local growth coalitions a thing of the past. Instead, while internal regional alliances will remain crucial, local power is inevitably flowing to those who can use the new information technology to deploy global power for local control of resources.

As people begin to consider alternative national and global policies that could return power to less elite forces within regions, it will become all the more critical to understand the policies that have shaped the economic implementation of Internet technology. Understanding the contingent decisions that were made in the past will make possible consideration of new alternative technological and economic regimes for the future.

2

How the Federal
Government Created the
Internet, and How
the Internet Is
Threatened by the
Government's Withdrawal

In a remarkable turn of societal imagination, many conservatives have begun picturing the computer age as the rejuvenation of small-scale entrepreneurial capitalism against the institutions of the nation-state. Alvin Toffler has talked about "demassification"; George Gilder has cited the "quantum revolution"; former House Speaker Newt Gingrich has promoted decentralization of government to local regions. A steady stream of conservative analysts have argued that new technology has made government's role, especially the federal government's role, irrelevant and even dangerous to the healthy functioning of the economy. Even the *Economist,* a magazine with an early enthusiasm for the Internet and usually a somewhat more balanced eye, has described the success of the Internet as the "triumph of the free market over central planning. Democracy over dictatorship."[1] The new conservative view: the private sector is the font of technological and economic innovation; moreover, the federal government should get out of the way and leave economic

41

development to the private sector, working occasionally with local governments promoting innovation and job creation locally.

Repressed in this bit of economic mythmaking is the key role that the federal government played in each step of the growth of the computer industry and in the birth and formation of the Internet. Further, this mythmaking ignores the fact that left to private industry, much of the computer technology would never have come to market and, in the case of the Internet, the result would have been less innovative and less of an economic engine for growth. In fact, it is unclear whether the integrated communication and information exchange that is the hallmark of the federally created Internet would have even come forth from the private visions and competition of industry. In the 2000 election, Al Gore was lambasted by political opponents and the media for having suggested that he, as a mere politician, might have played a role in the creation of the Internet. This political pummeling occurred despite the fact that key Internet technology innovators noted his crucial role as a political supporter of funding for the technology over the years. However, the mythmaking surrounding the Internet was so strong that Gore's justified claims (far more justified than most political claims) were made to seem bizarre.[2]

In this exploration of the nature of regional economies and the impotence of local government in the face of private industry, it is important to highlight the real power and need for national government involvement to ensure not just equity but also economic rationality. And the initial creation of the Internet highlights this contrast in a dramatic way.

The Internet is in many ways the product of central planning in its rawest form: planning over decades, large government subsidies directed from a national headquarters, and experts designing and overseeing the project's development. The government not only created whole new technologies to make the Internet a possibility; it created the standards for forms of economic exchange of information that had never before been possible.

The comparison has been made at times to the interstate highway system. However, the analogy would hold only if employees of the federal government had first imagined the possibility of cars, subsidized the invention of the auto industry, funded the technology of concrete and tar, and built the whole initial system. Karl Polanyi has argued that the underpinnings of the rise of industrialization required extensive government involvement—from labor market regulation to creation of international currencies—just to make the launch of industrialization possible. In a similar fashion, it is clear that a Promethean role was required for government to make information networking an economic reality.

The government achieved a technical success where private industry would never have sustained the R&D over the decades required. Where private industry was working to create proprietary standards that would have stunted innovation and blocked shared networks, the government sustained and expanded open standards that encouraged the maximum sharing of resources. This allowed other public institutions such as universities to develop a stream of free, quickly distributed software that enhanced the network and allowed continual and rapid innovation.

Beyond the technical success of the Internet, its development illustrated how federal planning and investment was a key to the success of the private economic explosion of activity surrounding the Internet. The government not only developed the hardware and software required, but also trained most of the initial engineers and entrepreneurs who would launch businesses around the Internet. Almost every major company's electronic networking endeavors have their roots in former government employees, and many were direct spin-offs from the government itself, in the form of either agencies or federal contractors. It has been precisely "big government" involvement in technology that has worked in symbiosis with entrepreneurial energy to create much of the technological and economic innovation of the last decades. By creating public domain systems and providing public domain technology, the federal government has helped avoid a world of purely proprietary technology dominated by monopolists that would have left little room for the entry of small start-up companies or their associated innovation.

It is worth remembering that even as the Internet is being taken for granted, a very different vision of the "Information Superhighway" was being promoted by private industry early on. It was a vision of proprietarily owned information networks with such companies as America Online, or AOL, and the Microsoft Network, or MSN, negotiating with cable companies and telephone companies to deliver exclusive and incompatible information services to the consumer. The headlines in 1993 were not about the Internet and software companies like Netscape but rather about mergers and financial deals between those who controlled the cables to the home. All assumed that those who monopolized control of the physical hardware connecting homes and businesses would reap monopoly profits in selling information services.

Fortune magazine described the ultimately unsuccessful merger of the cable company Tele-Communication Incorporated (TCI) and the telephone company Bell Atlantic this way in 1993: "It was the bold stroke of two captains of industry bent on securing their share of whatever booty washes ashore when the interactive age finally arrives. . . . When the dust settles, there will probably be eight to ten major operators on the highway, some earning their way mainly by collecting tolls for the use of their networks."[3] At the national policy level, the priority lobbying of companies was ensuring that a variety of billing options be built into the hardware of what was then designated the National Information Infrastructure (NII) to guarantee proper measurement and payment of services.[4]

In many ways, this private vision harked back not to the original federal highway system but to the first transit system that crisscrossed the nation's land—the railroads. And in fact, that historical legacy gives some sense of what a privately designed (if not always privately funded) system would have looked like. In the 1840s and 1850s, the first large railways were built, usually with incompatible track widths where trains entering the same city could not switch directly to another company's rail track. This was not accidental but a deliberate strategy by merchants sponsoring one railroad to avoid having another company (usually sponsored by merchants in a rival city) siphon off freight. It would take decades before the gauges of different train companies were all standardized and freight could be easily transferred from line to line for longer distances. Even as such standardization was achieved by the 1880s, giant railroad companies sought to create competing railway systems with power over enough territory to control the flow and pricing of significant portions of freight against competing systems, becoming the first major oligopolies in the U.S. economy.[5]

This vision of toll roads laboriously tying together incompatible proprietary systems was the dominant vision of the Information Superhighway, with the Internet often being dismissed as a toy for academics. The vision of monopoly carriers was so strong that rival companies called for a Justice Department antitrust investigation when Microsoft launched its own online Microsoft Network (MSN) as part of its new Windows 95 operating system. The fear was that as the gatekeeper of the operating system, it would establish an unfair advantage over other proprietary systems and gain exclusive control of the distribution of content to many if not most consumers. However, the proprietary MSN system introduced in August 1995 was summarily dumped four months later in favor of making all content and access Internet compatible. Ironically, when the federal government brought its antitrust suit against Microsoft in 1998, Microsoft's Internet access division was one of the company's few endeavors that the government thought was largely irrelevant to its antitrust concerns.

What will be detailed in this chapter is how and why those open Internet standards had triumphed over the corporate vision of competing standards.

Steven Levy, columnist on technology and a longtime chronicler of the free-spirited "hacker" ethic, argued that "the Internet can never be merely another profit center in their dreams of empire. Their power is based on monopoly, on controlling distribution. But the Net is built to smash monopolies."[6] While Levy was a bit too optimistic, it is clear that the technical foresight and the economic engagement of the government Internet planners had done what Microsoft's rivals had not been able to do and created a system where smaller entrepreneurial companies had a chance against monopoly-oriented players.

The result of that government support has been an explosive boom in Internet-related industry. By reducing the cost of entry, the Internet has been crucial in ensuring that deep-pocket corporations could not lock up whole areas in proprietary systems. Where phone systems, cellular systems, satellites, and cable TV remain a competing mass of often incompatible systems, the Internet has forged an integrated system of data communication so compelling that every other technology is rushing to integrate itself into the Internet in order to take advantage of the free-flowing commerce exploding over its networks.

However, by the mid-1990s, even as the broader public began enjoying the fruits of the Internet, nurtured for decades by the government, these new commercial companies that were spawned by the new technology became a focus of resistance to a continuation of the government role in assuring standards and access. New software companies such as Netscape—staffed by individuals who, like Netscape's Marc Andreessen, had been funded originally by the government to create browser software—would fight to take control of Internet standards, while the many new ISPs would lobby hard for continual privatization of Internet backbones and regional access systems. In a few short years, in the first half of the 1990s, both the governance structure and integrated backbone system created by the federal government over decades was privatized.

The danger is that with public and private investment in long-term basic research falling, the burst of Internet commerce engineered by decades of public investment may be consuming these fruits of past investment while undermining support for future government investment and coordination for the next generation of innovation. Creaming profits from decades of public investment, most of these new companies have resisted any suggestion of obligation to the broader public. Ironically, most of these same companies that had complained about government's role would within a few years see a new danger in the merger and monopolization of the new technology. By 1997, Netscape and Sun Microsystems, among other companies, would be calling for antitrust investigation of Microsoft for its anticompetitive control of Internet standards, while ISPs and others would note the danger of new mergers placing control of potentially more than 50 percent of Internet traffic in the hands of one telecommunications company, WorldCom, as it swallowed up MCI. In a stunningly short time, the industry moved from one in which the government nurtured a hothouse of entrepreneurial development and innovation to one in which giant companies vied for monopoly control of standards and technology.

In this chapter I will outline this trajectory of federal planning and engagement, entrepreneurial explosion, government withdrawal, and the subsequent merger and attempted monopolization of standards and technology. In Chapters 3 and 4, I will explore how the regional dynamics of Silicon Valley, themselves largely supported by the federal government, would be the major counterweight to the proprietary monopolization appearing in the wake of the government's withdrawal from Internet governance.

The Origin of the Internet and the Technical Triumph of the Government's Role

The Internet came out of a whole milieu of government funding for technology born largely during World War II and the Cold War. The Internet would succeed because it advanced these key points in government policy:

- Long-range planning and investment for future technical and economic needs
- Promoting a professional network to guide the process that crossed from government to university to business research labs that would act in the broad public interest
- Focusing on open standards that were as inclusive of multiple technologies as possible
- Taking advantage of public space and volunteer energy, especially from universities, to create a stream of free, quickly shared innovations
- Building a critical mass of participants to make the network viable for a broad audience

Investment and Planning

Markets for new technology exist on the margin: on the margin of the income individuals have to spend, on the margin of the costs companies have to develop that technology for the marketplace, and usually on the margins of any interest by private industry. Innovation in production that has immediate cost-saving or quality-increasing results is the more typical focus for corporate R&D. For this reason, most of the high technology we take for granted today was not built by private-sector initiatives, but is thanks to government planners who could fund technology for the ten, fifteen, even twenty-five years—in the case of the Internet—needed to make the innovation cost-effective and standardized enough for the private market.

The modern computer industry was born out of the government-driven research of that earlier era. While private industry in recent years has pushed forward the almost inexorable process of making computers cheaper and faster, computers still largely follow the design blueprint created in the war and postwar period under government funding and planning. In Britain, researcher Alan Turing used early computer prototypes to break the German Enigma code and would subsequently lay out much of the theoretical foundation for artificial intelligence research. In the United States, the multitalented John Von Neumann would head government research, including on the Manhattan Project, that set computer design in the postwar period, so much so that most of today's computers are referred to in the scientific literature as "Von Neumann machines." Claude Shannon, an early theoretician of the use of Boolean logic in computer circuit design, would lead work on defense-funded projects at Bell Laboratories that were critical to computer design, until 1947, when he became a professor at the Massachusetts Institute of Technology (MIT), where he would carry out further work on information theory in computers. Another MIT colleague, Norbert Wiener, would outline the theory and challenges of what he first labeled cybernetics (he later rejected practical computer work, fearing its use in weapons of mass destruction).[7]

These individuals and their research were bound together by a haphazard set of government projects and programs put together in the frenzy of World War II and the early Cold War. It was in the 1950s, particularly under the psychological impact of Russia's Sputnik success, that the U.S. government sought to regularize its technological research for maximum success, both technologically and economically. Given the biases in the United States against government intervention, it seems inevitable that the engine for industrial policy would be defense related. For the same reason, even federal intervention in highway construction and public education begun in the 1950s would officially be done in the name of defense.

However, President Dwight D. Eisenhower's personal experience made him distrustful of the bureaucratic interests in the Pentagon, which led him in the late 1950s to support the creation of new institutions largely independent of the specific military branches. One example was NASA, which ended up with much of the day-to-day applied research of the military at the time, while a new agency called the Advanced Research Projects Agency (ARPA) was created to help coordinate overall R&D spending by the military. The National Science Foundation was created outside the Defense Department in the same period to help fund nonmilitary research, although it would always have a strong relationship with the science-based military agencies.

A key appointment at ARPA came in 1962 when psychologist J. C. R. Licklider was hired to head a behavior sciences office there, an office that would evolve under Licklider's two-year directorship into the Information Processing Techniques Office (IPTO), which would direct the original creation of the Internet. Licklider had become interested as a researcher at MIT in how people and computers could interact to augment human activity and had been pulled into the evolving discipline of designing computer interfaces. He had become distressed by the amount of time he was spending just obtaining the knowledge from different places that he needed to even begin thinking about a problem. His main research originally was on how the brain processes sound and perceives pitch, but the preparation of the required graphs for his work overwhelmed all his time. And he found the use of computers utterly alien to his needs as a thinker and researcher; as an alternative he proposed a radical design leading away from data processing and toward the support of human thought.

His pre-ARPA career became embedded in the military-scientific network of the day. As a researcher at MIT, he worked at Lincoln Labs, where much of the university's military research was done. Licklider was involved in the Whirlwind program, in which computers were designed for the Defense Department's ground-based antimissile program (known as SAGE, for Semi-Automatic Ground Environment). Whirlwind used the first computers from which individuals could get information through visual displays rather than through punch cards; Licklider became obsessed with a vision of "interactive computing" as an alternative to both the typical automation use of computers and to the idea of artificial intelligence where computers would think like people. He envisioned a new kind of computer-based library that could manage information more effectively to improve human thought, working off an earlier vision by Vannevar Bush, who a decade earlier had promoted the idea of using microfiche in a similar manner. In 1960, Licklider published a paper titled "Man-Computer Symbiosis" in which he envisioned using computers to assist people in processing data in ways never before imagined. To accomplish this, computers would have to communicate better with people, but would also have to allow people to communicate with one another more effectively. Around this same period, he began working at Bolt Beranek and Newman (BBN), a consulting firm largely staffed by MIT graduates and brilliant drop-outs and that was often referred to as the "third university" in Cambridge. BBN had recently obtained one of the first minicomputers, the PDP-1, made by the Digital Equipment Corporation (DEC), which gave Licklider even more time to learn different ways to interact directly with a computer.[8]

When ARPA director Jack Ruina wanted to computerize military command functions at all levels, Licklider's experience gained from being both a psychologist and an expert on computer interfaces made him an obvious candidate for the job. So in October 1962, Licklider became head of the IPTO office at ARPA. As director, he had as his principle focus expanding the possibilities of his vision of interactive computing, from computer graphics to improved computer languages to computer time-sharing—the use of the same computer by multiple users simultaneously. With a yearly budget of $14 million, Licklider's office had more money for computer funding than all other government agencies combined. It was now in the position to support computer labs around the country, often with thirty to forty times the budget to which they had previously been accustomed.

ARPA would fund six of the first twelve time-sharing computer systems in the country which in turn would help spark the advent of the minicomputer industry in the 1960s—crucial to the industry and the Boston-area regional economy then, but as crucial to the development of the Internet over the next decades. A key creator of the minicomputer, DEC, had been founded by a team of MIT researchers from Lincoln Labs and would maintain a close relationship with the university. MIT researchers would continue to work closely with DEC, and the ARPA-funded Project MAC at MIT—given $2 million starting in 1963, with larger increases in following years—would design most of DEC's breakthrough time-sharing minicomputer, the PDP-6, which would be a crucial new tool for networking efforts across the country.

In 1964 Licklider would recommend MIT graduate student Ivan Sutherland, a pioneer in computer graphics, as his successor, who in turn would hire as his assistant (and in 1966, successor) a NASA employee, Bob Taylor. Taylor would continue Licklider's vision by pushing forward major funding for Stanford Research Institute to create an "Augmentation Research Center" run by Doug Engelbart, a man whose thinking had paralleled Licklider's in his conception of interactive computing. (In Chapter 3 I will discuss Engelbart's center and its role in the Silicon Valley technology explosion much more extensively.) Taylor would fund the first full-fledged graduate degree programs in computer science at Stanford, Carnegie-Mellon, and MIT and help fund the first computer graphics degree program at the University of Utah.

As head of IPTO, Taylor was frustrated that even as he had the power of several different computer systems in his office, all had different computer terminals and could not communicate with one another. Out of this frustration came Taylor's dedication to funding a multicampus computer-networking project that would evolve into the Internet. He saw this as a top priority, especially since researchers around the country often needed to share expensive computer resources at locations far from their primary research location. In 1968, Taylor and Licklider co-published their vision for the Internet in a paper called "The Computer as a Communications Device."

At the same time in the early 1960s, researcher Paul Baran had begun planning how to build the technology necessary for the goal of networking computers. Baran worked at RAND Corporation, a company set up to monitor and preserve the United States' operations research capability; of concern to him was the survivability of U.S. communication networks in the case of nuclear war. Modeling his ideas partially on the redundancy of neural networks in the brain, Baran envisioned the movement from analog signals to digital signals that could perform in a comparable networked system of digital transmission. Instead of possessing a central switching node in which a wire between two points would be reserved specifically for sound signals for a conversation, such a system would be a "distributed network" with each node connected to its nearest neighbors in a string of connections, much like the child's game of telephone. More dramatically, messages would be broken down into parts, travel the network, and be reassembled at the opposite end in a system called *packet switching*. This would allow fuller use of all lines in the network instead of lines being held open from end to end for each message. Each node would keep track of the fastest route to each destination on the network (and be constantly updated with information from adjoining nodes) and help route information without a need for central direction.

RAND was enthusiastic about Baran's idea, but when AT&T was approached about its feasibility, AT&T executives dismissed the concept and even refused to share information on their long-distance circuit maps; Baran had to purloin a copy to evaluate the ideas which he and RAND were convinced were right. Based on RAND's recommendation, the U.S. Air Force asked AT&T directly to build such a network, but AT&T still refused, saying that it wouldn't work (although a faction of scientists at Bell Labs did support the idea). This may have been technical myopia on the part of these business-oriented executives, but it was an economically self-interested myopia, a result at least partly of the political box in which the federal government itself had put AT&T. Such a distributed network threatened (and today does threaten) the central economic assets of the telephone industry: central computers and central switches. On one level, AT&T's resistance highlights the fact that corporate research labs, providing the main alternative to long-term government funding of technology, rarely if ever invest in fundamental technology that will likely undermine the economic dominance they currently enjoy. Compounding the problem for AT&T was the fact that the first winds of the deregulatory attack on the company were blowing and, specifically, the company was increasingly barred from selling anything having to do with computers that were attached to the phone network. As will be detailed in Chapter 5, this emerging deregulation and its political division between the phone network and the increasing computerization of telecommunications would have perverse consequences both for technology and for regional economies.

The air force contemplated building the network by itself but it was bogged down in internal organizational problems and never got the project off the ground. (This failed project is the source of the myth that the Internet evolved to help the military respond to a nuclear strike.) In the meantime, British physicist Donald Davies had begun promoting a similar idea of a computer network containing "packets" of information. He soon learned of Baran's ideas and was encouraged enough to push for support from the British Post Office, which ran the telephone system in Britain. Without the political issues of deregulation, the state-run phone system in Britain just treated the undertaking as a simple demonstration project. In 1968, the first computer-distributed network was established, on computers that were all located in London, at the National Physical Laboratory, where Davies worked.

Taking off on Davies's example, ARPA began its own networking project. Larry Roberts, from Lincoln Labs, was hired to oversee the computer-networking project. There was a certain hostility from many East Coast universities to sharing scarce computing resources; as a result, the network started out at four West Coast sites: the University of California, Los Angeles (UCLA); the University of California, Santa Barbara; the University of Utah; and the Stanford Research Institute's Augmentation Research Center. The plan was to install a new computer at each site as part of the network: this would avoid direct incompatibilities in the network and allow each campus to focus on a separate interface between its regular campus computers and the local network computer.

In 1968, ARPA advertised a bid for building the subnet computers (which would be called Interface Message Processors, or IMPs). IBM and other big computer companies declined even to make a bid, saying that it was not possible at a reasonable price. As in the case of AT&T, this reaction partially resulted from the myopia of those grounded in older technology, but it was also caused by a self-interested economic fear of the new minicomputer technology, supported by the federal government, that was challenging the dominance of companies such as IBM. IBM and others rightly feared that networking would make many government agencies and businesses rethink the need to actually own their own mainframe computer.

While the defense contractor Raytheon almost won the contract, in the end Licklider's old consulting firm, BBN, convinced ARPA that its relatively small operation (six hundred employees) could do the best job. With its ties to MIT, including a workforce made up largely of MIT graduate students, BBN had a good case for implementing technology largely developed at that university. With a large ($1 million) contract and the guaranteed market the government was providing, it was able to take on a project it could never have carried out on its own.

By October 1969, the network connection between UCLA and Stanford was established, and within months, all four "nodes" plus BBN itself were online. By the time the network was demonstrated publicly for the first time at the International Conference on Computer Communications in October 1972, there were twenty-nine nodes in the network (dubbed at this point ARPA-NET) clustered largely in four areas: Boston; Washington, D.C.; Los Angeles; and San Francisco. The entity that would evolve into the Internet had been born.

Professional Network of Experts

Beyond long-term planning, a key to ensuring that the Internet would expand in a dynamic way was the creation of a network of public-minded experts who helped guide the expansion of the network. Despite odes to the unbridled "anarchy" of the Internet, this was a closely supervised anarchy directed to the specifications of government yet marshaling the broad professional, volunteer, and eventually commercial resources of the emerging computer elite. In many ways, the very skill of the government in marshaling those resources with a light hand is a source of the sometimes rhetorical amnesia over its role. The smoothness of the Internet's creation and the building of a broad consensus over its shape created so much legitimacy for its design that it was seen less as a creation of "the government"—in other words, "them"—than as a broad creation of society as a whole.

The development of the Internet is in this way a perfect illustration of the ideal of "embedded autonomy" described by Peter Evans, in which a Weberian bureaucracy at ARPA and other federal agencies operated under broad professional norms and where "individual maximization must take place via conformity to bureaucratic rules rather than via exploitation of individual opportunities presented by the invisible hand."[9] The effectiveness of the ARPA bureaucracy was maintained through support being marshaled from a broad range of other government agencies, from universities, and from the private sector, which eventually saw an interest in a broad networking environment for online commerce.

Licklider had actually started this professional network at ARPA in the early 1960s when he reached beyond traditional experts at federal agencies and national labs to gather an association of experts interested in communication technology. He oriented ARPA to the goal of establishing contacts with university researchers around the country, establishing what he presciently called the Intergalactic Computer Network, which helped connect researchers interested in computer networking.

When ARPANET was created, UCLA was funded to establish a Network Measurement Center to oversee the evolution of the network. Forty graduate students at UCLA, many of whom would become leaders in both the public and corporate Internet world, helped run the center and coordinate with other researchers in developing the standards for running the ARPANET. The new technology itself helped involve a nationwide group of researchers and graduate students in these deliberations to help mold the evolution of the Internet. This national body became the Network Working Group (NWG), which was expanded after the 1972 "debut" conference to become part of an International Network Working Group, set up to promote international computer networking. Management of Internet "addresses," critical for the decentralized packet-switching network, would be housed at the Stanford Research Institute (SRI) in an institution called the InterNIC.

ARPA would replace the NWG by a more formal Internet Configuration Control Board (ICCB) in 1979 to extend the participation in the design of the Internet to a wider range of members of the research community. This was especially important as the ARPANET expanded to include a range of other government agencies and bodies and evolved into the diversity of the emerging Internet community. The ICCB was later replaced by the Internet Activities Board (IAB), which used a set of ten task forces to include a wide range of experts in the evolution of the Internet. As the Internet was privatized in the early 1990s, the private sector (led in many cases by former researchers for ARPA and its Internet-related funded projects) created the Internet Society in 1992, and the IAB reconstituted itself as the Internet Architecture Board and joined the Internet Society.[10]

At each step of the Internet's development, ARPA and associated government agencies expanded participation to an ever widening set of experts and technological leaders who, in turn, would encourage others in their academic, scientific, community, or business realm to support the effective development of the Internet. Further, the continual movement of personnel back and forth from academic, government, and (eventually) business positions created a cross-fertilization of ideas and a loyalty to the emerging network over any particular organizational loyalty. What is left open to question (as will be described) is what the increasing privatization of the Internet's governance means for maintaining this same healthy "embedded autonomy" in furthering a public-oriented Internet.

Creation of Standards

Beyond funding the initial computers and wires that connected the initial sites on the Net, the federal government's guidance ensured the creation of a shared set of standards for communication. It not only created its own initial ARPA-NET, but also set standards that could integrate all sorts of computer networks together into what would become the Internet. This was crucial to the explosion of innovation and the economic value of the Internet over time.

The reason for this is embodied in what has been called Metcalfe's Law (named after Bob Metcalfe, an ARPA-funded researcher and the founder of the networking company 3Com). This law argues that the value of a network does not increase linearly with additional computers added to the network but instead can be measured as the number of users squared. This means that the simple act of integrating different networks which were previously incompatible is a recipe for explosive increases in innovation and economic value, since networks of networks are exponentially more valuable than the sum of their parts.

This is the reason why proprietary information systems such as America Online and the Microsoft Network lost out to the Internet. While those services were incrementally increasing the value of their systems through enhanced "content," the Internet was exponentially exploding in value as each individual computer network was integrated into the broader network of networks. And the key was that only the government had an interest in creating a nonproprietary system from which no monopoly rents could be collected. In a sense, the Internet is nothing more than those standards: any computer over any wire can conceivably be hooked up to the Internet as long as that computer knows how to communicate in the Internet's language of information exchange.

ARPA oversaw the first standard network protocol in 1971 to allow a person at one computer to log onto other computers on the network as if he or she were a local user. This soon evolved into the standard Transmission Control Protocol (TCP), which was complemented in 1972 by the file transfer protocol (FTP), which allowed individual files to be exchanged between different computers. In 1976, DARPA (with "Defense" recently added to its name) hired Vinton Cerf, who as a UCLA graduate student had participated in the launch of the ARPANET and was by then a Stanford professor, as a program manager to work with Bob Kahn, a former BBN manager on the ARPANET and now working at DARPA, to create a system for integrating the ARPANET into other computer networks. By 1977, they had demonstrated the Internet Protocol (IP), which could be used to integrate satellite, packet radio, and the ARPANET. From this point on, new networks of computers could be easily added to the network. In 1981, DARPA funded researchers at the University of California–Berkeley to include TCP/IP networking protocols into the university's popular public version of the UNIX operating system, thereby spreading the Internet standards to computers throughout the world. By the 1980s, the focus for standards setting was the Internet Engineering Task Force (IETF), part of the Internet Activities Board, which would bring together researchers, government officials, and emerging business researchers to maintain consistent standards on the Internet.

An example of the cross-fertilization of staff and ideas outside the government was the case of Bob Metcalfe and Ethernet, the latter an important tool for networking computers together to access the Internet. Bob Metcalfe had designed the interface that connected MIT to the ARPANET and, given MIT's sprawling computer network, had an early sense that the need to connect local computers together was as vital as the need to connect them more broadly. (Early on, most ARPA traffic at MIT was between MIT computers, not with other campuses.) Metcalfe was hired in the mid-1970s at Xerox Corporation's new Palo Alto Research Center (PARC), which was headed by Bob Taylor, the former IPTO head who had started the ARPANET project. (PARC itself will be discussed much more in Chapter 3). Metcalfe was doing DARPA-funded work while trying to figure out how to cheaply network PARC's experimental personal computers. Using inspiration from ARPA's project in using packet switching over radio, Metcalfe created the system called Ethernet to exchange information between computers in what would come to be called local area networks (LANs). Ethernet was crucial for the expansion of the Internet since local computers could be networked together and then connected to other networks using the TCP/IP protocol and local router computers. Xerox would start selling Ethernet as a commercial product in 1980 (and Metcalfe would found 3Com to sell networking technology), while PARC head Bob Taylor donated millions of dollars of Ethernet equipment to universities to help expand the use of networking on campuses.

In all these ways, DARPA helped shepherd open Internet standards into the 1980s and 1990s, when they would be employed to radically expand the network to a wide range of users. In the process, it was clear that the professional norms promoted by DARPA and the community of researchers were critical in keeping individual profit-taking from undermining those open standards. As one example, in 1973 then IPTO head Larry Roberts was hired by BBN to run a company subsidiary called TELENET that would maintain private packet switching networks. In coming to BBN, Roberts carefully deflected a bid by BBN to take over ARPANET privately. J. C. R. Licklider, who returned to ARPA from MIT to replace Roberts as head of IPTO, soon found himself in conflict with his old employer, BBN, who was refusing to publish the original computer code for IMP computer routers they had designed. Making matters worse, BBN was becoming more and more reluctant itself to fix software bugs faced by the system (no doubt preferring to concentrate on programming for their for-profit TELENET subsidiary). Licklider, in the name of the openness of the Net, threatened to hold up BBN's federal contract funds unless the firm released the code publicly. BBN did so, thereby enhancing the tradition of open codes in the development of standards.

At the Internet Architecture Board, one of the main task forces became the Internet Engineering Task Force, which by 1985 developed a series of smaller working groups that were the focus for Internet standards decision making. Ironically, as data networks spread in the 1980s, it was the government experts at DARPA and universities who backed the flexible, tested TCP/IP protocol, while big private companies such as IBM, MCI, and Hewlett-Packard adopted an untested, bureaucratically inspired standard created in international committees for their private networking efforts. Vint Cerf, who had been hired at MCI to build that company's message networking system, remembers, "So I had to build MCI Mail out of a dog's breakfast of protocols."[11] It was only with the technical dominance of the Internet that most private industry would convert over to the public TCP/IP standard.

Using Public Space and Volunteer Energy to Create Free, Quickly Shared Innovations

A crucial element of the success of the Internet was the fact that the public space of the network harnessed the energy of universities, from both paid staff and volunteers, to provide a continuous stream of free software to improve its functionality. The Net itself allowed any new innovation to nearly instantaneously ricochet across the nation, even the world, without the friction of the costs of either distribution or purchase. This "gift" economy allowed new innovations to be quickly tested and to gain a critical mass of users for functions not even envisioned by the creators of the system.

The origin of this ethic of shared software, called the "hacker ethic" before the term became associated with electronic vandalism, was largely born at the ARPA-funded Project MAC at MIT. Designing new software and sharing the unexpected results became part of the way of life of students, an ethic that would contribute to both the creation of the Internet and the launch of the personal computer revolution. Games such as Spacewar and Adventure were created and widely shared as hacker outposts appeared across the country in the 1960s and 1970s.

Probably the most pervasive example of this hacker ethic was the early use of the ARPANET for electronic mail. Not even planned as part of its design, e-mail was created by BBN engineer Ray Tomlinson in 1972 as a private "hack" to piggyback on the file transfer protocol. Under the tolerant supervision of ARPA, use of the Net for e-mail communication soon surpassed actual computing resource sharing. Stephen Lukasik, ARPA director from 1971 to 1975, saw the importance of e-mail for long-distance collaboration and himself soon began virtually directing ARPA from electronic mail and his twenty-pound Texas Instruments portable terminal that he brought with him everywhere. Partly because of Lukasik's own frustration in dealing with the stream of raw mail, IPTO director Larry Roberts himself wrote the code for the first mail manager software called READ, which was soon supplanted by the popular MSG which invented the reply function.

In 1975, the first e-mail list was established for discussing standards and etiquette in e-mail exchange and for promoting free speech in the sometimes uneasy relationship with a defense-funded network. Other e-mail lists, along with the broad USENET set of discussion bulletin boards, would soon follow.

The game of Adventure was ported (translated from one computer system to another) onto the Net in 1976 and became a huge attraction as people began downloading it from all over the country. Through that game and numerous others to follow, many people became hooked on computers, and these games helped establish the ethic of freely sharing software far outside the original hacker enclaves. Innovations in operating systems, software tools and games would all freely flow through the emerging Internet, a large number of them available for free and encouraging continual new improvements from users.

What brought the Internet into its own, though, was the Gopher software developed at the University of Minnesota in the early 1990s. Building on the existence of individual Internet sites from which files and programs could be retrieved after one logged into a particular computer over the network, Gopher was a piece of software that could be used to create personalized lists of files from computers all over the Net and allow computer users to retrieve any file chosen from the list. With this innovation, the Internet became one giant hard drive that could be organized and presented to a particular set of users in whatever way made the most logical or aesthetic sense. Gophers sprang up on computers run by governments, universities, community organizations, and businesses beginning to stake a place on the Net. In a visual way, the Internet's vast resources could be presented and reached through Minnesota's "All the Gopher Sites in the World" Gopher site. For most commercial users of America Online and other service providers, Gophers were the initial contact they were given with the world of the Internet, and it created a hunger for the content they knew existed outside the proprietary walls of those commercial providers.

The next step, and one that brought the Internet into almost daily headlines, was the World Wide Web. The Web was initially designed at the European Laboratory for Particle Physics (CERN) in Geneva, Switzerland, to allow information to be shared internally among the sprawl of the CERN's computers—what would be designated as an intranet today. However, people quickly saw it as a useful way of sharing information between computer systems much in the manner of Gopher software, with the additional advantage of "hypertext" connections to internal parts of documents. In order to encourage its spread, Tim Berners Lee, the originator of HTML and the HTTP protocols, made sure that its code was available with no license fee.

In 1993, computer science students funded at the National Center for Supercomputing Applications (NCSA) located at the University of Illinois, Urbana-

Champaign, created Mosaic, the first major Web browser, which added graphics to the traditional text display. With almost unnerving speed, graphic-based Web sites exploded across the Internet as users downloaded the Mosaic browser needed to view them.

As Netscape, Microsoft, and others created commercial versions of the Mosaic browser, they were generally forced to give the software away to individual customers in attempts to sell related servers and other software to corporate customers. This commercial form of the Internet "gift economy" raised the danger that private companies would use such giveaways to distort standards for corporate profits (an issue we will return to) rather than to build on others' work, but it is a testament to the durability of Net standards that these standards have put restraints even on the ability of companies as powerful as Microsoft to fully undermine the Internet.

Building a Critical Mass of Participants

Despite the initial planning and funding of the technology, the Internet would not have become a mass phenomenon in the United States had the government not guided its expansion to the point where, with a critical mass of participants creating a diversity of free content, it became a viable alternative to the proprietary corporate systems. It might have remained a dynamic but limited tool for a small handful of researchers and elite government agencies—essentially the situation that existed until the mid-1990s in Japan and Europe until the Internet took off widely in the United States. Instead, the federal government helped democratize the system to a broad enough base that it developed an expansionary dynamic of its own.

Whereas ARPA had been the key government agency in funding the creation of the network and guiding its technical development, the National Science Foundation (NSF) played the most critical role in bringing the Internet to the masses—or at least the academic masses. The NSF had been born in 1950 to help promote basic science research, with a focus on the academic world. As early as 1974, an NSF advisory committee promoted the idea of a computer network to "offer advanced communication, collaboration and the sharing of resources among geographically separated or isolated researchers."[12] Nothing came of this proposal, but by the late 1970s, computer science departments were mushrooming. The ARPANET was threatening to split the computer science community into communication haves and have-nots; only 15 of the 120 campuses with academic computer science departments had ARPANET connections by 1979. Having an ARPANET connection became a crucial factor in whether a school could lure top graduate students and faculty. There was also the fear that high industry salaries combined with poor campus facilities was threatening to drain academia and leave no one to train the next generation of computer scientists. This was one of the first examples of information-networking technology threatening to centralize resources, both intellectual and economic, in fewer geographic places rather than decentralizing them.

The proposed solution was the Computer Science Research Network (CSNET), founded by six campuses in 1979. At first, the campuses were planning a private system with no connection to the ARPANET, but the NSF pushed for a proposal that gave universities the options of full ARPANET connections, a more limited connection, or an e-mail only connection based on the resources available to each campus. The NSF agreed to manage the network for the first two years (passing it onto a university consortium after that) while putting $5 million into the CSNET project to get it off the ground. By June 1983, more than seventy campuses were online, and by 1986, nearly all the computer science departments in the country along with a number of private computer research sites, were online. The NSF then turned to funding other campuses without computer science departments to become part of a broader network. Approximately thirteen hundred of the roughly thirty-six four-year institutions in the United States were assisted at a cost of about $30 million. Even as the NSF has phased out much of its participation in the Internet, institutions of higher education, including technical and community colleges, could apply for twenty thousand dollars in connection assistance; however, this would only be a small portion of the total cost, since a school had to provide access to all qualified faculty and students (although this often came from other state and federal government funds universities could tap). But the grants acted as a good leverage to build broader access to the network.[13] What is surprising in many ways is how little (relatively) was spent on this assistance, especially when compared with the tens of millions spent on proprietary networking in the private sector during the same period, and how much bang for the buck resulted. Of course, the very nature of open standards meant that every expansion of the network would so increase its value that institutions would be eager to pay a portion of the costs themselves to be attached to this ever more valuable resource.

At the same time, NASA, the Department of Energy, and other agencies were funding several special-purpose networks that combined with the ARPANET and CSNET, officially became designated the Internet as the TCP/IP standard tied them together.[14] The same forces that drove adoption of TCP/IP in the United States helped bring their adoption internationally, thus encouraging the international Internet community.

The most visionary step taken by the NSF came in 1985 when it built a backbone data connection between five supercomputer centers recently established. The NSF agreed to build and pay for the backbone only on the condition that the regional centers build regional community networks to assure broad access to the backbone and the other regional networks. In 1987, the NSF awarded a five-year, $14-million grant to Merit Incorporated—a consortium of eight state-supported colleges and universities—to manage this backbone and link all regional networks to the supercomputing sites and the overall backbone.[15] This led to an explosion in growth from ten thousand Internet sites in 1987 to one hundred thousand hosts by 1989.[16]

The NSF not only funded the backbone but also helped fund the participation of universities not currently on the CSNET along with funding the start-up of the nonprofit community networks. With the free NSF backbone provided as a common good, these community networks expanded rapidly. Out of this initiative came the initial regional network connections to the Internet—NYSERNET in upstate New York, BARRNET in the Bay Area, CerfNet in San Diego, SURANet in the Southeast, NorthWestNet in the Pacific Northwest.[17] These regional access networks would be available not only to universities but also to participating nonprofits and businesses who wanted to communicate with the broad technical network on the Internet, although commercial traffic over the Net was barred by NSF rules.

The Merit consortium would subcontract work for the backbone network to MCI (whose data-networking unit had been set up by Vint Cerf, one of the creators of the IP standard) and to IBM. As interest in the Internet grew as the decade of the 1990s opened, the NSF authorized Merit, MCI, and IBM to create a new partnership, Advanced Network and Services, Incorporated (ANS) to manage the backbone and in 1991 allowed ANS to create a for-profit subsidiary to provide businesses a backbone service unfettered by the NSF ban on commercial users. The idea was that a portion of the funds paid by commercial customers would pay for upgrading the backbone and the regional networks carrying commercial traffic. By 1991, there were 350,000 computers connected to the Internet, encompassing five thousand separate networks in thirty-three countries.[18] The number of computers on the Internet would continue to explode exponentially.

At the same time, the NSF allowed a number of other private companies (themselves derived from other federally funded networks) such as UUNET and PSINet, to form the Commercial Internet Exchange (or CIX) to compete in connecting business customers to the regional Internet services. There were real problems created by this privatizing of the Internet, but the immediate result was a continual increase in the number of those using the Internet for data networking.

Economic Triumph of the Government and the Pitfalls of Privatization

In the first part of this chapter I have outlined the technical and social reasons for the triumph of the government-backed Internet. However, the importance of the government was in far more than just its creating the Internet itself as a public good that the private sector could then exploit. The relationship was much more integrated between government and the private sector, with the government being the direct source of much of the economic activity and even the creation of the companies that became the innovators in the new cybereconomy. Mitch Kapor, the founder of the software company Lotus, has argued, "Encouraged by its successor, the rapidly expanding government/academic National Science Foundation Network (NSFNet), the commercial internet . . . represents the natural development and expansion of a successful government enterprise."[19]

Conversely, as the government has privatized large parts of the Internet, the issue has been raised of whether a key source of technical and economic innovation is being undermined in the country. In addition, the dynamics of private commercial Internet companies show the very specific regional dynamics of economic development that, if not balanced by continual federal government involvement, will merely reinforce regional winners in the economy and accentuate the class divide within regions. The following two chapters will illustrate that regional dynamic in the development of the Internet, specifically in regard to the Bay Area. The rest of this chapter will emphasize the overall role the federal government has played in making the Internet-related industry possible and the pitfalls of that privatization.

The economic role of the federal government in building the Internet industry has been in three broad areas:

- Training people with the technology skills necessary to create and staff the new markets
- Helping to directly create most of the initial Internet companies, either by providing the technologies and staff, funding start-ups through government contracts, or encouraging direct spin-offs of government institutions.

• Creating the Internet as the overall framework for innovation and expansion of the commercial sector

Training

Any theory of the economy that assumes that market demand drives innovation faces a major theoretical problem: if the response demands new skills to create the product, what is the process by which those previously unneeded and unused skills in society come under the control of an individual company? Some incremental increase in skills can derive from experience within the firm and established skills are in some cases passed on through workplace-based apprenticeship programs, but for any skills requiring years of training or previously unheard-of skills, that training is inevitably derived from sources outside the firm. Without that exogenous source of skilled workers, companies would never have been able to respond to the market opportunities represented by new technologies such as the Internet.

Many of the earliest programmers in the nation came out of the Defense Department's 1950s Semi-Automatic Ground Environment (SAGE) program, which was designed to defend against manned bomber attacks. While its military usefulness has been questioned, the program was crucial in training the first generation of computer programmers before almost any academic computer science curricula even existed. Out of the SAGE program was created the Systems Development Corporation (SDC)—a spin-off from RAND to write the SAGE software—which seeded programmers across other companies. To give some sense of the imbalance in private versus public training, in 1954 all the computer manufacturers *combined* provided just twenty-five hundred student-weeks of programmer training, while three years later SDC was providing ten thousand student-weeks of computer training alone. By 1959 there were eight hundred SAGE programmers and by 1963, SDC had forty-three hundred employees and six thousand former employees were in industry. Xerox Data Systems is just one early example of a company that was heavily seeded with government-trained SDC alumni.[20]

In the case of the Internet, the creation of firms has been an endless parade of government-trained researchers staffing key new Internet companies, from ARPA's Larry Roberts being hired to set up BBN's TELENET data networking, to Vint Cerf running MCI's data-networking project, to Marc Andreessen from the National Center for Supercomputing Applications (NCSA) being hired to build Netscape. As companies have looked to ramp up new industries, government agencies and government-funded university research labs have been a constant source of employees for cutting edge industries.

The Government Origins of Early Internet Businesses

Just as most of the initial employees in early commercial Internet companies were federally trained, a tremendous number of the initial Internet companies began life as government contractors or literally as private spin-offs from government entities. This fact exists for a simple reason: until technology development is fully mature, private industry has little ability to afford the years of development. Technological maturity is usually dependent on several engineering iterations in which important insights become public and no single company can recoup investment on anything but the last iteration of research that finally commercializes the technology. Given the social benefit and public ownership of most of these developmental stages, no single company can usually recoup the costs of research development.[21]

And since large near-monopolistic companies are the only entities other than the government that can come even close to profitably recouping such research costs, the existence of government investment and contracts has been the main route enabling smaller start-ups to even have a chance against larger players. The government has supplied the technology and a guaranteed initial market for many such start-ups over the years, allowing them to become stable companies and able to compete with their bigger competitors. There has been a strong symbiotic relationship between the federal government and entrepreneurial companies seeking to enter the marketplace.

When AT&T, IBM, and other companies refused to build the envisioned data network back in the 1960s, it was left to ARPA to find a company willing to take on the task. BBN, largely staffed by alumni of government-funded labs at MIT, saw a government contract as a chance to become a bigger player in the data world. It would later use the technological expertise developed in work on that government contract to launch the private TELENET data network system which was eventually bought out by Sprint. Chapter 3 will contain a more detailed discussion of how a range of Silicon Valley firms such as Netscape and Sun Microsystems developed as a result of technology and staff funded by agencies such as DARPA, converting that last cycle of innovation into commercial success.

What is remarkable is that as the Internet was privatized, essentially all the major companies who would take over running the broad physical architecture of the Internet would have deep lineages as offspring of government initiatives.

Obviously, the core ANS backbone built by the Merit university consortium with MCI and IBM was initially funded in 1990 by the government and its for-profit subsidiary thrived in the commercial world as the official backbone of the Internet. In 1994, America Online purchased the ANS commercial subsidiary along with its customer base for $35 million. America Online was attempting to move from being a proprietary online service to being a premiere Internet service provider and had previously depended on other services for Internet access.[22] With ANS's national network for Internet access, America Online quickly surged past consumer access competitors such as CompuServe and Prodigy to enlist more than 10 million customers by 1997.

One reason that ANS was sold to America Online in 1994 was that MCI was using its involvement in the ANS project to develop its own national backbone Internet service. Having hired IP co-developer Vint Cerf to head its data-networking division, MCI would take over much of the traditional university and government Internet service as the NSF ended funding for the NSFNet backbone service. Most of the regional service providers set up by the NSF were using MCI by 1994, with almost 40 percent of Internet traffic being carried over MCI wires.[23] Even with the end of the NSFNet, MCI's relationship with the government would not end. In 1995, the NSF would launch the very high speed Backbone Network Service (vBNS), known also as Internet II, to create a new-generation IP network connecting the original supercomputing centers around which its original backbone had been built. With a five-year, $50-million contract from the NSF to work on the project, MCI would once again have a strong advantage in developing cutting-edge expertise on the government's dime.[24]

Early in the 1990s, two direct government spin-offs had become the most direct competitors with the NSF backbone for the Internet. As commercial traffic began to appear around the regional non-profit Internet providers created by the NSF, the federal government helped spawn two commercial backbone providers who could connect Internet traffic nationally: Performance Systems International Net (PSINet) and UUNET. These companies would be the prime instigators of the Commercial Internet Exchange (CIX), with the aim of creating a commercial system of Internet traffic exchange to compete with the NSF backbone. PSINet was formed by a subset of the officers and directors of the New York area's regional NYSERNET community provider. By 1995, it had Internet access points in 157 U.S. cities and had expanded internationally to Canada, the United Kingdom, Israel, and South Korea.[25] In addition, a company called UUNET Technologies started life as a spin-off from a Department of Defense–funded seismic-research facility but became one of the key backbone providers of national Internet services. After a short alliance with Microsoft, UUNET was bought for $2 billion in 1996 by MFS Communications, which in turn was purchased within a year by WorldCom (itself soon to merge with MCI) as the core of its Internet service in competing in the global telecommunications battle.[26]

Another early, strong competitor in the Internet backbone service market was Sprint. Much to its later regret, GTE sold off its 50 percent share in Sprint for $1.1 billion to its partner United Telecommunications (which took on the Sprint name) in 1991 just as the data-networking market was beginning to explode. Sprint had a long history in data networking, which is unsurprising, since GTE had in the 1970s purchased BBN's TELENET service, which became the core of Sprint's original data service. With ARPA-trained talent from TELENET, the SprintNet service would make the company a leader in private data networking during the 1980s and positioned the company well, as Internet competition heated up, to become one of the largest carriers of Internet traffic by the mid-1990s.

Having largely retreated from the data-networking scene with its sale of TELENET in the late 1970s, BBN in the early 1990s aggressively reentered the Internet marketplace with its purchase of many of the NSF-created regional service providers across the country. In 1993, BBN purchased NearNet, the regional Internet provider for its New England home region, which made it the Internet service provider for universities such as MIT, Harvard, and Yale along with companies and institutions such as Massachusetts General Hospital, Raytheon, Polaroid, and Lotus. With its purchase in 1994 of the Bay Area Regional Research Network (BARRNET), which serviced Stanford, the University of California–Berkeley, Apple Computer, Hewlett-Packard and NASA's Ames Research Center, BBN had quickly become the largest provider of local Internet access across the country.[27] When in 1995 it acquired the regional provider for the southeast, called SURAnet, BBN was the Internet service provider for more than one thousand leading companies and educational, medical, and research institutions nationwide. At the time, nearly 50 percent of all local business Internet traffic was going through BBN, although that number would soon fall.[28] GTE, having sold off its interest in SprintNet years earlier, reentered the Internet marketplace by purchasing BBN in 1997 and combining BBN's Internet operations with its own nationwide fiber network to create a major Internet backbone competitor. In 2000, GTE spun its Internet operations off as a new company called Genuity, which had $1.3 billion in sales as a unit by that year.[29]

One of the only new companies to enter the market as a major Internet backbone provider in the late 1990s was Qwest, which jumped into the Internet backbone marketplace with assets derived from government contracts dating back more than a century. Qwest, a spin-off from the Southern Pacific Railroad, has a unique asset in the form of railroad right-of-ways to lay backbone fiber nationwide, rights that derived from when the first national railroad "backbone" was created when the federal government funded the creation of the Intercontinental Railroad. Southern Pacific had used those railway land assets to help GTE build Sprint in the mid-1970s, but having sold its interest in Sprint in the mid-1980s, Southern Pacific revived the use of its railroad assets to begin investing again in telecom services as the Internet took off in the 1990s. Once spun off, Qwest was able to finance its own nationwide wiring through having MCI, WorldCom, and GTE pay it billions of dollars to lay their fiber alongside its own. Using those profits, by 1998 it acquired LCI International, then the fifth-largest long-distance carrier, and then completed a $48.5-billion merger in 1999 with US West, a local Bell company, to create a telecommunications giant with $17 billion a year in revenue, 29 million customers, and 64,000 employees.[30]

Despite its late start, Qwest quickly joined the older companies in taking advantage of federal technology spending to help build its own innovation and research capability. As an alternative to the MCI-controlled vBNS "Internet II" research backbone, a university consortium in 1997 granted Qwest a $500-million five-year contract to help set up an alternative high-speed backbone called Abilene. Qwest also received a $50-million contract from the Department of Energy to upgrade its telecommunications system to superfast speeds—the goal being five hundred times the then current top speed—between the Energy Department's federal contractors, universities, research centers, and more than four dozen department laboratories.[31]

All these examples highlight the extent to which the private sector has depended on government contracts for developing both the innovative technology and initial markets needed to move into broader commercial markets after the technology was developed.

The Internet as Base for Innovation in the Private Sector

Despite the foregoing comments, what is most remarkable about the government's role in the Internet is not the specific investment in technology and training or the synergistic relationship with specific Internet companies, but rather the government's creation of the broad framework for commercial innovation. This went far beyond traditional conceptions of industrial policy in leveraging new industries or supplying the raw materials of innovation (whether research or people); it fundamentally shaped an electronic marketplace based on innovation centered on common standards rather than monopolistic divides between proprietary systems. The Internet is so ubiquitous now that it is hard to conceive of a world where we would need to access different software and different phone numbers to obtain the different kinds of services we might want, but that was the world being constructed by private industry. And it would have been an extremely stunted electronic marketplace purely because of the difficulties of integrating different services for the consumer. The advantages of electronic access would have been greatly diminished, which would have held back demand not only for Internet-related goods, but also for other related computer items that benefit from an expansion of an integrated networking world.

Karl Polanyi argued half a century ago that "[t]he road to the free market was opened and kept open by the enormous increase in continuous, centrally organized and controlled interventionism."[32] The reality is that the Internet is no accident—but neither was it a technological inevitability. It was the product of the U.S. federal government, in association with other nations' experts. This entailed a process of guiding its evolution and demanding that its standards be open and in the public domain, and that its reach be extended broadly enough to overwhelm the proprietary corporate competitors.

It is under such an open system that small companies can create Internet-related software products and know that these will be compatible with other products, given the pervasiveness of the standards. The fate of the companies that were building Microsoft's proprietary network—dropped by Microsoft and left with a useless set of products when Microsoft switched to Internet standards—reveals the shadow life of companies that depend on the whims of corporate standards. The open standards of the Internet and the easy distribution of products ensures that new companies have the ability to at least attempt to take on established players without the technology itself being used as a block against them.

This is critical for a whole range of information-based industries that Stanford economist Brian Arthur has argued are governed by the law of increasing returns for investment rather than the traditional decreasing returns of earlier economic models. The argument (which Arthur submitted as part of a legal brief to the Justice Department in the argument against Microsoft's original proprietary system and its incorporation into its Windows 95 operating system) is that because of a range of built-in advantages for early innovators, companies that attain initial control of a market have a massive advantage over latecomers.[33] Because business customers for software demand compatibility with other products they use and because they have to invest training time to use the initial product, those customers are often reluctant to change products, so early entrants to a market often have an overwhelming advantage in holding onto their market dominance.[34] By securing a degree of compatibility across all programs and cutting distribution costs, the Internet mediates against the worst monopoly effects of this increasing-returns effect.

The Losses from Privatization

However, as the federal government withdraws from its active intervention in ensuring open standards, competition, and investment in new technology, there is a worry that the larger corporate players could increasingly undermine the dynamism of innovation that is derived from the Internet.

Falloff in Basic Research

One prime concern is that private industry is eating the seed corn of research investments made over decades while society fails to replenish the technological commons with new basic research. Both government and corporate spending on long-term basic research is stagnating in both the private and public sectors. With the end of the Cold War, the Pentagon has cut back on research spending and civilian agencies are not taking up the slack.

Since the early 1980s, total private and public investment in R&D has declined as a percentage of national wealth, a trend that (if health sciences are excluded) accelerated in the 1990s. Once the dominant supplier of R&D—providing more than 50 percent in 1980—the federal government has reduced its contribution, to just 30 percent, with little indication of an upturn. Since government has always focused more strongly on basic research, as opposed to the last iteration of market implementation that dominates private-sector R&D, this decline in the government role reflects a sharp decline in basic research spending as a percentage of gross domestic product (GDP).[35]

Additionally, private investment in basic research has been declining. Traditional sources of basic research in the private sector are falling prey to increased competition. The 1984 divestiture of AT&T led to a smaller Bell Labs, and recent deregulation changes have led to even more reductions in basic research at both Bell Labs and the Baby Bell research centers. Bell Labs was recently divided between companies spun off from AT&T, and what remains emphasizes research that can be rapidly developed into immediate commercial products. Xerox's PARC laboratory has increasingly followed the same pattern. As business grew in the 1990s, corporate America increased its spending on new-product research, but basic research stagnated at $5.5 to $6.5 billion per year throughout the 1990s, dropping significantly as a percentage of overall revenue.[36]

Throughout the computer industry, R&D as a percentage of sales has been brought down by fierce competition between low-price vendors, such as Dell and Gateway, to ride on research conducted by others. This trend is extending to the telecommunications industry. While Microsoft runs a well-funded re-search center—mainly emphasizing new ways for people to interact with com-puters—it still invests nearly 99 percent of its $2-billion research budget on elaborations of existing software or testing rather than on long-term basic research. Instead of doing basic research themselves in the United States (with the attendant gains for training and spin-offs within the country), U.S. corpo-rations are increasingly focusing on monitoring research in other countries. In this spirit, American companies have increased their R&D spending *overseas* threefold since 1980, the Commerce Department reports.[37]

In the wake of Internet privatization, what remained of U.S. government funding for high technology was consolidated in a program called the High Performance Computing and Communications Initiative (HPCCI), which was mandated by legislation passed in 1991. This included funding for agencies including ARPA, the Department of Energy, NASA, NSF, the National Insti-tute for Standards and Technology, EPA, and the National Institutes of Health. The total budget for the program was $1.1 billion in the 1995 financial year, although the "bandwagon" effect of the Internet has encouraged, under this umbrella, the funding of programs that have little to do with research involv-ing computing and networking.[38]

The most dramatic, positive successor computer-networking research proj-ects were the Internet II projects discussed earlier in which the NSF and uni-versity consortium established new super-high-speed backbones for projects needing higher bandwidth and more reliable connection than the Internet itself could now provide. As an example, Abilene, one of the networks that makes up the Internet II project, operates forty-five thousand times faster than is possible with the average modem. This network, researchers hope, is the testing ground for applications ranging from "teleimmerson" conferencing, television-like Net broadcasts, and virtual laboratory–based visual collaboration.[39] While encouraging unto itself, such an extension of Internet technologies required little imagination and, in a time of overall decline in basic research investment, may just be absorbing funding from the more "blue sky" research projects—such as the Internet itself was originally—that increasingly have few sources of funds from either the private or public sector.

Privatization of research has other negative effects, as patent law has been extended to software and even business methods in an unprecedented manner, thereby hobbling new innovation under the threat of endless patent litigation. In the words of Web originator Tim Berners-Lee, "In fact, [patents] are a great stumbling block for Web development. Developers are stalling their efforts in a given direction when they hear rumors that some company may have a patent that may involve the technology. . . . Large companies stockpile patents as a threat of retaliation against suits from their peers. Small companies may be terrified to enter the business."[40] In place of a government-funded commons of research encouraging new innovation and competition based on the results of that research, a situation has arisen in which monopoly control of specific innovation is retarding both new development and competition out of a fear of litigation.

Weakening of Government Support for Standards

The end of the Cold War has threatened the back-of-the-hand industrial policy that has long been embedded within the Defense Department and associated agencies that were funded under the rubric of national security. Without that protective ideological gloss, a pervasive reaction threatens to fatally undermine ongoing support by the federal government in support of networking standards and extended access by all members of society.

Many observers date the research-and-policy retreat by the federal government to the departure of DARPA director Craig Fields in 1989 during the administration of the first George Bush. Bush officials objected to DARPA's "industrial policy" in funding a variety of research projects in conjunction with industry, including high-definition television and other technologies that were in the intermediate stage between conception and commercial production. His ouster was followed by that of other DARPA officials such as the agency's manufacturing head, who testified before a congressional panel that he was told that he focused too much on strengthening the industrial base at the expense of defense. This was all seen as a retreat from the engagement by the federal government in moving technology from the laboratory into effective development in the economy.[41]

It may have been a symbolic coincidence that 1989 was also the year that DARPA shut down the remaining nodes of the old ARPANET and merged them into the NSF backbone (itself to be turned over to the private sector a few years later), but the general direction of government withdrawal from active governance of the Internet was clear. Along with beginning to privatize backbone Internet service (which would be completed by 1995), the NSF began turning more and more functions over to private companies as profit-making ventures.

Since the creation of the original ARPANET, management and the allocation of IP addresses had been carried out under contract by the InterNIC at SRI. In 1992, the InterNIC was handed over to a for-profit company called Network Solutions, Incorporated (NSI), located in Virginia. In 1995, NSI was allowed to begin charging one hundred dollars to register any customer on the Internet for any two-year period. With revenues projected at $72.7 million in 1998, there was increasing controversy over a private company's making money off a monopoly, while other companies were buying catchy names and reselling them to other companies often for thousands of dollars with no compensation to the broader network. Even as debates raged over alternatives, such as competitive address registries, the more fundamental issue remained of how to deal with the endless policy issues that kept popping up, from trademark disputes to the security of the system, as multiple databases would be used to manage the network. In 1998, the federal government passed control of these functions and certain domain name disputes to a new nonprofit entity called the Internet Corporation for Assigned Names and Numbers (ICANN), but this further privatization of these Internet governance functions did little to stifle the inevitable disputes over control of scarce domain names and who would profit from their assignments.

Why the Feds Withdrew from Standards on the Internet

So why the withdrawal in the first place? The retreat of federal involvement has been based on a combination of ideological opposition, private industry desires, and the disappearance of a stable government bureaucracy able to assume the role of regulator. This is leaving Internet development increasingly in the hands of self-interested companies with little support for the "information have-nots" who have little chance for a voice in the future shape of the network, since its direction will have been largely decided outside any political arenas where they have representation.

The ideological assault on federal involvement in further developments of the Internet is strongly related to the end of the Cold War and the withdrawal of the "national security" fig leaf that had covered much industrial planning by the federal government since World War II. It was probably not a coincidence that technology-policy promoter Craig Fields was fired from ARPA in the same year as the fall of the Berlin Wall. While the Clinton administration made some gestures in asserting a public interest in the development of what it called the National Information Infrastructure (NII), privatization proceeded apace under it. What limited funds the Clinton administration allocated for encouraging community and local-government development of the Internet was vociferously opposed by conservative Republicans and, with the Republican takeover of the Congress in 1994, those funds were initially zeroed out and subsequently sharply limited, even as local need for the funds exploded with the expansion of the Net. After the 1994 elections, the National Institute of Standards and Technology at the Commerce Department and its Information Technology Laboratory feared criticism for doing standards work that private firms were doing. So it focused on tests, metrics, and other measures of standards rather promoting them directly.[42] The Telecommunications Act of 1996 further dismantled regulations that might have been used to require lifeline Internet service or other features of local utility structures, which will soon be a thing of the past as cable, telephone and Internet service providers cannibalize one another's markets in targeting the upmarket consumers.

Private industry had significantly benefited from government spending on the Internet in the period when the Net was not commercially viable and the government was the main market for Internet-related computer services. However, as a private market for Internet services appeared to emanate from the structure of the Internet, private industry has seen a vibrant public sector as a threat to their control of information markets. Companies that had started life as extensions of the government saw the opportunity for independence and extremely high profits as the government's role receded. As Peter Evans notes in his work, the success of government intervention in nurturing new economic sectors is often rewarded by the creation of a private sector interested in blocking further government action.

Similarly, the success of the private sector helped fragment and undermine the ability of an independent government class of policy experts to successfully work in the public interest. Partly this resulted from the outside hostility of ideological opponents and of business, which politically wanted to curtail the power of the public sector as the private sector succeeded commercially. With Defense Department involvement in high technology under assault and Republicans trying to abolish the Commerce Department, where most of the NII programs had been coordinated in the Clinton administration, there was little chance for a long-term view by public servants watching their political backs.[43] As important, though, was the way private opportunities for profit turned many of the engineers and ARPA employees from public servants into representatives of the private companies now pushing for limiting the federal role. From Bob Metcalfe, who became rich through founding 3Com, to Vint Cerf, who has become a major spokesperson for MCI, the founders of ARPANET who initially cultivated the ethic of freely sharing information and software were now fighting for profit share and private ownership of intellectual property.

The Threat of Monopoly and Government Reaction

With the shutdown of the NSF backbone in 1996, the nineteen regional networks located around the country were largely squeezed out of existence as they either were bought out by companies such as BBN or faded out under the onslaught of the larger corporate players. With the government no longer enforcing the automatic sharing of Internet traffic routed to people in different parts of the network, the large corporate players were forced to draw up "peering" agreements—commitments to automatically share electronic mail and Internet traffic—to carry each others' traffic, but this often left smaller players out of the game.[44] The very complication of pricing "free" information transit in an unregulated system, in which everything depends on individual agreements to share traffic, increasingly overwhelmed the system. By the middle of 1996, private Internet backbones were getting clogged, not so much because of lack of capacity but because individual companies were seeing little incentive to upgrade their own systems to transmit traffic for others. In the absence of either clear property rights over shared information traffic or comprehensive regulation, the Internet was developing classic symptoms of the "Tragedy of the Commons" as people increasingly referred to the "World Wide Wait."[45] In fact, the creation by research labs of the Internet II for their own use largely reflected the increasing unreliability of, degradation of, and obstacles to upgrading the core shared network of the original Internet under the demands of privatization.[46]

   With nonprofit regional service providers squeezed out, this has eliminating a local shared resource for comprehensive and equitable coordination to bring other nonprofits, schools, and other entities into the Internet (an issue we will return to in Chapter 6). Individual schools and government entities have, in most cases, been forced to turn to commercial service providers in an ad hoc way without the planning that extended the Internet to meet the needs of successive public institutions in its earlier stages. There is fear that the federal abandonment of a regulated expansion of capacity is leaving the way open for phone companies or cable companies to justify expanding capacity for elite customers at the expense of their local ratepayers. At the same times, private Internet backbones are being built to assure speedier information transit for selected customers, threatening the open equality built into the framework of the Net.

The Threat of Monopoly on the Internet: WorldCom

Even as some sang odes to the anarchy and competition brought about by the privatization of the Internet, the rise of corporate monopoly in the late 1990s began to threaten a new order for the networked world that would undermine the most basic values that had built the Internet. The liquidation of smaller companies by larger ones had begun when America Online swallowed ANS, GTE swallowed BBN, and a host of other mergers created larger and larger corporate players. Up to a point, this was rationalized as a shaking out of competition and the creation of companies large enough to deliver the integrated services desired by customers.

But with the bidding between WorldCom and GTE for ownership of MCI Communications—with WorldCom emerging victorious—many saw a threat of monopolization that could undue the Internet. WorldCom had already grown into a dominant player in Internet traffic; beyond its purchase of UUNET, it had initiated a complicated deal with America Online in 1997 in a $1.2-billion takeover of the CompuServe online service. Not only did World-Com acquire CompuServe's network services data-networking unit, in exchange for giving CompuServe's residential customer base and $175 million to America Online, WorldCom acquired ANS from America Online.[47]

When MCI agreed to be acquired by WorldCom for $36.5 billion in late 1997, it was merely the most recent in a wave of mergers and consolidations that had occurred in the wake of the 1996 Telecommunications Act. However, by combining UUNET, ANS, CompuServe's network services, and MCI's Internet unit, along with a number of smaller Internet companies, WorldCom was now bidding to create a company that would control up to 60 percent of all Internet traffic in the United States.[48]

Opponents and even supporters of WorldCom declared that a monopoly was the likely result of the merger. Everyone from competing phone companies such as Bell Atlantic, to Ralph Nader to unions such as the Communication Workers of America worried that this was not merely the case of a big dog threatening others by its size, but that the nature of Internet privatization would give the merged entity a stranglehold on crucial points of access to the Net.[49] As long as the NSF was exerting strong regulation of the Internet, the temporary dominance of companies such as ANS as a backbone or BBN at the local level made little difference since all parts of the network were required to share traffic without discrimination. However, for a entity like the Internet dependent on cooperation as well as competition, the new lack of regulation became an invitation for a company to emerge that would use its size to extort monopoly prices as the price for cooperation.

Part of the NSF privatization in 1994–95 had been the creation of key "network access points," or NAPs—switching platforms where Internet traffic is transferred between different networks. While not the only interconnection points, the NAPs are the fastest and most efficient spots for transferring data or for independent local ISPs to hook up to the Internet. An assortment of different companies had been given contracts to operate specific NAPs, but with its merger WorldCom, would control five NAPs, including the two dominant ones: San Jose, in the heart of the Internet explosion; and Washington, D.C., which had become the main connection for traffic going to Europe. Bell Atlantic would file its opposition to the merger, arguing, "As owner of five of the NAPs, WorldCom has the ability to influence the terms by which traffic is shared not only between its network and other networks, but among other networks as well. An ISP cut off from the WorldCom NAPs is in dire straits."[50]

This threat was not academic. Earlier in 1997, WorldCom's UUNET had announced that it was terminating peering agreements with all networks other than its fellow backbone providers such as Sprint and MCI. Such peering had been the Internet's underlying strength from its inception, because its very decentralization was based on cooperation in making sure that each message got to its ultimate destination regardless of who owned which network. The most prominent challenge to UUNET's decision was David Holub, the president of Whole Earth Networks, which ran a venerable electronic service and ISP called the WELL. He argued vociferously and publicly that since the WELL was already paying UUNET to place its computers physically at the NAP, UUNET should not be able to use that ownership to extort more money for the right to transfer information as well. Holub's public fight so frightened the WELL's board of directors that they fired him, marking the new fear in the

ISP community of angering the large backbone providers and NAP owners who now dominated the Internet.[51]

Ironically, even as WorldCom and other companies were ending peering agreements with smaller ISPs, they were collaborating with those same ISPs to demand that the Federal Communications Commission (FCC) maintain the essentially free interconnection to the local phone infrastructure that had been granted to Internet carriers when it was an academic and public entity. Even as WorldCom was charging local ISPs for access to its infrastructure, it was able to exploit the local phone infrastructure for Internet traffic while paying nearly nothing for its upkeep.

Adding to criticism of the merger were fears that WorldCom not only was seeking a stranglehold on the Internet but also was uninterested in extending its benefits or broader telecommunications services to average consumers. WorldCom vice chairman John Sidgmore stated in late 1997 that WorldCom was eventually going to abandon MCI's plans for expanding residential access to telecommunication services: "Our strategy is not in the consumer business." With WorldCom's focus on business customers and its key strangleholds on network access, the earlier vision of cooperative regional networks promoting broad access to the Internet was increasingly giving way to a situation in which global multinationals were using control of regional chokeholds to serve elite customers and their own bottom line, mostly at the expense of the average citizens who paid for the Internet's creation in the first place.

The problem is that in the wake of privatization and the NSF's withdrawal from active governance, the general competition between Internet companies had already led to Internet traffic congestion and disputes that had raged out of control of the informal structures for governance that had been left in place. Many conservative commentators who had once attacked government telecommunications monopolies now praised WorldCom as the savior of the Internet through consolidation and rationalization. George Gilder, the right-wing guru of technology, rapturously praised the merger: "Mr. Ebbers [World-Com's CEO] will be the salvation of the Internet. . . . Like John D. Rockefeller and Michael Milken before him, Mr. Ebbers has shown the magic of entrepreneurial vision and guts. . . . [Only Ebbers can] build a Net that can bear the burdens of continual exponential growth."[52]

To the undoubted disappointment of Gilder and the partial relief of consumer advocates, antitrust authorities allowed the merger to proceed but only with strong conditions, focused especially on avoiding Internet dominance by the combined entity. Both the U.S. Department of Justice and the European Commission regulators required that MCI's Internet business, valued at $1.75

billion, be sold off to Britain's Cable and Wireless, thereby bringing a new major player into the Internet backbone and services market.[53] Yet two years later, the large backbone companies still maintained their cozy free peering exchange with one another while often refusing peering to smaller ISPs with no government regulation or oversight to constrain their actions or avoid balkanization of network interchanges.[54] This added to the general connection issues throughout the Web with larger backbone providers imposing periodic intentional blackouts of Net traffic on smaller rivals.[55]

In the absence of a strong government presence that could impose basic rules of cooperation and equal access, the same debates on monopoly and on equity of access for consumers would continue, especially as new issues concerning high-speed broadband access became more urgent as the decade ended, bringing questions about Internet access right on a collision course with the debates on general telephone regulation that had raged from before the AT&T breakup in the early 1980s. (We will return in Chapter 5 to the dynamics of deregulation and the regional politics of those deploying local telecom infrastructure in shaping regions in the global economy.)

The Threat of Monopoly on the Internet: Microsoft

If the threat of monopoly control of the physical wires undergirding the Internet could be partially deflected by traditional antitrust action, a similar effort using the blunt instrument of antitrust law would cause a much more public firestorm in targeting Microsoft's monopoly threat of controlling the intangible standards of the Net. While the rise of the Internet has been a challenge to the desktop monopoly that Microsoft built in the 1980s, the withdrawal of strong government-backed standards also made it an opportunity to expand its reach to a degree that had been impossible before. With electronic commerce exploding throughout the 1990s, the Internet became the decisive realm of computer competition.[56]

The threat to Microsoft was obvious: with its tradition of open computing standards connecting computers of all kinds, the Internet looked ready to make proprietary operating systems for individual machines an anachronism. As the Internet broke into the national consciousness in 1994 and 1995, it appeared that millions of computers were connecting to one another with Microsoft having nothing to say in the matter. The rise of Netscape and a host of other new Internet companies seemed to promise a new era of competition including a whole new cast of companies. The final nail in Microsoft's coffin seemed to be the failure of its proprietary Microsoft Network service in the face of open Internet standards.

However, the opportunity for Microsoft that arose from the growth of the Internet lay in whether it could seize control of those standards, thereby reinforcing its control not only of the desktop also but of corporate computing, which is increasingly dependent on the Internet. Combining its traditional tactics of hardball agreements tied to its desktop monopoly, control of the programming tools used by most programmers, and an avalanche of technology and company acquisitions—amounting to more than one major technology per month by 1996 and 1997—Microsoft launched a bid for control of the Internet that was dizzying in its breadth and depth.

The most visible part of the battle for control of the Internet was the so-called browser war, the fight largely between Microsoft and Netscape Communications over which piece of software would be used by computer owners to surf the Internet. Netscape initially dominated distribution of browsers beginning in 1994, but it rapidly lost market share to Microsoft, and as the latter's Explorer browser was more and more integrated into Microsoft's operating system, most analysts expected Microsoft to take over this software market.

But browsers are more than a piece of software—they have a decisive impact on all standards for Web design. Browsers are the means by which any computer user may "read" information from a World Wide Web computer server that is situated at a distant place. If the dominant browser was designed not to read a certain kind of information—a kind of graphics, software effect, and so on—then Web page designers would be loath to use that kind of information or technology, while they would tend to support software standards that are compatible with the dominant browser. And if you were a software company such as Microsoft selling Web servers, Web design, and an array of Internet commerce software, you would have an overriding interest in controlling those Web standards.

Unfortunately, the government's withdrawal from offering strong support for Internet standards had created the opportunity for Microsoft's bid for control. Belatedly, in late 1997, even as the FCC was exploring whether the WorldCom-MCI merger would undermine competition, the Justice Department launched its own investigation of Microsoft, culminating in the landmark lawsuit and court decision against the company in 2000.

But despite the decision (especially in light of whatever modifications or settlement that may have occurred by the time you read this), it is unclear what a blanket call for competition means in an institution such as the Internet, where cooperation on standards demands consistency throughout the network. In the absence of active government involvement ensuring such consistency, many people defended Microsoft's monopolistic practices as serving a purpose for customers. In a world of rapidly changing technology, Microsoft's monopolistic grip on standards and different market segments gives consumers some assurance of stability and interconnection between products. The always looming presence of Microsoft inevitably framed the alternative strategies pursued by competitors to assure vibrant standards; and these will be detailed in the following chapters.

Regional Production Districts as an Alternative to Monopoly?

If there was any alternative to monopoly in the wake of the privatization of the Internet, it was the vibrant cooperation of Silicon Valley's Internet firms. If opportunism and monopolization is the rule in global competition in the absence of strong government intervention, open standards and nonopportunistic competition seemed to be more embedded in the institutions and relationships of a specific region such as Northern California. It is clear that while national government planning and intervention were the driving force in the creation of the Internet, its expansion as a commercial enterprise was initially supported and defended against proprietary attacks by the range of Silicon Valley firms committed to its open standards.

In turn, the economic and technological strength of Silicon Valley has deep roots in both government spending and technical engagement over the years. In Chapter 3, I will outline in detail the ways in which federal involvement built the region's general technological strength and, in particular, how regional firms and the Internet expanded in a synergistic relationship between government intervention and private-sector entrepreneurial activity. This synergy had profound effects not only on the region but also on the global shape of the Internet economy. In that sense, government support for such regional economic networks can be seen as one more form of pervasive government intervention that deeply shapes the more general economy.

The question becomes how that regional cooperative model, built over decades with government support, survived the privatization of the Internet. In Chapter 4, I will explore the regional efforts to revitalize that cooperative model in the new era of privatization to support the Internet against proprietary alternatives. These efforts will be compared with the broad national experience in using such regional business-to-business consortia as an engine for economic and technological growth. A crucial question is, Even if businesses in a region benefit from such cooperative regional models, what are their relationships to broader sectors of the regional economy in promoting jobs and economic equity? If this regional community is one that benefits only an economic elite, it is a rhetorically shrunken conception of community that, in the absence of the original government intervention that expanded economic opportunity, may be little better for the majority of the population than the monopoly alternative.

Federal Spending and the Regionalization of Technology Development

What created the Internet, and why is Northern California at the center of its development? Nothing to do with the government, at least according to *Wired*, the fashionable technomonthly that promotes the Net as the embodiment of a new paradigm in human development, unshackled by government, scarcity, or even geography. Since its inception, its covers have featured a parade of cybermoguls from such hip antiestablishment outfits as CitiCorp and TCI Cable, all supposedly self-made men riding the wave of cyberspace benefited by nothing other than pluck and a vision of a libertarian world of freedom.

There is an almost charming Gatsbyesque quality to this denial of the past in plotting the brave new unregulated future of cyberspace. Less charming to many is the fact that so many people buy not only the magazine but also the ideology that is reflected in so many high-tech firms. Richard Barbrook and Andy Cameron of Britain's Hypermedia Research Center have labeled this viewpoint the "California Ideology" in a widely circulated article that notes the convergence of conservative social cutbacks with an ideology that forgets the defense-funded past of the Silicon Valley region. This ideology "promiscuously combines the free-wheeling spirit of the hippies and the entrepreneurial zeal of the yuppies" while ignoring the increasing inequality fueled by information technology.[1]

Responding to the article, Louis Rossetto, editor and publisher at *Wired*, denounced the "laughable Marxist/Fabian kneejerk" ideology that overlooked the contrast between Europe's "statist" support for technology and the process by which the United States had developed its technology, based not on government spending but on capital markets and free market development. "In point of fact," Rosetto argues, "it was the cutback in American defense spending following the Vietnam War and the subsequent firing of thousands of California engineers which resulted in the creation of Silicon Valley and the personal computer revolution."[2]

The levels of intellectual repression embedded in this sentence are astounding—from the myth that such companies as Apple Computer emerged from the garages of their designers like Athena from Zeus's head to the assumption that the concentration of engineering talent in the Bay Area was a fact of nature much like the pleasant climate. And it represses the fact that the very Internet boom that is *Wired*'s raison d'être, like the semiconductor technology that runs its hardware, are inescapably the products of government support in the United States.

At the most obvious level, government spending was the engine of economic and technological growth in the region, from federal contracts that built the intercontinental railroad, to defense contracts that spawned the first wave of electronics companies during and after World War II, to the spending on the Internet funneled through key Bay Area institutions such as Stanford, UC-Berkeley, and Xerox PARC. For decades in Silicon Valley, winning a defense contract was almost the only means through which a start-up company had a chance to enter technology markets against already established firms.

As critical, though, was the role of the government in fostering the long-term business and social networks that made the region's technological innovation possible. Political scientist Robert Putnam has noted that the social capital that fuels regional economic growth is not the result of short-term events but is part of long-term persistent patterns that evolve over lengthy periods of time and embed themselves within the fabric of a region. It was a synergy between federal government and local actors, aided by periodic infusions of federal cash, that slowly evolved the interconnected cooperative model of technological development that became Silicon Valley's hallmark.

Nowhere is this clearer than in the way the region benefited from and in turn contributed to the federal government's shaping of the open standards of the Internet, standards that in turn reinforced the region's dominance of the new Internet-related businesses. As was detailed in the preceding chapter, open standards allow dramatically expanded technological possibilities and economic growth, but the reality is that the benefits of such standards-driven growth tend to accrue to regionally concentrated firms in which collaboration can fully exploit the opportunities from open standards. Competing proprietary standards controlled by corporate titans can be directed from geographically scattered enclaves (such as Microsoft's Bellevue or IBM's Armonk headquarters), but the Internet's structure as an open system backed by the federal government had the seemingly paradoxical effect of reinforcing the importance of a handful of specific regions, especially cementing Silicon Valley in a lead role.

This chapter will outline how these three government-driven factors—spending, promotion of social networks, and support of standards—were crucial in creating Silicon Valley, which became in turn so crucial to the birth and expansion of the Internet. Starting with a short history of the role of the government in the emergence of Silicon Valley, I will move on to examine that history in the context of debates over regional economic development. In turn, that early technological development would help shape a network of firms and collaborative behavior that would attract new waves of public investment and private commercial development fueled by information networking and culminating in the explosion of Internet-related businesses this decade. While internal regional factors were not irrelevant to these results, the overall history makes clear how much regional development is dependent on national and global decisions outside the control of local actors.

Early History of the Valley: Building the Social Capital of Technological Innovation

If there is any clear idiosyncratic indigenous reason for the rise of Silicon Valley, it is indisputably the founding in 1887 of Stanford University, the font of both technology and engineers who would fuel much of the technological hyperactivity of the region. Yet Stanford's origin itself, though the university is a private one, is hard to separate from the government. Its founder and benefactor Leland Stanford was not only a U.S. senator and a governor of California but was also one of the four founders of the Southern Pacific Railroad, the economic power behind the throne of state government for decades until the Progressive movement of Hiram Johnson broke its hold in the early part of this century. And the fortunes of the Southern Pacific were made through the massive federal subsidies paid for the western leg of the transcontinental railroad. In fact, Stanford himself drove in the golden spike at Ogden, Utah, that completed the intercontinental link. Out of the profits of that venture, Stanford was able to endow the university he founded with $20 million, one of the largest philanthropic gifts ever made in his day, along with a seventy-two-hundred-acre farm near the redwoods of El Palo Alto.[3]

In this way, the federal government of a century ago, through its first economic linkage of the continent, indirectly invested in the founding of an institution that would in turn help electronically link the nation and world a century later. And from the beginning, Stanford University had the practical orientation to science and technology that would be its hallmark. Early on, the Bay Area was a hotbed of radio experimentation, and with the earthquake of 1906, a number of firms in the budding industry were encouraged to move south down the peninsula toward Palo Alto and the science center of Stanford. In 1909, a former Stanford student, Cy Elwell, traveled to Copenhagen to ask for the rights to manufacture a special telegraph system called an arc transmitter. On Elwell's return, Stanford president David Starr Joran was impressed enough with the idea that he invested five hundred dollars of his own money, and other faculty and local businessmen followed in supporting the founding of the Federal Telephone and Telegraph Company. Federal would go on to hire a succession of Stanford graduates for years until the company moved East in 1932 (to eventually be absorbed into ITT). The year 1909 was also when San Jose established the first commercial radio station in the nation.

In the meantime, the company would serve as a focus for the emerging radio and audio industry of the region. In 1911, the firm hired Lee de Forest, the man who had patented the first vacuum tube, along with producing three hundred other patents during his lifetime, and who had come West after being fleeced by his own company's board of directors. He stayed at Federal for only a few years but helped build the reputation of the region's companies. The sinking of the Titanic led in 1913 to the passage of the Radio Law, requiring radios on all commercial passenger ships, and that combined with military demands in World War I spurred the region's radio industry, which expanded rapidly.[4] Not for the last time, the mandating of technological standards by the federal government would benefit the region.

A second generation of electronic pioneers was being born in this period. Frederick Terman moved to Stanford in 1910 with his father, Professor Lewis Terman (famous for his co-invention of the standard IQ test), and Frederick soon joined the ranks of ham radio hobbyists in the area. He received a BA from Stanford and a doctorate from MIT. He had meant to pursue a career on the East Coast, but on a trip home he came down with tuberculosis and, through an initial part-time teaching load, ended up as a professor at Stanford himself. In 1924, he was made head of a new "radio communications laboratory," which would become a center for technological innovation on the West Coast. In 1932, Terman wrote a classic text titled *Radio Engineering,* and by 1937 he was a full professor in charge of the Department of Electrical Engineering.

Upset about graduates moving back East for jobs in the 1930s, Terman encouraged a number of his students to start up companies locally. Two of them, William Hewlett and David Packard, had graduated in 1934 and moved back East. Four years later, Terman recruited them back to the region with fellowships to complete electrical engineering degrees. For a master's thesis, Hewlett created a frequency oscillator, a device to produce controlled electrical signals at whatever frequency was desired. Terman personally lent the two $538 and arranged a one-thousand-dollar bank loan to get them off the ground in their own company. With Hewlett's wife baking the first painted panels in their oven, Hewlett-Packard was born as a company.

World War II and Its Aftermath Fuels the Explosion of Bay Area
High-Tech Firms

However, it was only the massive military spending of World War II and the postwar period that launched Hewlett-Packard and so many of the firms to follow into the stratosphere of growth. Hewlett would be called to active service in 1942, ending up in the development division of the War Department's special staff where he made key contacts for the postwar period. At home, Packard started filling government orders that by the end of the war had Hewlett-Packard earning $2 million a year with two hundred employees.

Another regional company that grew exponentially during the war was Litton Engineering Laboratories, which a former employee of Federal Telegraph had started when Federal moved East in 1932. Manufacturing glass vacuum tubes, Litton found himself running a million-dollar company by the end of the war as radar expanded the demand for his product. After the war, Litton incorporated and as a military contractor would grow to be a billion-dollar company by the 1960s.

The other important company that made its reputation during the war was that of the Varian brothers, Sigurd and Russell, who worked in the Stanford physics lab run by their collaborator, Bill Hansen, the developer of a lightweight radar tube called the Varian Klystron to help Allied aircraft fly in the dark. (Stanford would receive $2 million in royalties for its support of the brothers). While the brothers worked at Sperry Gyroscope during the war, they returned to the area in 1948 and founded Varian Associates, which would become a powerhouse in the electronics instruments business.[5]

Overall, the federal government pumped $35 billion into California between 1941 and 1945, and the state has continued to receive 30 to 40 percent of all Department of Defense R&D contracts since then. This spending not only built up electronics, but also created a broad-based manufacturing economy in the Bay Area. Of $5 billion in ships built during the war, $3 billion worth were built in the Bay Area alone.[6] The geopolitical need to take on Japan in the Pacific led the federal government to create, almost overnight, a radically expanded manufacturing base in California that would fuel growth for decades and create a ready local market for the electronic products of the emerging Silicon Valley.

During the war Frederick Terman had been recruited by Vannevar Bush to direct a massive research team in the East dedicated to jamming enemy radar. Recognizing a coming postwar wave of federal funding, Terman returned to Stanford after the war determined that it and other California institutions would get their full share of defense contracts. In writing guidelines for spon-sored university research, he argued, "The war . . . brought to the West the beginnings of a great new era of industrialization. If Western industries and Western industrialists are to serve their own enlightened and long-range inter-ests effectively, they must cooperate with western universities and, wherever possible, strengthen them by financial and other assistance."[7] Berkeley was also emerging in this period as a science powerhouse, riding partially on the reputation of Berkeley physicist Ernest Lawrence, who had won the Nobel Prize in 1939 for his development of the cyclotron, a mechanism to accelerate a stream of protons to break open the nucleus of targeted atoms.

But Terman did not stop at attracting research money; in collaboration with his contacts in the federal government, he developed a homegrown industrial policy to economically link the university to new high-tech endeavors. He began promoting the idea of using the extensive university land, which the original bequest barred from being sold, as an industrial park for high-tech-nology firms. Varian Associations, where Terman served as a board member, became the first leaser of land in 1951 when the company agreed to an annual sixteen-thousand-dollar lease for ninety-nine years on four acres—what would become one of the sweetest land deals in the Valley as prices rose in the region. When Hewlett-Packard set up headquarters in the park in 1954, it became the nucleus of the emerging Silicon Valley. By 1955, seven companies were in the park, thirty-two by 1960, and seventy in 1970. By the 1980s, all 655 park acres were leased, with ninety tenant firms making $6 million a year for the university. This was important not just because it linked Stanford to emerging high-tech firms (with more than 60 percent of Stanford's engineering faculty consulting with government or business over the years). Additionally, the unrestricted income from the leases could be used by Terman for his ambitious faculty recruitment goals to push Stanford to the forefront of academia.[8] The cluster of firms around Stanford would become a business community in which the companies' links to one another would become increasingly important over the years as technological collaboration became more and more important.

The other critical institution in the area was the Ames Research Center, established in Mountain View in 1940. It would later become part of NASA, and its presence in Northern California encouraged Lockheed to establish its Lockheed Missile and Space Company subsidiary in the Valley in 1956 in order to be close to Ames. When Lockheed set up its new Missile and Space Division in Sunnyvale in 1957, Stanford agreed to supply faculty members to train workers in return for Lockheed's helping build the university's aeronautical engineering department.[9] Lockheed would quickly become the region's largest employer. Added to this was state government support for the rising engineering talent of San Jose State University. The combination of Ames, San Jose State, Hewlett-Packard, Stanford, and now Lockheed became an irresistible force in attracting new research centers to the region. Throughout the 1950s, the electronics giants began opening divisions in the South Bay: General Electric brought its nuclear division to San Jose, Westinghouse brought its heavy equipment division to Sunnyvale, Sylvania built a plant in Mountain View, and IBM opened a research lab in downtown San Jose to design its first computer disk memory system.[10]

Silicon Comes to the Valley: Semiconductors and Defense-Driven Entrepreneurs

It was the end of the 1950s that would see the creation of the semiconductor industry, which would give the region its enduring public fame. At Bell Labs in 1948, a team of researchers, led by William Shockley, invented the first transistor to replace the large, bulky vacuum tube in electronic devices. The federal government immediately gave Bell Labs a contract to expedite development of the transistor. By 1950, when Bells Labs had figured out how to produce the transistor more commercially, AT&T was feeling the heat from an antitrust suit filed in 1949 (which would ultimately lead to AT&T in 1956 being barred from selling transistors and computers in the open market). In response, AT&T in the early 1950s widely licensed the technology and gave detailed seminars in how to manufacture the new transistors.[11]

Thus, the federal government not only funded the development of the technology but also ensured that it would be widely available in the public domain. Combined with defense funding that channeling money to far-flung regions distant from Bell Labs headquarters, the new semiconductor technology would facilitate high-tech booms in a number of regions far from traditional manufacturing regions in the East and Midwest. One of the first companies to take advantage of this new technology was Texas Instruments, which grabbed a top Bell scientist and in 1952 developed the first silicon version of the transistor for sale to the military. By concentrating on supplying the military, Texas Instruments quickly became the largest merchant supplier of semiconductor devices, although because of the open licensing from Bell Labs, more than twenty-six companies were competing with it by 1956. Because the military was funding its own experiments in miniaturizing transistors, it was clear that a rich prize of military contracts would exist for the firms that delivered a cheap, small integrated transistor.

For the future of semiconductors in Silicon Valley, the key event was the return in 1955 of William Shockley to his boyhood home in Palo Alto, where his father had taught mining engineering at Stanford. With his return he would attract much of the initial talent to the region, which would set in motion an industrial explosion in semiconductor production in the region.

Shockley's effect on the region was truly a surprise, especially for the physicist himself, who had expected to become a millionaire but instead saw the entire leadership he recruited to the region quit and form their own company. Shockley was described as a "genius, but a real prick" to work for, and eight of his top employees left to form Fairchild Semiconductor with $1.5 million in support by Fairchild Camera and Instrument Corporation in New Jersey. In 1959, Fairchild was given the contract to supply transistors for the Minute Man I missile program's guidance system. Robert Noyce took the helm of the new enterprise and it was his invention of the integrated circuit that same year (along with Jack Kilby of TI, who shared the patents) that would make Fairchild's fortune.[12] The military, having paid premium prices throughout the 1950s, and thereby funded the integrated circuit's evolution into a mass production item, began buying integrated circuits in a volume that would make the industry boom. Considering that IBM refused to use integrated circuits in its 1960 model 360 series of computers, the military support was critical for the new industry. In a RAND study on the effect of the military on the semiconductor industry, Anna Slomovic has written, "Two government procurement decisions were responsible for moving integrated circuits into large-scale production. In 1962 NASA announced that its prototype Apollo guidance computer would use integrated circuits. Shortly thereafter, the Air Force announced the use of integrated circuits in the Minuteman II guidance package."[13] With nearly 100 percent of integrated circuit purchases in 1962, the military's initial support was crucial in supporting the iterations of development that would expand the market for commercial applications throughout the next decade.

In the 1960s, the federal government would indirectly push integrated circuit technology into the consumer market when it began requiring that televisions be able to receive UHF channels, channels that traditional vacuum tubes often failed to receive reliably. Fairchild built a prototype television using solid state circuits that helped convince manufacturers to buy their integrated circuits. Much as legislation requiring radios on ships after the Titanic disaster had led to an initial technology boom out of the region, so too did this new standards mandate by the federal government help spur the expansion of commercial sales of the Valley's semiconductor products tied to solid state technology.

As the 1960s progressed, Fairchild's employees would begin leaving to start up their own companies in the hot emerging market for semiconductors. Partly because of the corporate parents' refusal to spread stock profits among the employees and partly because of the immense profits to be made by running their own companies, even the eight founders would flee Fairchild. By 1967 the energy of Fairchild was gone, the company replaced by a slew of new firms, including National Semiconductor, Advanced Micro Devices, and Intel, this last led by Noyce himself. The Defense Department contributed to this splintering of Fairchild through its policy of requiring a second source for all computer chips; with second-source contracts to be had, technological startups had ready markets once they secured a bit of financial backing. With Intel's invention in 1972 of the 8008 microprocessor—a device to combine all needed circuits for computer processing on a single chip—the first computer product was created for which the military would not be the first major customer. Its successor, the 8080, produced in 1974, would become the basis for the rise of the home computer industry in the Valley, while the density of engineering talent and electronics production would become an almost irre-sistible magnet for a whole range of computer start-ups in the next two dec-ades.

As will be detailed, the key concepts and technologies driving the commer-cial personal computer industry—from basic design to the computer mouse to icon-based operating system—were the products of government-associated labs at places such as Stanford or Xerox PARC. Another key part of govern-ment support in the region was the continued expansion of education and training of local talent to staff these new ventures. While Stanford and Berke-ley get most of the press attention, as important was the evolving state and community college system in California (itself enjoying the benefits of ex-panded federal tuition support for students). By the 1970s, San Jose State Uni-versity was training as many engineers as either Stanford or Berkeley, and six surrounding community colleges offered technical programs that were among the best in the nation. Community colleges, particularly, would work with local businesses by contracting to teach private courses for their employees, even holding courses at company plants to enable employees to attend after hours.[14] These evolving education to business links were further reinforcing the economic networks that were so crucial to collaborative innovation in the region.

Coincidentally, just as the computer industry found its first mass commer-cial market after decades of dependence on government R&D and purchases, ARPA was expanding government support for computer networking and the Internet that would launch the next computer revolution two decades later.

Augmentation Versus Automation: The Internet Origins in Silicon Valley

If the earlier waves of technology in the region were driven by the geopolitics of defense, the networking wave of computing that eventually exploded in the 1990s derived partly from defense support as well, but sprang from the most visionary reaches at ARPA rather than from direct weapons procurement. The defense and semiconductor industries in the Bay Area made it a magnet for attracting both funds and top talent for the energy unleashed by ARPA, and in turn, the energy of open standards unleashed by and guided by the federal government would spur the growth of companies in the region that would be in a position to dominate the global Internet-related industry when ARPA's work came to fruition decades later.

One of the earliest threads in the origin of the Internet dates back to an

Atlantic Monthly article published right after World War II and written by Vannevar Bush, the prominent MIT researcher who defined much of the vision for government support of science in the postwar period. In the article, he discussed the problem that many scientists were unable to keep up with all the knowledge they needed and proposed a machine using microfilm he called the "memex" to enhance the powers of human memory and association. One of his most important readers was Doug Engelbart, a young sailor who read the article as World War II was coming to an end in 1945. Moving beyond microfilm, he became fixated on the idea of augmenting human capacities through use of computers to better link people both to stored ideas and to other people working together in groups. He contrasted his vision of using computers to "augment" human knowledge with the more mainstream focus on the "automation" of individual tasks.[15]

Coming out of the navy, Engelbart would work at the Ames Research Center from the late 1940s until 1951, when he would leave for a doctorate at UC-Berkeley. He eventually arrived at the Stanford Research Institute in 1957, a creation of Frederick Terman's dedicated to defense-related research and assistance to West Coast businesses. There he would receive a small initial grant from the U.S. Air Force Office of Scientific Research to pursue what Engelbart called an "augmentation laboratory" to explore how people and computers could share knowledge. He would publish his ideas about using computers not just for the traditional computation and data processing in which they were currently engaged, but also as a tool that could act as an adjunct to individual creativity and as a link between human beings. While broadly ignored, his ideas paralleled those in J. C. R. Licklider's article "Man-Computer Symbiosis," which had led to Licklider's recruitment into ARPA. In 1963, Bob Taylor, then at NASA, pushed through new funding for Engelbart's project and in 1964, having moved over to ARPA, he was able to fund Engelbart's project with a million-dollar time-sharing computer plus half a million dollars to staff and run his proposed laboratory. Out of the Augmentation Research Center (ARC) would come an array of researchers who would go on to become leaders of their own research teams at universities and commercial R&D divisions in the region and across the country.[16]

Considered an odd duck by his SRI colleagues, Engelbart was suddenly running what could arguably be called one of three most important computer labs in the country—the others being MIT, where explorations of time-sharing computers were already spawning revolutions in the minicomputer industry, and Stanford's Artificial Intelligence Lab (SAIL). Engelbart was building on the tradition of interactive computing largely funded at MIT by a yearly $3 million ARPA grant focused on time-sharing called Project MAC. It was MIT hackers at Project MAC who largely designed DEC's breakthrough PDP-6 time-sharing minicomputer and would spend endless hours making it perform tasks beyond the expectation of any of its creators.[17] Probably the most important precedent for Engelbart was Ivan Sutherland's SKETCHPAD program, which initiated the manipulation of computer images, allowing users to resize and manipulate graphics. And it was Sutherland who had been hired to run ARPA after J. C. R. Licklider left and who put Bob Taylor in charge of ARPA's Information Processing Techniques Office, from where he would develop the ARPANET project.

Engelbart worked from Sutherland's precedent to concentrate on the manipulation of text and ideas. ARC would combine hardware design, "building the tools" that would in turn support an "intellectual workshop" to create the breakthrough software needed for Engelbart's vision. Working with seventeen colleagues and going through three rapid cycles of hardware revolution, by 1968 he was ready to publicly demonstrate the results at an engineering conference called the ACM/IEEE Joint Computer Conference in San Francisco. And the results stunned the audience.

Hooked up by microwave communication to the computers back at SRI, Engelbart would demonstrate the array of tools developed at ARC: the first "mouse" used as an input device, a windowing environment that could rapidly switch between a menu of information sources and models of information, word processing on screen. None of these had ever been seen before, and in an age when most programmers were still interacting with computers through punch cards, the idea of word processing was a revelation.[18] What was demonstrated was only the showiest example of a set of tools developed to facilitate communication and shared-information-based work among intellectual collaborators. ARC was already using text editing to share common data through hypertext storage (the method of linked pages used in the World Wide Web) and ran an electronic mail communication system with dedicated e-mail distribution lists among the researchers—all this years before these innovations would come to the ARPANET. ARC would also pioneer videoconferencing years before it was developed commercially.

It was also in 1968 that Bob Taylor, now in charge of the ARPA office developing the Internet, would publish with J. C. R. Licklider an article titled "The Computer as a Communication Device," in which the authors outlined a vision of ordinary people using new tools to share information, arguing that "when minds interact, new ideas emerge." Taylor and Licklider would cite ARC as the prototype of their vision.[19] And it was in 1968 that Bob Taylor asked Engelbart to make ARC the Network Information Center (NIC) for the ARPANET. Engelbart saw the ARPANET as the perfect vehicle for extending his vision of distributed collaboration, so in 1969, SRI would become the second computer on the ARPANET. At the NIC, Engelbart would help identify and organize electronic resources on the Internet for the easiest retrieval. Until 1992 (when the NIC functions were awarded to other companies), the function of the NIC at SRI would include administration in assigning IP network addresses and domain names for all servers, essentially creating the yellow pages for the Internet.[20]

Surveying the initial implementation of the ARPANET in a speech in 1970, Engelbart could already envision the evolution of the networked community in which "there will emerge a new 'marketplace,' representing fantastic wealth in commodities of knowledge, service, information, processing, storage, etc."[21] In the early 1970s, ARC began collaborating with business managers in envisioning the office of the future. However, when ARPA cut off funding to ARC in 1975, the project rapidly shrunk down to Engelbart plus a lot of software. SRI had separated from Stanford in 1970, and SRI had never been that excited about Engelbart's vision in the first place. SRI preferred a vision of "office automation" to replace, not augment, workers. So SRI sold the software to a company called Tymshare, and Engelbart went along to help sell his software as part of Tymshare's general office automation service. Engelbart would eventually become director of the Bootstrap Project at Stanford, still pursuing his dream of using technology to enhance human potential.[22]

What is startling about Engelbart's achievement, and often ignored because of the institutional liquidation of ARC, is how many of the conceptual computing breakthroughs and initial implementations were achieved by his team. To name just a few that are critical to the networked economy:

- pioneering distributed electronic mail and e-mail lists five and seven years before ARPANET
- implementing word processing a decade before it began to appear in offices
- designing the mouse as an input device sixteen years before Apple introduced it to the world
- creating a windowing environment twenty years before Microsoft did so
- envisioning hypertext-linked documents in a distributed environment a quarter-century before CERN and Mosaic created theirs

All this was paid for by the federal government thanks to the vision of Bob Taylor and ARPA. As important for Silicon Valley, this federal investment would contribute to making the region a magnet for new visionary talent and a wellspring of the networked economy.

Xerox PARC: Making It Personal

When Xerox CEO Peter McColough announced his intention in 1969 to make Xerox *the* "architect of information" for the future, the attraction of the already existing Silicon Valley companies and the research energy of Stanford, especially the ARC project and Stanford's Artificial Intelligence Laboratory, were irresistible. Xerox had recently bought a computer company called Scientific Data Systems, whose breakthrough time-sharing computers had been largely designed by UC-Berkeley students funded by ARPA.[23] When Bob Taylor was hired in 1971 to run the computer research lab at PARC, he knew whom to recruit. Many of the earliest PARC researchers would come directly from the nearby ARC lab or from BBN, which had build the initial ARPA computer network. But Taylor's experience in organizing annual meetings of major ARPA researchers and hosting a separate annual conference for key graduate students involved in ARPA projects allowed him to locate the best minds across the country. The establishment of Xerox PARC in Palo Alto became a critical bridge in pulling together in one place fifty to sixty of the hackers and visionaries who had been working in separate streams of research for the past decade.[24]

If the attractions of Silicon Valley would pull people to PARC, there were significant factors pushing researchers there as well. Federal research funds were becoming increasingly militarized as the Vietnam War dragged on, so funds for pure computer research were drying up. The congressional passage of the Mansfield Amendment forced ARPA to provide specific justification for most computer projects, thereby ending the open-ended hiring of hackers at places like MIT that had fueled much of the creative innovation there.[25] On the flip side, many computer researchers with antiwar views found their employment by the Defense Department more and more unacceptable; many were looking for alternatives, and Xerox became a top option. Bob Taylor himself had left ARPA in 1969, disgusted with the war after being sent repeatedly to Vietnam to deal with the controversy over the army's "body count" reports.[26] He spent a year at the University of Utah, but to him, as to others with antiwar views whom he would recruit, Xerox looked like a golden opportunity to continue the computer and networking research to which he was devoted. Perversely, even as military demand was powering employment in

Silicon Valley, the militarization of research was also helping the region by driving a cadre of antiwar hackers from the East Coast to the West.

Out of this concentration of talent would be produced most of the personal computer technologies and design concepts that exist today. Applying the concepts developed at ARC, PARC researchers in 1974 created the Alto personal computer, a machine that would include a mouse, graphical icons, and a windowed environment. Given the commitment of ARC and ARPA alumni to collaborative computing, the PARC researchers created Ethernet data connections between all their computers to allow file exchange and electronic mail. In a different section from the computer lab, PARC also created the first laser printers, usable by any computer on the network, and pioneered what is called client-server computing to maximize the computing power of all computers on the network. PARC-designed software for the Altos included word processing, page layout, and electronic mail management, all using graphical user interfaces for ease of use. In summary, PARC had created the modern personal computer and office system well over a decade before any company would commercially apply their ideas.[27]

PARC's work was fueled by support from government-funded institutions, both through direct funds for a small number of projects and, more important, through collaboration-inspired breakthroughs. The most important, as described in the preceding chapter, was the use of the ARPA-funded ALOHA-NET radio-network system as a model by ARPA alumni Bob Metcalfe in designing Alto's Ethernet networking technology. Until ARC disappeared, PARC researchers continued to work with colleagues still at that institution, although PARC's focus on personal computers led to a greater and greater divergence in vision. Where Engelbart at ARC was focused on more centralized systems that would enhance groups that were working together in universities and industry, PARC became more and more focused on maximizing the power in the hands of each individual computer user.[28] In a sense, this is symbolic of the tension and the splits in approaches in the computer industry that would persist for the following twenty years. And with the commercialization of the Internet, as will be noted at the end of this chapter, the fundamental debates were replayed in discussions about personal computers versus network computers.

PARC not only contributed many of the central technologies and concepts for the personal computer revolution; it would also contribute many of the key company leaders who would commercialize those concepts. Alan Kay, a key Alto designer, would become chief scientist at Atari and later a senior researcher at Apple. Steve Jobs would visit PARC in 1979, and his guide that day, Larry Tesler, would join Apple in 1980 to work on the Lisa and Macintosh projects, which would first bring graphical user interfaces to a wide computing market. Bob Metcalfe would leave to set up 3Com to sell Ethernet technology to network offices and university computers, a critical prelude to broad participation in the Internet.

More indirectly, PARC and the rest of the Palo Alto–based labs helped foster an environment of shared information that helped spawn many of the earliest computer companies. PARC programmers would contribute programming time to community computer projects, many of them run by countercultural and antiwar activists who saw technology as a new tool for liberation. In turn, when the Intel 8008 chip came to market, a number of those engineering-minded community activists, notably Fred Moore and Lee Felsenstein, would in 1975 organize the Homebrew Computer Club, a legendary collection of students and engineers who would come together to hash out many of the ideas needed to assemble personal computers. Meeting biweekly in the auditorium of Stanford's Linear Accelerator Center, local hackers would mix with researchers from PARC and Stanford's labs, sharing the ideas that would birth many of the personal computer companies in the Bay Area.[29] In many ways, the Homebrew Computer Club deserves its mythic reputation, since it symbolizes the convergence of the region's political networks, embodied in radical activists such as Felsenstein, with the government-funded technology institutions of the defense establishment. With this overdetermined social network blooming in the region, if it had not been Homebrew members Steve Jobs and Steve Wozniak, others from that area inevitably would have led the personal computer revolution.

One crucial PARC defector from the Silicon Valley region was Charles Simonyi, the man who had designed the breakthrough Bravo word processing programs for the Alto. After meeting Bill Gates during a trip to Seattle, he would join Microsoft in 1981 as the company's director of advanced product development. At that point, Microsoft's business was essentially translating computer languages for use on personal computers, its premiere product being Gates's translation of the BASIC language (itself the creation of Dartmouth professors working on an NSF grant back in 1964). It was Simonyi who laid out the business plan for Microsoft that would promote its expansion into databases, spreadsheets, and of course, Simonyi's specialty of word processing. Microsoft Word, designed by Simonyi and a Xerox protégé, would become the keystone of Microsoft's push into software applications. Simonyi would describe himself as "the messenger RNA of the PARC virus" at Microsoft.[30]

Xerox itself never really successfully commercialized the early ideas generated at PARC. Partly this is because of the company's separation of its research and manufacturing divisions that analysts such as Florida and Kenney have so roundly criticized. But the reality was that the research being done at PARC was intentionally noncommercial. The PARC researchers well understood that computer technology was delivering double the power every year, so they were designing not for what was commercial in the mid-1970s but for what would be needed and used in the offices of the 1980s.

What the experience of PARC and Stanford's ARC labs shows is the indispensability of public investment both to the advance of basic technology and to regional economic innovation. PARC has been lauded for its critical contributions to the proliferation of computing. And every large corporation has vowed never to make the same mistake Xerox made in funding it in the first place (and have cut back their long-range research budgets to prove it). From Xerox's financial point of view, PARC's computer lab seemed like a costly mistake—although the laser printer, developed in a different section, was a large moneymaker for Xerox. But from general society's perspective, however, it was an astonishingly fortuitous mistake. In a time of cutbacks in nonmilitary research, Xerox PARC prevented the dissipation of the community of ARPA and MIT researchers. By maintaining ties to the national network of ARPA researchers and to the local university institutions while being embedded within the private economic innovation of Silicon Valley (itself the product of previous iterations of government support of R&D innovation), PARC would help set the stage for much of the explosion in the personal computer revolution in the Silicon Valley region. The United States needs to use ARC, PARC, and their embeddedness within the ARPA network of public-oriented researchers as a model for public investment in research.

Even as Xerox retreated from its commitment to pure research at PARC in the late 1970s and early 1980s, its legacy, and continued DARPA investment in the region, would lead to a new birth of Internet-driven industry. This new wave of networking companies in the region would not only absorb the technology of networking, but by actually promoting the ethic of open standards embodied in the Internet would also position the region for economic dominance as the Internet came to replace proprietary networks in the 1990s.

Before we turn to how and why Bay Area firms benefited disproportionately from the Internet, it is worthwhile to revisit some of the theoretical debates on the impact of the government on the high-tech industry and regional development.

Did the Government Create the High-Tech Industry?

There has been a vibrant debate about whether government support was the critical factor in the success of high technology in the United States and even whether its role has been destructive in a number of instances. As noted briefly in the first chapter, an overwhelming result of research by analysts such as Ann Markusen and Stuart Hall is that defense spending played an overwhelming role in the regional location of high-technology firms. Without defense spending, the existence of research universities had little effect on high-tech employment, a fact that has left vacancies in any number of attempted university-based "science parks."[31]

Not that universities were irrelevant to high-tech growth. Quite the contrary. E. J. Malecki found that university R&D was critical in combination with other factors. When the government funded self-contained government labs for defense or other government research, it rarely attracted complementary private industrial R&D. Where defense spending went either to large labs or to programs using mostly large companies, as in the space program, those outfits were usually so self-sufficient that few spin-off companies were generated. However, defense spending concentrated in regions with leading universities such as California and Massachusetts had significant agglomeration of high-tech firms.[32] While the reasons for this clustering around universities is not completely clear, at least one obvious reason is that whereas government labs and space programs hold onto employees for long periods of time, universities "fire" waves of employees in the form of graduate students, who, with the availability of defense contracts or a milieu of government-funded technology, often can be involved in start-up firms after graduation.

Some researchers, among them Richard Florida and Martin Kenney, have argued that military spending has been a drain on innovation, diverting companies from continual innovation in bringing products to market by directing toward military needs.[33] There is little question that using the military as the main tool for industrial policy in the United States has led to a waste of potential engineering talent, but it is less clear that much of that engineering talent would have been there at all without the existence of military R&D and a military market for the initial products of high-tech innovation. After World War II, money from the Department of Defense plus that from the Atomic Energy Commission accounted for 96 percent of all research money to universities—the key source for training new engineers.[34]

In the 1950s, the federal government accounted for two-thirds of all computer-related R&D and still accounted for more than half of R&D in the mid-

1960s. The two most cited commercially funded breakthroughs, the integrated circuit and the microprocessor, were products of companies, such as Texas Instruments and the Silicon Valley firms, that were immersed in that publicly funded milieu of technological innovation. Parallel teams at Texas Instruments and Fairchild created the integrated circuit within months of each other, and they repeated this breakneck competition in creating similar microprocessors within months of each other a bit more than a decade later. This reflects the fact that most commercial technology is short-term innovation based on publicly funded basic research. Robert Noyce himself argued that at best, his talent may have pushed forward developments by a year or so, noting that "these were ideas whose time had come."[35]

While the home computer boom commercialized technology and spread it widely over the past two decades, until the Internet (itself again the product of careful federal support) came along, the industry was producing its designs on the basis of general models drawn from decades-old government-funded projects. Paul Allen, Bill Gates's partner in creating Microsoft, began devoting a chunk of his wealth (more than $100 million) to a think tank firm called Interval founded in 1992 (and quietly shut down in 2000) precisely because Allen felt that short-term engineering and commercialization has not been able to deliver the breakthroughs of a generation ago. "Everyone out there," Allen has said, "is doing things in a one- to three-year time frame, and so we're trying not to do that."

Even in the commercialization of computers, the government played a strong role, from mandating the ultrahigh frequency (UHF) standards that created the first major mass-market demand for integrated circuits to the antitrust pressure on AT&T. Through the latter, the government made sure that the technical knowledge of transistors was broadly enough available in the public domain to allow a proliferation of competitive firms involved in the production of semiconductors.

The deeper criticism of the role of defense spending on R&D is not whether it was responsible for much of the technological breakthroughs of the past half century: the record is relatively clear that is was. Rather, the question is whether that policy indirectly hampered day-to-day "process" innovation in high-tech manufacturing. Florida and Kenney argue that the very emphasis on technological breakthroughs has left U.S. companies at an international disadvantage in the competition to efficiently produce and qualitatively improve already existing products. Where once industrial-research labs worked hand in hand with company manufacturing facilities, the availability of military R&D contracts caused the "decommercialization" of industrial R&D with

manufacturing facilities oriented to commercial markets and industrial research aimed at producing technological breakthroughs. With half of the big
corporations' R&D paid for by the Pentagon and with the accompanying cost-
overrun budgeting, such R&D was the ultimate safe bet for corporate profits.
However, this helped create a split in corporate culture, between the segment
making cutting-edge products for the military and that making cheap products
for the public. Scientists were separated from production, and manufacturing
came to be seen as a "profit center" where time and resources were not to be
wasted on developing innovations in production techniques.

This functional separation soon led to a geographic separation; in the words
of Florida and Kenney, "R&D grew further and further estranged from factory
production, as companies moved their manufacturing plants to new low-wage,
non-unionized locations and then relocated their R&D facilities to suburban
campuses."[36] They cite Xerox PARC as the poster child for the failure of corporate R&D because of its separation from the main manufacturing culture of
Xerox. This left its breakthrough technological innovations ignored within
Xerox, so start-up companies such as 3Com commercialized its Ethernet networking system, and the Apple Macintosh raided its mouse and icon software
system.

But Florida and Kenney's analysis of Xerox PARC just reinforces the essential difference between the dynamics of R&D for the commercialization of
technology by individual firms and the more public-oriented research that is
needed for the fundamental technological breakthroughs that power whole
regional and national economic transformations. And to whatever extent the
separation of research labs from manufacturing facilities hurt U.S. industry
nationwide, that separation benefited the Bay Area tremendously, given the
disproportionate placement of such centers there. So once again, national policy, however unintended, had strong effects in shaping the strength of Silicon
Valley's regional economy.

The Regional Economic Effects of High Tech

When Silicon Valley is viewed in a national context, its pattern of development, while unique in some aspects, still follows broad trends in regional
economic development among defense-driven areas. The historical military
presence in the Bay Area has already been discussed, but it is worth emphasizing that even through the 1970s and 1980s, military spending continued to flow
to the region (just as it had nationally in a range of fast-growing regions

around the country even during the military cutbacks of the 1970s). Even as some of the semiconductor firms found more and more commercial markets, the largest employer in the region remained Lockheed Missiles and Space Company, and by the early 1980s, it was receiving new billion-dollar contracts for its Sunnyvale operation. Other major defense contractors included FMC Corporation of San Jose, which was doing $1 billion of military business a year in that period; Ford Aerospace; GTE Sylvania; and ESL Incorporated. In 1983, as the Reagan defense budget ramped up, Santa Clara County had direct defense contracts of $4 billion on top of the large indirect purchases of its electronics by military subcontractors in other parts of the country.[37]

In a different vein, a number of commentators and researchers have noted the way the serendipitous birth in Palo Alto of two pivotal figures, Frederick Terman and William Shockley, were crucial to the emergence of the region as a high-tech center. M. J. Taylor has argued that location theories ignore the basic fact that a majority of new firms appear in their founders' hometowns.[38] While this individualistic view of regional organization history forces a certain respect for contingency, it has to be examined in the context of the long-term environment that produces innovators interested in a particular area of production and the resources available that make it possible for any business to appear. Terman's and Shockley's early interest in technology had been stimulated by a South Bay environment alive with technical innovation that was stimulated economically by the federal government in World War I. This interest was reinforced by massive World War II federal spending that would cement their desire to stay within the region. In the same way, the appearance of Microsoft in Seattle had much to do with that environment being powered by Boeing and the federal government in World War II and the aftermath of technical excitement. This would infuse a young Bill Gates with excitement about technology, and Seattle would be a region offering the technical support that would bring him back there when he established his own firm.[39] This reinforces the fact that economic development is almost impossible to formulate as short-term economic incentives; rather, is ultimately shaped by national and global economic forces that, as Robert Putnam observes, usually have long-term effects and often unexpected effects on specific regions.

The focus on Frederick Terman as a person also helps to obscure a focus on Terman's role, namely as a conduit for national industrial policy driven by the Department of Defense. The fact that so many unelected "civic boosters" like Terman ended up being the conduits for industrial policy in the postwar United States has helped contribute to a certain degree of collective amnesia about its existence, while also creating the impression that much of the defense-driven suburbanization of industrial development (and the consequent abandonment of many inner-city areas) was a natural phenomenon rather than a policy-driven result. Frederick Terman the person can be placed in the pantheon of indigenous Silicon Valley heroes, but Frederick Terman as the coordinator of national research and defense funds makes clearer the national priorities that helped give birth to and sustain the region's technological innovation and economic growth.

The other major explanation given for the birth of high technology in the region is the availability of venture capital. Much of the discussion of venture capital is intertwined with debates about the policy of lowering the capital gains tax, since those who advocate the role of venture capital cite quite impressive variations in the availability of venture capital tied to the capital gains tax rate. In 1969, the sum of $171 million in private funds was dedicated to venture capital firms nationally. Then the tax law in 1969 raised the capital gains tax from 28 to 49 percent. By 1975, only $10 million in new money was dedicated to venture capital. In 1978, lobbying by Silicon Valley and other high-tech firms led to capital gains taxes being scaled back to 28 percent. A decision in 1979 by the Department of Labor allowed more pension fund money to go into venture capital. In 1981, the capital gains tax was further decreased to 20 percent. By 1982, $1.4 billion in new money was dedicated to venture capital.[40] While impressive when seen as raw numbers, however, the birth of critical firms such as Atari, Apple, and a range of others during the mid-1970s when formal venture capital was scarce shows that venture capital was hardly the determinant of innovation and business success in the region. The fact that continued innovation and the expansion of venture capital would follow the 1986 increase of the capital gains tax to 28 percent also throws doubt on the causal link of tax rates and business expansion. Other analysts have argued that venture capital is more a causal result of defense funding infusions into the region combined with the region's already existing role as a banking center.[41] And many business leaders, including Andrew Grove and Gordon Moore at Intel, have questioned whether what they termed "vulture capital" hurts or helps the region, since it often merely encourages the dismemberment of solid firms as employees churn through different ventures looking for the quick profit.[42]

The interaction of venture capital and defense spending highlights the fact that Silicon Valley evolved not based on internal dynamics (although there were some important indigenous forces) but because of the region's link in a global system of research, production, distribution, and financial markets. Researchers David Gordon and Linda Kimball see the specific strength of the

Silicon Valley region as nonreplicable without those broader global linkages; otherwise, attempts to subsidize such high-tech centers merely create "illusory Silicon Valleys" based on limited branch-plant technologies with little chance of generating the diversity of energy that drives innovation in the Bay Area. They go so far as to argue that "objective factors" generating indigenous high-tech agglomerations matter little and that analysts should concentrate on the "comprehension of a region's social and economic position within the corporately organized integrated circuit of global production."[43]

How true this statement is overall will be explored in the following chapter in looking at how the Internet as a tool and an economic force changed business relationships between firms within Silicon Valley and with multinational firms globally. What is clear is that the Internet did strengthen the Bay Area's focus as the center of technology nationally and globally. Where proprietary information services would have thrived in a diffuse set of geographical enclaves, the very openness of the global Internet has had the seemingly contradictory effect of reinforcing the prominence of a few regions as coordinators of the technology and standards. The Bay Area assumed this role because it was so focused a product of the same public investment that created the Internet in the first place. In turn, its businesses shaped themselves to support that emerging Internet and were therefore positioned to benefit most dramatically as it expanded geometrically in the 1990s.

Sun Microsystems and Open Standards: Making Virtue a Commercial Necessity

Probably no single private company contributed more to the sustaining of the open standards driving the Internet, or benefited as much from the adoption of those standards, as Sun Microsystems, a seller of high-performance computers. Sun would enter, then dominate, the market for what were being called workstations, relatively inexpensive stand-alone machines aimed at engineers, that were beginning to replace time-share minicomputers. Started in 1982, Sun would be one of the fastest-growing companies in history, making the Fortune 500 within five years and clearing $1 billion in revenues in six years.[44] By 1995, the company would sell 1.5 million high performance computers used as the core systems for networking in government, universities, finance, and engineering. And from the first day of operation, every single Sun computer was shipped with hardware and software designed to be hooked up to the Internet. It was on Sun computers that much of the Internet would be networked in the 1980s and it was on Sun workstations that the first Web browser, Mosaic, would be designed.[45]

Sun's commitment to open standards reflected the company founders' emergence out of the milieu of Bay Area graduate students immersed in the ARPANET. When Stanford M.B.A.'s Scott McNealy and Vinod Khlosa teamed up with Stanford student Andy Bechtolsheim, who had developed a new high-performance computer using off-the-shelf components, it was natural for them to adopt UNIX, the popular university operating system, as the operating system for their new computer. And it was natural for them to bring in as a co-founder Bill Joy, the premiere UNIX and ARPANET programmer at UC-Berkeley in the late 1970s and early 1980s.

When Sun entered the emerging market for workstations, their main competitor (aside from established minicomputer makers such as DEC and Hewlett-Packard) was Route 128–based Apollo, the start-up company that in 1980 had pioneered the workstation market. Apollo was committed to proprietary standards for both hardware and software, a traditional strategy in the computer industry. Such a policy guaranteed Apollo high profits on each machine, first from sale of their own technology (as opposed to Sun's having to pay others for their components) as well as a continual stream of profits from licensing fees from proprietary software.

Anna Lee Saxenian and others see the adoption of open standards by Sun as a cost-cutting move to break into a market that would never have trusted a proprietary system developed by a company run largely by graduate students: "When Sun Microsystems pioneered open systems in the mid-1980s, it was largely making a competitive virtue out of an economic necessity. As a start-up, Sun lacked the financial resources to develop the broad range of new technologies needed for a computer system."[46] While it is true that Sun took advantage of the diversity of firms in Silicon Valley to quickly and cheaply produce their initial workstation design, it is important to understand that far beyond making a virtue out of a necessity, Sun's founders saw the virtue of open computing standards as a centerpiece of corporate strategy, even as a religious imperative. And making that virtue of open standards a necessity for all computer customers was the key strategy used by Sun to defeat rival companies using proprietary systems.

The commitment to UNIX's open standards on the part of Sun's founders, especially Bill Joy, derived from the fact that UNIX was the first operating system to be developed independent of specific hardware, thereby giving users and programmers more freedom from the dictates of hardware designers. It could be "ported" to different machines, thereby allowing the same program to run on completely different hardware. UNIX was created at Bell Labs from research it had initially done in collaboration with MIT's Project MAC. Since in the late 1960s, AT&T was being barred from most commerce involving computers, AT&T widely licensed UNIX, mostly to universities. UNIX was especially popular with ARPANET programmers working on a wide variety of computers because they needed to created an integrated set of software tools for managing their emerging network. UNIX had developed into a number of lackluster variations, so in the late 1970s, UC-Berkeley researchers (largely funded by DARPA) developed an improved version that was dubbed UNIX 4.1 BSD (Berkeley Software Distribution). Bill Joy, the lead programmer in the Berkeley UNIX effort, was again funded by ARPA in 1981 to create a new version of UNIX including TCP/IP networking protocols.[47] With a minimal $150 license fee, Berkeley seeded its UNIX version with its Internet protocols throughout the university world.

Commercial versions of UNIX, however, were splintered between various incompatible proprietary versions. Far from UNIX being a widely used standard in business that Sun could just hop a ride on, Bill Joy and the Sun team had to create a standard and sell private industry on the gospel of open computing. They took a number of steps so that the UNIX sold on Sun's computers would be seen as a real standard. Sun gave away the BSD UNIX and TCP/IP networking software with every computer they sold. When Sun developed a Network File System (NFS) in 1984 that enhanced network computing by making it possible to share files between different computers, they did not try to sell this advance as normal software. Instead, they licensed it to the industry for a nominal fee and even published the specifications for the software on the Usenet electronic bulletin board so anyone could construct an alternative to the NFS if they wanted to avoid the license fee. Usable on DOS, VMS, and other operating systems, it was a vital advance for networking and increased trust by customers that Sun would be an honest guardian of the open standards it was promoting on its hardware. Another key step was made in 1985 when Sun approached AT&T, recently allowed back in the computer industry, and worked out an agreement to merge Sun's Berkeley UNIX with AT&T's System V, further enhancing the public view of Sun's UNIX as *the* standard.

The first sales of their initial workstation model, the Sun-1, had nearly all been to universities, which, not surprisingly, liked the UNIX standard promoted by Sun. These sales had given Sun both the technical feedback and the credibility to help them sell their second version, the Sun-2, to a number of commercial buyers, notably resellers in the engineering and computer-aided design (CAD) field. Based on its commitment to networking, Sun had made some inroads into finance and automated-manufacturing markets as well. However, by the beginning of 1986, Sun was still a relatively small ($115.2 million in 1985 sales), if significant, player even in its own workstation niche within the computer industry. Apollo still dominated the workstation market and both Hewlett-Packard and DEC had introduced new workstations of their own in 1985.

What moved Sun into dominance of the workstation market and into a position to take on even higher-end-computer markets was the federal government's decisive move in 1986 in support of Sun's UNIX standard. By the mid-1980s, the federal government was faced internally with a mess of different computer systems that needed to be networked together. Because of the close ties of the Department of Defense to university researchers (largely fostered by ARPA/DARPA), the federal government already had an affinity for UNIX. So in 1986, the government passed regulations that no company could bid on any government computer contract unless its system offered UNIX as an option. This gave Sun a huge advantage in securing a large slice of the $500-million, five-year National Security Agency (NSA) contract then under bid. Sun's and AT&T's version of UNIX was now the benchmark for selling to the government and university markets (along with many private industry customers who would follow the government's lead in standards). This was reinforced in 1988 when the air force declared DEC's proprietary version of UNIX, called Ultrix, ineligible for government contracts.

With the NSA contract under its belt, Sun in 1986 passed Apollo as the top seller of workstations. Within a few years, Sun would have more than 50 percent of the federal workstation market, accounting for 20 percent of the company's sales.[48] And by 1991, it would add forty state and local districts to its list of customers as all levels of government adopted open computing standards and often passed regulations similar to the federal government's.[49]

Apollo along with other workstation makers would do a complete turn-about in 1987 and beginning promoting its "open computing" UNIX systems, but as in the case of Sun's other rivals, it was too late. Sun would use the advantages of close relations with a range of suppliers and collaborators in Silicon Valley (and increasingly around the world) to maximize its advantages in a world of standards that it and the federal government had helped establish. Sun would help promote standards for windowed environments within UNIX and continue to encourage networking standards of all kinds, advertising its own computers with the slogan "The Network is the Computer."[50] This standardization would not be complete, however, especially as rivals such as

DEC and Hewlett-Packard (which absorbed Apollo in 1989) committed themselves to a combined rival UNIX standard. This division over UNIX standards
in the early 1990s would come back to haunt all the workstation makers as
Microsoft and Intel began invading their turf.

Sun's most radical move came as the company, working with computer
scientists at UC-Berkeley, developed a new higher-power computer hardware
design called SPARC (for Scalable Processor Architecture). As soon as it developed the processor, it began licensing it to other vendors in July 1987. Never
before had a computer manufacturer unveiled its CPU (central processing
unit) architecture for the price of a license and a royalty. Working off an
operating standard while selling advanced hardware was one thing; allowing
clone companies to attempt to beat you on price on your own hardware was
something else. Sun, however, not only licensed the technology, but also supported the creation of an independent SPARC Vendor Council to set standards, including competitors such as Toshiba. Just as creating the UNIX
standard had encouraged a number of other Bay Area firms (notably Silicon
Graphics) to jump into the workstation market in the late 1980s, Sun was
betting that the sacrifice of some sales within the SPARC market would be
more than compensated by an enlarging of the overall demand for SPARC
machines, especially as Sun pushed deeper into higher-end computers.[51] This
would encourage a host of companies, mostly in Silicon Valley, in innovating
new chip designs around SPARC-compatible standards.[52]

It is crucial in evaluating Sun's commitment to open standards to understand that its success was not inevitable. An open standard *if established* does
give a significant advantage to a company able to collaborate with the myriad
of technologically innovative firms in Silicon Valley. But, counter to Saxenian's
argument, the establishment of that standard in the first place was not based
on an economic determinism of companies driven by the need to respond to
rapid product cycles and technological complexity.[53] Rather, it depended on
Sun's advantage in riding a standard saturated across the country's universities
under the guidance of DARPA and on the federal government's destruction of
the viability of competing proprietary systems through its completely blocking
itself off to those alternative systems. It was in that national context and application of federal public policy that Silicon Valley's industrial-district-style innovation would win out over Apollo and DEC in Massachusetts.

For a counterexample in which proprietary standards did quite well in the
1980s, one need only look at the success of the "Wintel" duopoly managed by
Intel and Microsoft. Where Sun spent the 1980s and 1990s licensing specifica-

tions of its hardware to anyone who wanted it, Intel spent the same period in court suing anyone, especially Advanced Micro Devices, who might try to duplicate the workings of its microprocessor. And while Sun worked relatively hard to establish trust among customers and software vendors as an honest steward of the UNIX system standard, Microsoft spent the same period being cursed by software vendors and investigated by antitrust lawyers. By the mid-1990s, the result was that Intel had driven nearly every competitor out of the personal computer processor business (with a small resurgence by rivals late in the decade), and Microsoft had come to dominate every software category it chose to enter.

The difference was that in the personal computer market there was no pre-existing government-backed standard to adopt and the government never used its purchasing power to punish proprietary standards. And as the government pulled back from enforcing strong UNIX standards in the early 1990s, what was dubbed the "UNIX wars" led to a splintering of standards between different companies, opening the way for Microsoft itself to move into the corporate computing environment. In this sense, it is clear that the roots of success of the collaborative system of production in Silicon Valley are not indigenous but are extremely dependent on a supportive national system of standards. The virtue of open systems, therefore, much as virtue in general society, pays off for its practitioners only when society supports its practice and punishes the "immoral."

With Microsoft increasingly dictating proprietary standards, production of personal computers was based on little cooperation, so it spread to far-flung enclaves around the country and the world, from Texas to South Dakota to South Korea. While UNIX companies such as Sun did negotiate many international contracting deals, the production and design of Sun's workstations, as with competitors like Hewlett-Packard and Silicon Graphics, had to remain rooted in the collaborative environment of Silicon Valley as standards and innovation moved forward without any one company controlling the evolution.

As we will return to at the end of this chapter, it was only as personal computers were integrated into the open standards of the Internet that Microsoft was forced to slightly adjust its tight control of its operating system to more open standards, "embracing and extending" the Internet, in the words of Microsoft executives. Not coincidentally, this came at a time when Intel processor power began to approach that of lower-end workstation servers, pitting Microsoft directly against UNIX-based systems for the first time in

competing for the networking market. Unfortunately, in the absence of continued government support for open standards, this seems more like the triumph of the false penitent feigning virtue until the retreat of the law in order
to win out in the end.

Cisco and the Commerce of InterNetworking

If Sun had to struggle to establish the standards that would help sustain the
rise of Internet networking, Cisco Systems would hop the explosion of those
standards to become in 2000, just before the NASDAQ collapse, the most
valuable company on earth.[54] With a name less known to consumers than
those of many of the other computer titans, Cisco supplied the technology at
the heart of the Internet—the computer "routers" that direct the data traffic
from one local network of computers to another, both within a large company
or across the Internet. Cisco's routers, successors to the original IMP computers set up by ARPA in the 1970s, now account for more than 80 percent of
routers connected to the Internet.[55]

As with Sun, Cisco's success came from its origins steeped in the culture of
the Internet in the Bay Area. Its founders were Stanford academics Leonard
Bosack and Sandy Lerner, husband and wife, who were also the computer
systems managers for, respectively, the computer science and business schools
in the early 1980s. The two systems were unconnected, so Bosack set out to
develop a device to allow the two local networks of computers to communicate
and share data. What he created was the router, a combination of software
and hardware that could inexpensively forward data packets from one computer site to another. Stanford had been funding its own networking effort, so
it was indifferent, even hostile, to Bosack and Lerner's efforts. Bosack and
Lerner probably would not have launched their own company, despite this,
but Stanford then refused to allow them to make similar routers for friends at
Xerox PARC and Hewlett-Packard. Enraged, they left their jobs.

No established firms were interested in their technology, so in 1986 they
mortgaged their house, maxed out their credit cards, and persuaded friends
and relatives to work with them, with the incentive of generously shared stock.
But what really made their business possible was the ARPANET, which allowed them to cheaply let other Net engineers know about their new product.
Within a year, Cisco was selling $250,000 worth of routers a month to univer-
sities across the country with no sales force and no paid advertising of any
kind. (Cisco did not buy its first advertisement until 1992.) Cisco would do

$1.5 million worth of business in its first fiscal year, which ended in July 1987, even turning an $83,000 profit. Technically in violation of ARPANET guidelines prohibiting commerce on the Net, Cisco was the first major company to build a market from scratch on a foundation of direct Internet marketing. Of course, it helped that Lerner and Bosack came out of the same milieu of the Internet as the university engineers doing the buying and that the NSF and other federal agencies were supplying much of the cash during this rapid expansion of the Internet at universities.[56]

It also helped that most networking companies were concentrating on the fight to network desktop computers, in a situation in which a whole mess of proprietary hardware and software systems were battling it out. The original Ethernet desktop networking technology faced new competition from token ring and ARCnet technologies, with each of the technologies splintering into internal differences. Computer companies across the country were fighting to lock customers into proprietary desktop networking systems, even as they ignored the emerging market for networking the networks based on standard Internet protocols.[57] Cisco began finding corporate customers who needed to connect far-flung local networks of desktop computers or who needed their own connections to the emerging Internet for commercial users.

Like Sun, as other companies entered the wide area networking market, Cisco made sure it had the top technology in order to dominate router networking, but it was the universal standards of the Internet that made technology, not control of proprietary standards, the measure of success. This is in sharp contrast to the fate of 3Com, a technological leader in desktop networking, which nearly foundered after it allied with Microsoft in the late 1980s in a fight over proprietary standards for desktop networking. By 1991, Microsoft had abandoned the fight in that round of network standards (returning a few years later with its Windows NT system), forcing 3Com into massive layoffs and the abandonment of almost all desktop networking. 3Com survived but only by committing itself to competing in the open standards of the wide area networking market as had Cisco.[58]

Cisco continued to grow rapidly (minus its founders, who left in 1990 after a conflict with the new management brought in by venture capital investors) and maintained its commitment to staying at the cutting edge of technology. And like Sun, which was happy to port its expertise to whatever processor its customers demanded, Cisco readily adopted new networking technology if customers wanted it. If it did not have the technology in house, Cisco bought the companies that did, spending billions beginning in 1993 on takeovers of both start-ups and established businesses in technologies in which they lacked the expertise. This included Cisco's $4 billion acquisition of its Silicon Valley neighbor StrataCom, a leader in new high-speed switching technology, to expand its tools for networking clients.[59]

With this growth, Cisco's CEO, John Chambers, has made Microsoft-like statements about Cisco's software becoming "a de facto standard" and that the company's purpose "is to shape the future of the network industry the way IBM shaped the mainframe and Microsoft did the desktop."[60] Even as competitors like 3Com and Bay Networks complained about the danger, the reality was that unlike with Microsoft, every Cisco product has to work with every other product using established Internet standards. When Cisco invents new ways to improve on those standards, it has to submit them for approval to the Internet standards boards before customers will accept them.[61] While many see such boards as losing authority (as the case of Netscape would soon show), much to Cisco's chagrin, the company has continued to have to fight for dominance the way it has been doing from the beginning: offering the best product based on Internet standards. And it has met ongoing stiff competition from the competitors it has not yet bought out, many of them, like 3Com, riding the same Silicon Valley Internet roots as they did.

Destroying the Village to Save It: Netscape and the Web Standards War

It was with the World Wide Web that the Internet broke into national consciousness and where Netscape Communications would become the central Bay Area firm around which a slew of new Silicon Valley companies would form. But unlike Sun, which rode public UNIX standards to rapid growth, Netscape began its life with a direct assault on the original government-based standards created by the National Center for Supercomputing Applications (NCSA). In this, Netscape would play a three-cornered game both against the NCSA and against Microsoft, who it knew would quickly be coming out with its own controlled standards. Netscape's initial success would be based on the virtual withdrawal of the government from any serious intervention on behalf of Internet standards. Netscape would develop a distinctly regional economic strategy built around standards, drawing on the expertise of a wide range of Bay Area companies in a fight that would start over standards for Web software but would explode into a battle for the very architecture of future computer hardware and software.

As described in the preceding chapter, the initial Web browser, Mosaic, was created at the University of Illinois at Urbana–Champaign where the NCSA was located. The NSF had officially funded the NSFNet backbone of the Internet to link five major supercomputing centers, including NCSA, and NCSA's software development group had concentrated for years on high-performance information-sharing and collaboration software. Even before Mosaic, the NCSA had back in 1985 created software "clients" for PCs and Macs, called Telnet, to allow people to access and use computers connected to the Internet as if the user were locally based. A different computer center at Illinois was responsible, as well, for the popular Eudora client for electronic mail on PCs and Macs. The NCSA had worked to create a graphics-based collaborative tool for sharing documents called Collage, so it was natural for them to form a team to develop a graphics-based version of the Web HyperText Markup Language (HTML) protocols created by CERN in Europe.[62] The result of this forty-member team was Mosaic, first introduced on the UNIX platform in January 1993, with Macintosh and PC versions introduced in August 1993. Copyrighted by the University of Illinois, Mosaic could be downloaded for free by individuals and by companies wishing to use the Internet for internal communications.

However, the NCSA did not want to become a help desk for commercial applications, so in August 1994, the University of Illinois assigned future commercial rights for licensing NCSA Mosaic to Spyglass, Incorporated, a local company created by NCSA alumni to commercialize NCSA technology. The goal was for university researchers to continue developing longer-term technology and standards to be incorporated into browsers, while Spyglass would help license the technology to companies addressing immediate customer needs such as support, speed, and security. Spyglass began widely licensing Mosaic to computer companies, including IBM, DEC, AT&T, NEC, and Firefox Incorporated, who was working to integrate Mosaic standards into Novell networking software for the personal computer.[63]

With the licensing agreement requiring Spyglass to closely share technology with the NCSA, Spyglass president Douglas Colbeth noted that the benefits of the commercial version of the viewer, dubbed Enhanced NCSA Mosaic, would be a "stable and standard" product across multiple computer platforms.[64] And the combination of license fees to the university and the potential economic development benefits to the surrounding community made it appear that the Urbana-Champaign region might be about to experience the same government-technology-driven boost that Stanford and Berkeley had given to the Silicon Valley region. "This is the classic example of technology transfer that

Congress envisioned in setting up the supercomputer center in 1985," argued Larry Smarr, director of NCSA at the time.[65] Stable standards and technology transfer to the community made it appear that Urbana-Champaign would be appearing on the map as an upstart technological rival to Silicon Valley.

But it was not to be. Watching Mosaic from the Bay Area, Silicon Graphics CEO Jim Clark, a veteran of the workstation standards wars, understood how much money could be won if a company could take control of the standards of this new Internet tool. So Clark left his company and set out to destroy Mosaic and replace its government-backed standards. He met with Marc Andreessen, a member of the Mosaic team who had been hired at a Bay Area Internet security firm called Enterprise Integration Technologies. Out of that meeting in April 1994 was born Mosaic Communications Corporation (later to be called Netscape). With Clark putting up the capital, Andreessen recruited five other Mosaic team members from NCSA to design what, in house, they called Mozilla, the Mosaic-Killer. In six months, Clark's team had created a powerful browser, which the team called Netscape, with easy-to-navigate features, and which loaded graphic images faster than NCSA's Mosaic. But Netscape did something else—it included the ability to display text formatting that did not even exist in the HTML standards embedded in the NCSA Mosaic browser. This meant that Web pages designed to work with Netscape would not be readable by all the other Mosaic-based browsers. This would encourage people to use Netscape browsers and, as Netscape developed them, would encourage Web designers to pay Netscape for the server software that developed Web pages using their modified standards. It was in this later market of selling Web design tools costing from fifteen hundred to fifty thousand dollars where Netscape intended to make their money.[66]

And then Clark and Andreessen compounded their fracturing of the NCSA standard by giving their version away over the Internet. The University of Illinois had demanded that Clark's company pay for a license before selling their version. Clark later said that he refused because the university was demanding an ongoing per-copy royalty: "I didn't tell them, but we had intended to allow people to download it, and they were going to charge me. The amount varied, but nothing is innocuous when you're talking tens of millions of people."[67] The point of the licenses by Illinois had been, along with collecting a little revenue, to control the standards and make sure that the only free version available was the official NCSA standard. Netscape would essentially "dump" its version onto the Internet, thereby undercutting the rest of the commercial browser companies who couldn't duplicate Netscape's actions because they were fairly paying per-copy license fees. So Netscape, being the sole enhanced commercial browser flooding the Internet, was able to destroy NCSA-led standards and take over standards creation itself.

Unlike with Sun Microsystems, where the government would decisively support open government-based UNIX standards, the federal government did nothing to support NCSA's standards. Other companies and analysts would immediately condemn Netscape's actions as a monopolistic move.[68] The government, however, made no investigations into possible monopoly practices, brought no lawsuit alleging intellectual property infringement, presented no announcements that the federal government would use only NCSA-approved codes in government Web sites or that it would refuse to buy any Web servers (namely, Netscape's) based on such nonstandard formatting; no signal came from the government that they would oppose Netscape's takeover of the standards. Instead, the University of Illinois, after a bit of public grumbling, threw in the towel. They signed an agreement with Clark in December 1994 that allowed Netscape to be sold without a license for the minor concessions that the word *Mosaic* be removed from the firm's title and that no mention of Mosaic be made in marketing the browser.[69] Given the moves toward privatization of government Internet functions in recent years, the failure of decisive policy is not surprising. Criticism had already been leveled against the University of Illinois and NCSA for their commercial relationships,[70] and, in the context of December 1994, a month after Newt Gingrich's antigovernment message had stormed to a majority in Congress, government officials probably had even less appetite for defending the wisdom of government regulation of standards.

In a perverse way, Clark and Netscape would justify their destruction of the government standards based on the expected weakness of the government in defending them. They predicted that Microsoft would soon use its dissemination of the operating system to take control of standards if Netscape did not do so first through free distribution. Argued Clark:

> At some level, standards certainly play a role, but the real issue is that there is a set of people, a set of very powerful companies out there, who don't play the standards game. For the standards game to work, everyone has to play it, everyone has to acknowledge it's the game. Companies such as Microsoft aren't going to sit around and wait for some standards body to tell them, You can do this. If your philosophy is to adhere to the standards, the guy who just does the de facto thing that serves the market need instantly has got an advantage.[71]

And once Netscape had taken control of the standards from NCSA, in order to gain trust in its management of the Web standards, the company drew on many of the same collaborative practices and resources in the Bay Area as Sun had in gaining trust in its stewardship of the UNIX standards. In fact, Sun itself would become a key partner in the alliance against Microsoft over standards for interactive aspects of the Web. Crucially for Internet commerce, Netscape worked closely with Enterprise Integration Technologies (EIT), Andreessen's old firm, to agree on security protocols for online transactions, working through an initially regional consortium called CommerceNet (much more about that organization in the following chapter). Netscape would build the possibility of "plug in" architecture into its browser, so an explosion of new firms could easily follow Netscape in Internet distribution of new products that could instantly be incorporated into individuals' desktops. Netscape, having seized leadership of Web standards, would continue to work with the old Internet fellowship of engineers embodied in the Internet Engineering Task Force (IETF) and the more recent World Wide Web Consortium (W3C) based at MIT and run by Berners-Lee, who came to MIT in late 1994.

And when Microsoft entered the game, with its own Internet Explorer browser to appear on every Windows desktop, the grumblings over Netscape's occasional forays into proprietary advantage would lessen as the alternative fear loomed that Microsoft would take over the whole computing world. Having come late to the Internet, Microsoft initially directly licensed Mosaic browser technology from Spyglass in December 1994—a license netting Spyglass about $13.1 million. But when Microsoft began giving its browser away at the end of 1995, the rest of Spyglass's licensing revenue (amounting to $20 million) disappeared as the browser war settled into a two-company fight between Netscape and Microsoft.[72]

In the end, Netscape would argue that the beloved public village of standards was threatened by Microsoft, and that Netscape had only destroyed the village in order to save it. And if saving the village made Jim Clark's Netscape worth $4.2 billion (the price America Online paid to purchase it in 1998) and snatched leadership of Internet development away from Illinois back to Silicon Valley—well, this was just returning leadership of Internet-based computing to the region that government support had made the leader in the first place. However, Netscape's ultimate failure and the rise of companies such as America Online, based in Washington, D.C., emphasized the new pressure on the Silicon Valley model of innovation.

But beyond the economic battle of the browser war was a more fundamental divide over the future of computing. In the region that birthed the personal computer, these Bay Area Internet-driven companies would make the replacement of the personal computer by "network computers" the path for moving information processing off the desktop into the realm of network servers—not coincidentally the bastion of Silicon Valley technical dominance.

Network Computers and Massive Parallel Processing: The Bay Area Challenges the Personal Computer

By 1996, the evidence was clear that the Internet was birthing a very different technological strategy for Bay Area firms. Building on the infrastructure of client-server workstation firms, including Sun, Hewlett-Packard, and Silicon Graphics; networking firms such as 3Com and Cisco; and a slew of software firms, a new explosion of Internet-based companies were appearing in the Bay Area region. A quarter century of investments by the government in the Internet were coming to fruition, and with a disproportionate share of those funds and expertise having been channeled to the region, Silicon Valley was in a position to reassert its authority over the direction of technology.

In May 1996, a new alliance of Netscape, Sun Microsystems, Apple Computer, IBM, and the software company Oracle announced the specifications for a new standard of computing, the network computer. In many ways, this proposal was the long-time culmination of the Internet's direction, with four key Silicon Valley companies (with IBM essentially tagging along) working to ride their long-time expertise in order to retake control of the computing world. Instead of processing power being concentrated on the desk with software individually installed on each computer, the network computer would be an inexpensive customized processor in a box that would access CPU power over the Internet, downloading software "applets" as needed from a centralized server. Priced under one thousand dollars (and it was hoped as low as five hundred dollars), such network computers would be more economical in both homes and offices than the traditional personal computer. What these companies proposed was nothing less than to kill the paradigm of personal computing that the Bay Area itself had introduced nearly two decades before.

Microsoft was officially invited to join the group, but declined for obvious reasons, since the other clear goal was to destroy Microsoft's and Intel's duopoly control over the home and office desktop. If most computing power was be located in the network, there would be no need to constantly upgrade to a faster (read Intel) processor; instead, offices or homes could simply subscribe to higher levels of processing power as needed (or as developed by server designers). Similarly, the network itself would be responsible for upgrades in operating systems and application software, so there would be no need to constantly buy upgrades of (read Microsoft) operating systems and applications.[73] The overall goal, as Sun had argued in advertising, was to make the network the computer, thereby maximizing the resources available to each user and allowing constant upgrading of those resources as technology advanced.

In many ways, this was a return to the vision of Doug Engelbart at ARC, a vision that the Xerox PARC researchers had abandoned in favor of putting maximum processor power directly in the hands of individual users. In the mid-1970s, the absence of a broadly available Internet made such a movement to personal computing the natural development, but historically, personal computers may be seen as a two-decade-long detour. Seeing the connection between their direction and Engelbart's vision, Sun and Apple have been prime funders of Engelbart's nonprofit Bootstrap Project, where today he continues the research he started decades ago on collaborative computing.[74]

However, each of the companies involved in the Network Computer (along with the many more who would immediately sign up in support) had more than a theoretical interest in replacing the personal computer. With its entire business strategy tied up in the Internet, Netscape had the most obvious stake in preventing Microsoft from using its dominance of the personal computer to in turn dominate other Internet-based tools. Apple and IBM had both, in their own way, lost the personal computer market to Microsoft. Despite four million computers being sold worldwide by Apple in the previous year, Microsoft's dominance of the desktop was clear enough that even Apple loyalists were pronouncing the company on its deathbed.[75] With Microsoft's control of the Windows operating system, IBM had been reduced to being a clone maker for a computer once described as being "IBM-compatible." With no room to innovate, given Microsoft's absolute proprietary hold on the operating system, IBM's R&D annual budget of $6 billion was nearly useless in the personal computing market.[76]

Sun Microsystems, in many ways the linchpin of the alliance, saw the face-off with Microsoft in nearly religious terms. With Windows NT machines increasingly penetrating the network-server market, which that had traditionally been Sun's stronghold, Microsoft was emerging as a clear and present danger to the UNIX standards that Sun had worked so hard to establish. Some of this could be blamed on bitter battles that had partly fractured those standards in the early 1990s, but much of Microsoft's march was based on the reality that Windows was on so many desktops, increasingly combined with

Microsoft's Explorer Web browser, so that using Windows NT on the central server became more and more attractive for many companies. It seemed that 1996 would be the year that NT-based server sales would pass UNIX-based sales.[77]

Sun's weapon against Microsoft and the linchpin of the network computer was Java, a unique programming language invented at Sun that could run the same program on any operating system platform—thereby making Microsoft's control of its operating system irrelevant. Most important, Java programs are designed to run seamlessly across networks: whereas ordinary programs take up megabytes of hard-drive space on conventional PCs, little Java applets can be delivered as needed to the desktop. With Netscape's incorporation of Java into its browser, the first effect of this was to make Web sites more interactive, but the broader effect could be to harness the Internet to deliver the day-to-day programs people use without a full-scale computer being needed to run them.

To accomplish this goal, Sun used every lesson it had learned from its experience with UNIX. It licensed Java for the asking to any company for nominal fees. It began developing a complete operating system based on Java called JavaOS. It launched a subsidiary called JavaSoft whose purpose was less to make money than to work with all licensees to keep the Java standards stable. Along with IBM and other companies, Sun pooled together a $100-million venture-capital pool call the Java Fund to seed start-ups, using the language.[78] By the end of 1996, almost every major company, except Microsoft, had agreed on a single set of Java standards for all software, a level of standards agreement Sun had never achieved with UNIX. Even Microsoft reluctantly licensed Java to develop applications because of customer demand[79] (although Microsoft would be sued by Sun for violating the license by modifying the language in order to undermine those standards). Respect for Sun (and fear of Microsoft) was evident in 1997 when the International Organization for Standardization made Sun the official guardian of Java standards, a role usually given to independent associations rather than individual companies.[80]

Aside from helping Sun achieve a "halo effect," as one Sun executive put it, this public-minded standards creation was designed to hold off Windows NT and allow Sun to focus the computer industry on deciding which company could design chips and machines that would best run the Java software created around those standards. With a variety of new SPARC processors in production that were designed to maximize Java processing speeds, possibly by three to five times the speed of existing platforms, including Windows NT, Sun and its allies charged into the competition against Microsoft.[81]

While network computers competing head to head with personal computers were slow to enter the market, given the slowness of most home Internet connections, the real promise of network computing and Java soon began appearing in a raft of "smart" consumer appliances, from cellular telephones to stereo equipment to set-top boxes on cable television systems.[82] In early 1998, Cable and Wireless agreed to use network computer technology in cable set-top boxes to be installed in 7 million homes and businesses in the United Kingdom, Australia, and Hong Kong.[83]

For companies managing hundreds, often thousands, of desktops, the idea of moving maintenance of software from the desktop to central computers became increasingly irresistible. In many cases, companies moved the software outside the company altogether, hiring firms to provide software as needed over company networks connected to massive computer "server farms." In a real sense, the entire computer, telecommunications, and consumer appliance world was increasingly being described as a race between Silicon Valley's model of open networking standards and Microsoft's proprietary Windows maneuvers.

In some ways, the most interesting booster of the network computer model was the database software company Oracle Computer, whose CEO, Larry Ellison, was the main instigator of the alliance. At the simplest level, he was there on the stage announcing the network computer because he feared that Microsoft, was moving dangerously far into his company's territory. Oracle was the second largest software company in the country, but had rarely competed directly with Microsoft since Oracle made software for every type of computer *except* the personal computer. With Microsoft moving increasingly into the server arena, an assault on Oracle's software by Microsoft was inevitable.

However, Ellison had much more than a defense strategy up his sleeve. For Ellison the interest in the network computer was in the computers and data that users would be accessing at the other end of the phone line. Ellison had become a billionaire by developing fast relational databases in the late 1970s to assist corporations and government in more easily organizing vast amounts of information (his first contract, typically, had been for the federal government, in this case the Central Intelligence Agency [CIA]). He had used IBM's own research to get a three-year jump on Big Blue, as IBM is known, and unlike IBM, which initially produced their databases only for their own machines, Ellison wrote versions of his software for every brand of mainframe and networked server possible, ending up dominating the database market. He never gave a thought to the Information Superhighway, considering it only when British Telecom approached the company in 1993 to help run an experimental interactive television service. Unexpectedly, Ellison realized that being a database creator put Oracle smack in the middle of the interactive age. "Better to be lucky than smart," he deadpanned at the time.

But Ellison was more than lucky; a few years earlier he had invested at least $60 million of his own funds to take control of a struggling Northern California supercomputer firm, nCube, which specialized in massively parallel processing, linking thousands of microprocessors. Like most supercomputer makers, nCube had built its business selling to government labs and universities, but Ellison saw that a commercial market could be opened up for these machines if he linked Oracles's software to such an endeavor. In 1988, he decreed that all Oracle software would be written to run on nCube supercomputers. nCube was able to score sales to corporate customers, among them BMW and Shell Oil, and ended up capturing 65 percent of the market for massively parallel systems in Japan. Other deals were soon inked with Bell Atlantic for interactive television experiments in the United States.[84] And as the Internet took off, Ellison saw the potential to sell nCubes bundled with new high-end Oracle multimedia Web servers for large enterprises needing to manage large numbers of users accessing a range of multimedia information.[85] As broadband access took off, nCube moved itself out of the hardware business into selling the server software needed to manage multimedia data on such computers.

What was at work here was not only an assault on the personal computer market but also on the lower-end server market, which Microsoft was invading as well. And Ellison was just following Sun and other Bay Area workstation firms in pushing the technology toward supercomputer levels. The dilemma for mainframes had traditionally been that despite their reliability and central control, crucial for managing large databases, the relative costs of computing power had been much cheaper on smaller machines, thereby pushing systems toward networks of smaller machines. However, new research, driven by government funding, had been using massively parallel processing to bring down the costs of supercomputing and allow much more flexible computing levels, responding to economic needs based on what has come to be called "Scalable architecture." Supercomputer research had always depended on federal government support and procurement, but after cutbacks in the late 1980s, federal support for supercomputing began skyrocketing in the mid-1990s. Partly to support simulated testing of nuclear weapons (given the Nuclear Test Ban Treaty prohibiting real testing), the Department of Energy's Accelerated Strategic Computing Initiative (ASCI) had its budget ratcheted up from $85 million in 1996 to $600 million per year by 2001.[86]

Following long-standing trends in the Bay Area, Silicon Valley firms were following government funding of new technology into new markets and, in the end, a new paradigm of computing. In 1996 Silicon Graphics acquired Cray Research, the premiere supercomputing firm, and Sun bought out key technology from Cray as well that year. Both built on Silicon Valley relationships to leverage themselves into competition for the large "enterprise" level computing markets. Intel also held a large share of the supercomputing market and won the first multiyear pact, worth $45 million from the Department of Energy's supercomputing initiative to push for its technology. IBM won the second pact, worth $93 million, but Silicon Graphics entered into the big leagues with a funding agreement with the Department of Energy as well.[87] All these technologies would in turn be applied to the commercial markets, which will have an increasing thirst for expanded computing power.

And Ellison was working to dominate the Internet database markets that were emerging and that would be dependent on advances in supercomputing technology. The need to keep pushing the technology was clear. Although many analysts concentrate on a lack of wired bandwidth to the home as the major barrier to multimedia Internet industries, the inability of central computer servers to manage the media information demanded is equally serious. To put this in perspective, one of the largest commercial databases in existence by 1993 was American Airline's Sabre computer reservation system, taking up one hundred gigabytes of computer memory (roughly one thousand times the memory of a typical desktop computer in the year 2000). However, the same amount of computer memory could only manage an interactive online library of fifty movies. It was this reality that made Ellison see the design of supercomputing servers combined with broadband Internet access as the next big market opportunity.[88]

And at the most visionary level, Ellison during the 1990s had promoted what he called Oracle's Alexandria Project, the name evoking the ancient Greek library in which the goal was to contain all the world's published works. The project, built on the dream of using computers to change the way human knowledge is gathered and stored, would bring together unfathomably large multimedia databases including books, art, films, and news coverage. Needing memory ten thousand to one million times larger than that of databases Oracle currently sold, such databases would act as centers for global commerce and learning while enabling radically new forms of collaboration.[89] In this vision, the network computer was the first step on the road to a new computing future.

Conclusion: The Engineer's Lament—Technology and Its Discontents

The Alexandria Project would be the fulfillment of Doug Engelbart's dream of augmented collaboration, itself inspired a half century earlier from Vannevar Bush's vision of a device to rapidly access the ever escalating information faced daily in our modern lives. It was Vannevar Bush who contributed mightily to the postwar federal government support of the R&D of the computer hardware that would make possible Engelbart's laboratory at SRI in the 1960s, as well as subsidize the vibrancy of innovation in the region. In turn, Engelbart's work would advance the tools for networking that would help attract a whole new generation of innovators to Silicon Valley. With that base of innovators, continued government subsidies of the open standards of the Internet would help "lock in" technological leadership by the region's innovators, who themselves so much the product of that public investment, would push forward technological change based on the standards that they had themselves helped develop. To this day, the federal role in funding research in the region (a yearly total of $14 billion for California) supports technology strength in the Bay Area.[90]

Yet Engelbart today has doubts whether, despite the ever expanding number of computers, this technological change is really in the end improving the economic and social lives of most people in society. In one interview, Engelbart cited an MIT study that showed technology having dramatic effects on productivity in specific work situations yet having little, and even negative, effects on productivity overall in society. This reinforces his half-century-long frustration that so much of the focus in implementing new technology has been on automating existing tasks rather than on using the technology to augment human capabilities in truly new ways.[91] Engelbart's lament in many ways echoes his regional predecessor, Lee De Forest, who would write later in life that the result of his invention of the vacuum tube and radio technology was the debasement of culture and the prostitution of his technological child: "What have you gentlemen done with my child? He was conceived as a potent instrumentality for culture, fine music, the uplifting of American mass intelligence. You have debased this child, you have sent him out in the streets . . . to collect money from all and sundry, for hubba hubba and audio jitterbug. . . .

Some day the program director will attain the intelligent skill of the engineers who erected his towers and built the marvel which he now so ineptly uses."[92]

Engelbart's and De Forest's regret reflects the fact that just as the invention and development of technology is driven by economic and political imperatives often independent of general welfare, so too the implementation of that technology reflects those broader social forces, often at the expense of both social uplift and equality.

In the following chapter I will explore how the implementation of Internet-related technology in the Bay Area has reflected the social and economic imperatives that have driven its production and how the social forces driving its production in turn have shaped the overall political economy of the region. In discussions of creating a "Smart Valley" in the region, political and business leaders have echoed Engelbart's hopes of translating the technology into a broader-based change in economic relationships, but, as we will see, its implementation has more reflected the social realities of economic inequality both within the workforce and between areas within the region.

4

Business Cooperation and
the Business Politics
of Regions in the
Information Age

In early 1992, Silicon Valley faced an economic crisis where job growth since the mid-1980s had been lagging behind the national average. Fed by both defense cutbacks and a sense of foreign and domestic competition against their high-tech products, business leaders in the region created a new organization called Joint Venture: Silicon Valley, its goal "a community-wide effort . . . to construct a rational blueprint for the continued economic vitality of Silicon Valley."[1]

With the publication of a commissioned report, *An Economy at Risk,* along with a jammed conference of technology leaders—with more than one thousand attendees—that was held by the new organization in the summer of 1992, the organization became the focus for business revitalization within the region. Belatedly, government leaders were also encouraged to become involved, with San Jose mayor Susan Hammer becoming co-chair of the organization in 1993. Not only would Joint Venture work to strengthen existing high-tech organizations such as the San Jose Chamber of Commerce, the Santa Clara Valley Manufacturing Group, and the American Electronics Association, it also unleashed a host of new working groups and a range of new incorporated organizations. One of these spin-off organizations was Smart Valley, its vision being to "create an electronic community by developing an advanced information infrastructure and a collective ability to use it."[2]

Smart Valley, in turn, would become the catalyst for a host of initiatives and new organizations focused on using advanced communication technology to strengthen the regional economy. For most in Silicon Valley in the early 1990s, the Internet was seen not as a focus of commerce itself but rather as a new tool for regional revival, of the creation of electronic communication bonds that would bring about more efficient production within the region. In doing so, Smart Valley was following the model of Singapore, a "smart island," and the "science cities" being planned throughout Japan, themselves in turn trying to improve on the model of regional technology development in Silicon Valley itself.[3]

Two major initiatives of Smart Valley, CommerceNet and the Bay Area Multimedia Technology Alliance (BAMTA), were in turn incorporated as separate organizations with greater yearly funding than that of their parent (thereby adding to the explosion of interlocking business organizations spawned by Joint Venture). CommerceNet, initially chaired by John Young of Hewlett-Packard and involving almost every major computer company in the region, was established with an early vision to make business-to-business networking over the Internet more feasible, while BAMTA saw its mission as uniting together the diffuse and often noncooperating artistic, software, and hardware talents of the Bay Area in creating new multimedia technologies over networked technology. All these initiatives seemed to be fulfilling the original mandate of Joint Venture to strengthen Silicon Valley vis-à-vis other regions.

However, both these spin-off initiatives rapidly went through transformations that downplayed their specifically regional nature in favor of allegiance to a national and even global perspective on the industry. CommerceNet had allowed a few non–Bay Area companies and federal agencies in its membership from early on. Such outside members proliferated so quickly that Commerce-Net began talking of creating regional subgroupings around the country, which was followed by the acceptance of foreign member companies, including Japanese companies, revealing a rather startling change from the nationalist anti-Japanese stance of the Valley in the 1980s. While the bulk of members still hailed from the Bay Area, the organization began to publicly downplay any specific regional focus. Similarly, BAMTA very quickly expanded its membership nationwide and globally and soon officially wiped the words *Bay Area* from its name, preferring "Broad Area Multimedia Technology Alliance."

What did this transformation mean? For companies that established the network and remained heavily involved, it did not necessarily mean a loss of loyalty to their home base in Silicon Valley. However, it did reflect the technological, social, and economic forces buffeting the high-technology economy and how these institutions in turn played real roles in reinforcing those same globalizing economic forces. Further, these changes reflected the specific alterations that led the Internet from being seen mostly as a high-tech tool for regional revival back in 1992 to being perceived as a global industry unto itself by 1994 and 1995. And it reflected the new dynamics of high technology in which global initiatives are almost impossible to separate from the regions from which those initiatives are launched. It is in the latter sense that, paradoxically, the abandonment of regional language by these Bay Area initiatives reflects the intense importance of the region, not its absence.

What Is a Region in the New Economy?

It is clear that as a regional space, Silicon Valley and the Bay Area are very specific products of national forces, especially the investments of the federal government in technology in the postwar period. Yet even as some of those investments were withdrawn (although not to the extent that some believe), Silicon Valley and other such regions continued to loom large in the geographic space of the global economy. Despite the odes to "virtual corporations" and the rhetoric proclaiming that global communication technology made irrelevant the location where one did business, real companies in the specific regional economy of the Bay Area were playing a disproportionate role not only in guiding the new technology of the information age but also in leveraging that technological dominance into global power over finance, media, and even politics in the new economy.

Yet if regions mattered as much, if not more so than in the past, in shaping business organization, the globalization of work for poorer, less skilled workers continued to accelerate. The irony is that those families most bound by poverty and ties to specific spaces are the ones most at the mercy of the ravages of globalization, while the most cosmopolitan engineers and professionals seemed to find solid community in the nexus of business and living spaces in regions such as Silicon Valley. Business leaders increasingly found that far-flung global alliances were most easily built on the foundations of trust in such regional spaces, the key being that those least dependent ultimately on a specific space end up with the most power within a region. The fact remains that the economic action of technology innovation is overwhelmingly local, but the power of corporations to choose their venues is global and outside the control of local actors who desperately try to negotiate with these global partners.

As described in the first chapter, a range of theorists (Saxenian, Piore and Sabel, Best) have explained this regional agglomeration by citing the economic tradition of Alfred Marshall and the idea of industrial districts. Instead of the traditional neoclassical economic view that firms can be treated as isolated units seeking the lowest-cost production area, these new views identified the milieu of production to be as important as the individual firm. Such industrial districts create a shared employment pool along with a flexible set of production partners that can be easily rearranged as needed for new innovation or changes in the market.

However, Bennett Harrison and Robert Reich, among others, emphasize that even in such industrial districts, multinational corporations are the driving force in economic decisions. Corporate webs use the flexibility that is present in regions to maximize the effectiveness and profitability of their local offices and factories while using local knowledge to enhance global empires. They promote the collaboration of symbolic analysts in local regions to create the key information assets for global competition, while outsourcing low-skill production jobs to the array of regions, which have low labor costs.

Yet there is another factor shaping regional economic space that is becoming increasingly important, namely, the establishment of agreements on information standards, which are increasingly crucial for economic growth. The creation and maintenance of standards is essential to capitalist accumulation, but the means for achieving those standards create tensions that shape economic geography in distinct ways. At the most basic design of technology, government standards have become increasingly crucial, but the dynamics of corporate power inevitably pushes for privatization of those standards.

On the one hand, the preference of individual corporations is often for proprietary standards that they themselves control, since that route allows the company controlling the standards to collect monopoly rents, the primary example of this today being Microsoft. However, such proprietary standards narrow competition and innovation, while often leaving incompatible technologies littering the landscape. The alternative is for the government to promote open standards supported broadly by a range of private-sector companies, which while foregoing monopoly rents on proprietary standards gain by a much broader expansion of the overall economic pie because of the innovation stemming from open standards.

Creating such standards can open up entirely new markets, which grow more quickly and explosively if consumers of technology products have faith that they are not buying into a dead-end proprietary technology. But such open standards, which require continual interaction between all partners in a rapidly changing technology field, will fail unless the social networks among companies are strong enough to assure each one that they will get a fair piece of the overall economic returns from cooperation. Such social networks can theoretically be built between companies in widely disparate locations, but empirically their success is much greater when companies have the multiple interactions and the shared social capital derived from sharing the same regional space.

The kicker is that while open standards, such as those around the Internet, may have originated with a strong regional character, they must also generate globalized marketplaces and globalized systems of alliances for ultimate success. This is one of the roots of the contradictions between localized cooperation, highlighted by analysts such as Saxenian, and the multinational corporate direction observed by Harrison and Reich. And in the absence of government involvement, needed to continually replenish the local social capital and legitimize the standards, it is unclear whether those open standards can long hold out against the opportunism of global corporations defecting from cooperation in favor of proprietary or semiproprietary approaches.

In this chapter I will use the development of the Smart Valley initiatives to explore, first, the process by which the Internet changed from being merely a tool for "smart networking" in a region to being a focus for a whole new industry, even a whole new kind of marketplace, that would radically modify the mission and goals of regional actors and regional development in the Bay Area. Not only was the Internet a new marketplace for innovative goods produced in the region for global consumption; the Internet itself began to modify past industrial practices and industrial links dependent on pre-Internet modes of communication. Thus, the projects initiated by Smart Valley would help undermine the original regionally based goals of integration envisioned in its creation. At the same time, these developments revealed the basis for regional cooperation in the new information economy: the collaboration of top-level engineers and their corporate leaders in setting standards that lock in economic dominance for the region.

Second, I will step back to put these developments in the broader context of the politics of business collaboration, both across the nation and historically in California. Even as specific production relationships globalize at an ever more rapid pace, there has been a countervailing trend of ever increasing business consortia relationships within regions, most often with government assistance to help stabilize the relationships between businesses. Precisely because of that globalization and the increasing disconnect between firm self-interest and the need to create "public goods," both technical, as in the case of basic research, or social, as in the case of standards, multifirm business collaboration has become crucial within the new high-technology economy. And beyond specific public goods, the emergence of the Internet as a radically new marketplace has required (and strained) new forms of business cooperation and trust within regions to ensure its proper functioning and stability. The chapter will examine the crucial role of standards in the economy and discuss whether the open collaborative standards around the Internet developed in the Bay Area region can stand up in the battle with Microsoft. In regard to the Bay Area itself, collaboration through the Internet has begun to highlight how this new technology is allowing the region to try to leverage global power in a range of economic spheres from finance to politics.

Third, I will outline how this new global form of cooperation at the regional level leaves out many communities within the region. In a "politics of abandonment," the focus on the regional needs of high technology increasingly focuses the provision of public goods narrowly for the elite while leaving only public squalor for those communities on the periphery. This reflects the reality that the cross-class collaboration of industrial production that had stabilized regions in the past has given way to the new politics of business, in which collaboration between engineers and corporate leaders is the new basis for regional development.

The Politics of Regional Revival: Conceiving a "Smart Valley"

When Joint Venture was launched in 1992, the high-tech executives involved saw the Internet as merely one of a range of networking tools available to spark a regional revival. Initially, the program was focused as much on private local area networks (LANs) and other proprietary electronic networks as on use of the Internet in order to link businesses, schools, government offices, and health care facilities in the region. With plans for dedicated fiber-optic lines to run this local technology, the emphasis was in many ways on uses of the technology to lock in regional relationships, much as dedicated technology had locked in relationships between many individual companies.[4] The main vision of electronic networking was that "[The] foundation of communications can enable innovations in regional services that will assist Silicon Valley to improve its appeal and capacity to retain and grow enterprise. The Valley has the ability to become a leader in telecommunicating and in high speed, high bandwidth networking of companies, suppliers, and individuals."[5]

In this, Smart Valley was like a number of regional information-networking projects around the country in the early 1990s. In Arizona, a collaborative project involving Arizona State University, Digital Corporation, and the Times Mirror Corporation sought to reinforce the same kind of regional collaboration of local industry using Times Mirror's cable technology along with Digital's Ethernet networking technology. "We've tried to work with some major manufacturing companies and their suppliers to begin to look at what utility these companies can gain by being connected across a metropolitan area network," said Dave Rosi, Digital's director of video and interactive services, back in 1994. Defense contractors such as McDonnell Douglas were the first customers in the region who saw an advantage at that time in encouraging smaller contractors to conform to a single regional standard.[6] Clearly, both Digital and Times Mirror Cable saw advantages in tying that standard to their own technology.

The Route 128 region followed up on Silicon Valley's initiative with an online network called the Commonwealth Exchange (originally called CommerceNet East until Silicon Valley folks objected), which was explicitly designed to avoid having Silicon Valley get a jump on Boston's high-tech corridor. Funded jointly by the state of Massachusetts and the Massachusetts Telecommunications Council (MTC), an organization of technology companies in the Boston area, the Commonwealth Exchange was envisioned as a system to recruit employees, place orders, and collaborate regionally. One initial focus was using "groupware" software from the Boston-based Lotus Corporation, called Notes, popular on proprietary networks for collaboration, to connect government agencies, schools, and businesses working on new projects.[7] Like the initial conceptions of Smart Valley, use of the Internet was merely an option among more proprietary possibilities that would reinforce regional standards (and not coincidentally assist specific regional companies that were promoting those technologies).

Other initiatives in this period of the early 1990s focused less on specific regions than on whole industries in order to make U.S. production more efficient in international competition. The American Textile Partnership (AMTEX) was created in 1994. AMTEX was organized to link different parts of the textile industry, from fiber to retail, with eleven Department of Energy laboratories in developing technologies to recapture global market share— through better manufacturing processes, electronic commerce, and quick response to consumer demand. About eighty industry partners were involved, and annual funding by 1995 was pushed to $50 million, provided jointly by industry and government. Almost half of that funding was devoted to electronically networking the industry.[8]

It was in these industry-specific networks that the emergence of Internet standards as a business standard became more and more obvious. The auto industry had come to electronic networking early, with its suppliers using proprietary networks. Seeking to escape the costs and limits of such networks (usually called value-added-networks, or VANs, in industry parlance), Ford Motor Company announced in March 1995 that they would begin using Internet-based IP standards over a secure private version of the Internet managed by Ameritech Corporation to exchange automotive-design files with contractors. This followed the determination in 1994 by the Automotive Industry Action Group, the technical advisory organization for the automobile industry, to adopt IP as the communications protocol for communicating with parts suppliers. With parts suppliers often working with multiple companies all over the country, the use of a standard set of protocols became more and more important. Ford soon was working with General Motors and Chrysler to develop what was soon called the Automotive Network Exchange (ANX) to allow contractors to use Internet connections to link to the private network with encryption and security for sensitive data being exchanged.[9]

So it was in the context of competing regional networking initiatives, often involving the use of proprietary or company-specific technology, and national industrial networking initiatives, which were turning more and more to Internet standards, that Smart Valley and its spin-off projects were developed. Partly, the Smart Valley initiative was an effort to ride the new federal interest in information networking that emerged out of the 1992 presidential campaign and the call from Bill Clinton and Al Gore for creation of a national Information Superhighway.

Coming out of Joint Venture (itself partly funded by federal and local government), the Smart Valley initiative was funded primarily with corporate contributions, but its position helped in leveraging millions of dollars in federal funds for its other incorporated initiatives. Michael McRay, the program manager at Smart Valley, who worked extensively on helping to develop the CommerceNet initiative, described Smart Valley's role thus: "We are venture capitalists without capital. We help consortia find resources." After that point, Smart Valley stepped back to offer strategic advice, but each initiative would take on a life of its own.[10] In the case of CommerceNet and BAMTA, that life would prove to be very different from its sponsors' initial conception.

CommerceNet emerged out of the initial Smart Valley idea of creating local networks connecting firms to enhance production in the region. Backed by Silicon Valley firms, Stanford's Center for Information Technology and the Bay Area Regional Research Network (the NSF-sponsored regional Internet access network that was just about to be sold off to BBN), Smart Valley solicited a $6-million federal grant in late 1993 from the new Clinton administration's Technology Reinvestment Program.[11] This would be supplemented with $6 million in corporate and local government support over the initial three years. Sponsoring corporations would pay anywhere from five thousand to thirty-five thousand dollars a year to join CommerceNet, with many federal and local government agencies paying those fees as well.[12] Through this public-private partnership, the federal government hoped to see commerce on the Internet grow while the regional companies hoped to foster more regional growth.

Most of the staff involved in Smart Valley held to the view that activity in the region would trump the global effects of the Internet. Michael McRay at Smart Valley saw the cooperation stemming from CommerceNet as part of a long tradition of rivalry and market competition in the area. "Silicon Valley has a wonderful history of competition and collaboration," said McRay, explicitly citing Saxenian as a guide in Smart Valley's thinking about regional economics: "There's definitely a regional advantage. Geographically defined clustered industries have a real need to keep on their toes. They are blessed in having a cluster in the electronics industry. I don't see going on-line losing that regional advantage. It goes to that issue of trust. These very close business relationships are fostered by having a geographic proximity. . . . The Internet has played a key role in this cross-fertilization."[13]

McRay summed up the effects of the Internet this way: "While the Internet allows you to find out about things around the world, sometimes the most important function is to find out what is happening in your own backyard." In this view, the social networks were just as important in determining the future of a region as was its communications network. And if the Internet strengthened the social networks, it could only strengthen regional advantages.

CommerceNet would allow the further disintegration of companies to subcontract functions that are not within their "core competencies," with the expectation that this would more likely happen regionally than internationally.[14]

Randy Whiting became the first chair of CommerceNet's Sponsored Projects Committee, the day-to-day governance body for the varied projects initiated at CommerceNet. Whiting, a Hewlett-Packard manager at the time who would later become CEO of CommerceNet, described the initial mandate of CommerceNet as

> to automate the value chain and supply chain of electronics in the Silicon Valley. At the high end, we (at Hewlett-Packard) are more of a user than producer of electronics, so the idea was to bring together the main high-end consumers, and then we could deal with the whole wealth of contractors "downstream"—fab, consultants, etc. . . . We would be using the technology to improve the efficiency of that region. In a business-to-business relationship, the Internet was a good solution because there isn't a generalized infrastructure [of specific networking technology] in place. We had a regional imperative."[15]

BAMTA had a similar initial focus with the addition of tying in many of the graphics and software companies often located in San Francisco's "multimedia gulch" that had often been more peripheral to day-to-day business in Silicon Valley. Additionally, BAMTA had the extra impetus of the push by NASA's Ames Research Center to commercialize multimedia technology in order to justify its continued existence during a period when NASA was looking to cut costs. No group was bringing together the disparate parts of the multimedia industry in the region, so they decided to create BAMTA to deal with the cradle-to-grave production of networked multimedia products, from the creation of content to the tools for managing that content, to delivery to the consumer.[16] NASA agreed to fund the project with an initial grant of $5 million over two years, obtaining a commitment that the money would be matched five to one with private funds and in-kind donations. BAMTA quickly grew to forty-three companies focusing on a range of areas from the delivery of health care information over networks to the creation of what was called a Collaboratory, to test multimedia products in combination in a facility with state-of-the-art equipment, a special boon to smaller companies. NASA specifically supported three large projects aimed at air traffic control, reconstructive surgery assistance over networks, and networking high-technology laboratories.

NASA framed its involvement squarely in economic terms, both for its own bureaucratic self-interest and in its public interest goals. NASA wanted to make Ames a model for how NASA could partner with local industry in order to commercialize its technology. "The fundamental reason is that if you commercialize technology," argued Paul Kutler, the NASA director assigned to oversee the agency's involvement in BAMTA, "you can create economic growth and new jobs. The new mandate for NASA and other agencies is to commercialize their technology. Some have been in classified areas with state-of-the-art technology which were not in the public domain and now can be declassified and used to spur the economy, especially in heavy Department of Defense areas of the country." As to why NASA decided to invest so heavily in this kind of venture when they did, Kutler stated, "Fundamentally, it gets down to that there is world economic competition. In aerospace, it used to be McDonnell Douglas versus Boeing versus Lockheed versus Martin versus Hughes, and now it is Boeing and McDonnell Douglas versus Europe." In stark terms, NASA and other agencies saw sparking regional economic growth in new technologies as a nationalist economic strategy against competition from other countries' leading industries.[17]

As federal investment in both CommerceNet and BAMTA showed, high-technology centers could still attract more resources to reinforce their lead. And even if they were to lose the investment advantage in tangible dollars, many involved in Smart Valley saw its initiatives as helping to consolidate local advantage by reinforcing the intangible regional political and social connections that help the Bay Area thrive. Bill Davidow, a longtime Intel executive who had become a prominent venture capitalist and was an early financial supporter of Smart Valley, saw building "the intangible asset base in Silicon Valley" as the greatest accomplishment of the initiatives. Whether through ongoing projects or through Smart Valley lectures, he saw a local exchange of new ideas that the Internet cannot quite duplicate. "The Net can facilitate a lot of things, but there's something that happens in face-to-face interchange that isn't going to happen over the network; I may be wrong in that since I was in a generation that grew up without the network, but I suspect that over hundreds of years, mankind has evolved skills that are oriented to physical interaction that won't exist for the network. Maybe in the next million years, we'll have it down pat."[18] What all these comments show is that the originators of the Smart Valley initiatives assumed that electronic networking and the initiatives spawned by Smart Valley would facilitate local interaction to increase productivity and lead to the growth of production links within the region. However, while some of that no doubt occurred, what would become clear is that the globalizing effects of the Internet would be more dominant and the role of regional interaction would be much different from what was envisioned by Smart Valley's originators.

Creating a Global Electronic Marketplace

It was partly a testament to the rapid changes in the Internet and partly a testament to the recovery of the Bay Area economy (and the two were not unrelated), but the Smart Valley initiatives rapidly abandoned their focus on using networking technology to enhance production within the region and instead began to focus on using the Internet technology for marketing to customers, especially for marketing Internet products themselves. With a global market to conquer, most Bay Area firms saw an increasing need for global alliances, and endeavors like CommerceNet and BAMTA became prime vehicles for facilitating the alliances that would stabilize the Internet marketplace and create a bigger pie for all involved in the exploding networking industries. CommerceNet and BAMTA would, within a year of their establishment, officially abandon their emphasis on Bay Area firms and local collaboration, welcoming membership from companies all over the country and the world.

Many of the company executives soon recruited into Smart Valley projects became quite skeptical of claims of regional production links. Mack Hicks, a Bank of America vice president and the second chair of CommerceNet's Sponsored Projects board, leaned heavily toward the view that the Internet spelled the end of the limitations of geography. "We have to watch out here, since we think we have Silicon Valley—but there is no region per se." Hicks saw the whole CommerceNet project less as a regional industrial policy than as a font of experimentation whose effects would benefit all business.

> To quote Mao, let a thousand flowers bloom. Some people will use [electronic commerce] to contact others in the area. Others will use it to take from the area to go out into the world. . . . Any thought of boundaries—state, religious, age, sex—all go away. It is the ultimate promise of marketplace and personal diversity. . . . In an industrial society, cities were appropriate because of the regional nature of business and the delivery of services; in the information age there is no boundary for the delivery of services so there is no reason for regions. You want to compete in the international region."[19]

Hicks noted that Bank of America's chief regional rival, Wells Fargo, had been using the Internet to transform itself from a large regional bank into a major international player. With a good Web site and new strategies and services, Wells Fargo could now have an international presence without the costs of new branches in London and New York. So even as mainly local banks like Wells Fargo merged and downsized local employment, partly with the help of information technology, that same technology was helping them find new markets worldwide to replace less profitable customers previously served regionally. (Note that the next chapter will deal carefully with the effects of the new networking technology on banks and other noncomputer companies such as utilities with a traditional regional focus).

For most of the companies involved in CommerceNet, reaching customers in the global marketplace became a much higher priority than working more closely with suppliers in the region. CommerceNet's Randy Whiting argued that CommerceNet had started out with the mission to be a "regional marketplace" to make local companies more competitive, but the companies involved quickly decided that "the real value was not that I could deal with my suppliers more effectively but how could I reach my customers more effectively." With the emphasis on marketing, the focus was much more on relationships outside Silicon Valley, helping to push forward membership in the consortium for companies outside the region and internationally.[20]

BAMTA went through a similar transformation. Although with a later start than that of CommerceNet, its transformation to a global consortium was even quicker. Mason Myers, the Smart Valley staff member who helped get BAMTA up and running, argued that it became obvious quite soon that the Bay Area focus was more important to NASA than most of the businesses involved: "NASA is government and they feel bad about Moffett [the air base that closed down in the early 1990s] and they worried that Ames might be closed down." Bay Area businesses didn't want the words *Bay Area* in the name and soon had it dropped, leaving only the acronym. Not that those business leaders had no vision of Silicon Valley's role in the multimedia future, but they argued that a "Bay Area-centric" approach, as Myers derisively calls it, would miss the global business opportunities in multimedia networking. Still, as Myers saw it, in the end, "the Bay Area will do very well: you have everything you need, you have the technical folks, the creative folks, and the ad folks. The only other area that has it is New York, which doesn't have the technical folks, but that will be less important in the future."[21] But doing well meant hooking up with multinationals such as Kodak that were key to the multimedia future.

Whiting was a bit sardonic in seeing the whole Smart Valley effort as a case in which "it started with the academics at Smart Valley, EIT [a consulting firm that helped get CommerceNet off the ground], and Stanford who didn't come from the business side. . . . We came in and said screw it, let's look at the selling side." This led to big banks and multinationals joining in the rest of the effort, rather than many of the small contractors, as originally envisioned. In many ways, the Joint Venture/Smart Valley project was set up according to all the academic models (including explicit reference to Saxenian) of encouraging more "industrial district" style collaborations, but in the end the larger business leaders in practice saw less need to strengthen local production and more opportunity in exploiting local advantages in skills and technology to sell to the global market. The reality was much more in line with the model of industrial districts withering under pressure from global competitors skimming the skills and advantages of local regions.

The very expansion of the Internet helped fuel this focus on global sales, since Silicon Valley suddenly found itself at the center of the most hyped technology since the personal computer. Job growth arising from computer networking and related software production exploded in the region with employment at the top five computer-networking firms in Silicon Valley growing nearly 300 percent between 1991 and 1996.[22] Selling the idea of the Internet to a global market became a primary goal of those involved in the Smart Valley projects. Networking projects within the region came to be seen as less about improving productivity than about showcasing the technology for the global media. Networking firm 3Com's CEO Eric Benhamou became an important director on Smart Valley's board and was heavily involved in efforts to wire the public schools for the Internet. Benhamou saw regional networking as a launchpad for his company's global expansion. While he expected 3Com's employment to expand in the region, he clearly saw a diminishing percentage of the company's payroll as 3Com moved into new world markets.[23]

Beyond the hardware of networking, Silicon Valley saw an explosion of new software firms being linked to the Internet, with Netscape leading the pack of new firms soon to become the dot-com explosion. Throughout the early 1990s, even as defense-related companies and most hardware-based firms were losing jobs in the steep California recession, software continued to expand rapidly. Semiconductor and other hardware employment staged a recovery by 1994, but software job growth significantly outpaced new jobs in manufacturing in the mid-1990s as new Internet software firms exploded in the region and older software firms such as Oracle expanded with new Internet-related strategies.[24] The culture of innovation was important in these new software successes, but direct production links as originally envisioned by Smart Valley mattered much less in software than did expanding the global appetite for the region's new Internet software products.

The Internet Undermines Local Supply Networks

Beyond the disinterest in using the Internet for regional supplier networks, the specific projects that CommerceNet emphasized actively undermined much of the need for closer supplier-customer relationships. Partly, this reflected the abandonment of private electronic communication networks, which themselves had served in the 1980s and early 1990s to lock in relationships because of the costs of infrastructure investment. With networks based on public IP protocols, networking costs were significantly lowered and this allowed looser, more fluid, and more market-driven relationships with vendors. This in turn allowed, with the support of CommerceNet and other standards-setting groups, an electronic marketplace for subcontracted goods to emerge and to begin to replace more stable buyer-supplier relationships that had depended partly on the costs to gain information on alternative contractors.

At the most prosaic level, the Internet has become a cheaper alternative for corporate networks. Private networks had been based on being metered prices per kilobyte of data sent. Companies found that the twenty-five thousand to thirty thousand dollars per year it might cost for an Internet connection was less than the cost of using a private VAN for just ten hours of typical data use. More crucially, any private VAN connection limited a company's connections to other businesses and virtually excluded connections to individuals. Added costs for private high-speed "toll road" Internet access might push up costs, but the public network was still one-hundredth of the cost of older proprietary networks.[25]

Software costs for maintaining internal intranets or business-to-business extranets were also less than those of older proprietary networking systems because of the shared standards and flexibility of using Internet-compatible servers and support software. This led to explosive growth in Internet-based software; by the end of 1996, 64 percent of the top one thousand companies had intranets, a 30 percent increase from just six months earlier.[26] With standardization from IP protocols, the costs of making new supplier and buyer connections tied to electronic communication significantly dropped, thereby lessening the lock-in of relationships pertaining to proprietary software.

What is remarkable is how quickly the U.S. network infrastructure had abandoned the proprietary network choices of the late 1980s and early 1990s

that researcher Francois Bar, in a 1990 study, noted had created "the fragmented US infrastructure[, which] risks inhibiting flexibility in productive
arrangements among firms."[27] As mentioned earlier in the chapter, many
larger companies such as those in the auto industry were quite willing to
forego any monetary advantages of proprietary networks for the flexibility of
public standards. This opened up access to communication networks to a
much wider range of companies than could have afforded to lock themselves
into proprietary network systems. Similarly, in the 1980s large Silicon Valley
firms such as Hewlett-Packard had invested abundant resources in a private
electronic network for suppliers. While the company prized this network for
its security, it began to push for a more public system to more easily deal with
its ever changing production needs. And as that public Internet system took
form, as Hewlett-Packard's Randy Whiting noted, the company found that
instead of merely creating smoother supplier-buyer relationships, the Internet
was opening up the possibility of marketizing many transactions. This allowed
new vistas for direct electronic purchases from suppliers and new opportunities for electronic sales to Hewlett-Packard's own customers.

Information technology has increasingly been used to coordinate global
production networks, hold inventories down, and promote just-in-time delivery. The continued growth of Federal Express (FedEx) in the 1990s was largely
a product of this globalized information infrastructure, whereby it used online
tracking to increase customers' confidence in ordering parts from far-flung
areas; by mid-decade FedEx was calling itself less a transportation company
than an information-technology company. In one example, Silicon Valley's
National Semiconductor Corporation was using FedEx to slim its global ware-
housing system and to cut its total logistics costs from 3 percent to 1.9 percent
between 1993 and 1996.[28]

On the selling side, one key project at CommerceNet became the Cataloging
Working Group, which focused on how best to put company catalogs on the
Internet, the heart of what would become business-to-business networks, or
B2B networks, as they were soon called in the inescapable Internet jargon. If
specifications for specialty products could be precisely and flexibly presented
to customers, relations between buyers and sellers would need less time-
consuming human massaging for deals to be made. Instead, more subcontracting relationships could be "marketized" rather than there being a dependence on tight supplier-relationships to deliver precise goods. CommerceNet
organized a large group of company executives focused on this area, since the
payoff would be especially dramatic in high-technology areas where electronically cataloging the specifications needed for products could expand the

sources from which a company buys. Decreases in the transaction costs of finding such information could allow the substitution of more market relations for structured relationships with suppliers. The key for the working group was to create catalog systems that presented information with complex specifications in an easy-to-understand way.

By allowing much quicker searches through a variety of parts makers, such interactive catalogs would decrease the advantages of long-term relationships with subcontractors and also minimize the need for person-to-person meetings for working out detailed specifications. From a salesperson perspective, Randy Whiting saw the Web interface as operating in a way remarkably similar to the way that transactions have traditionally occurred through the intermediary of salespeople. What salespeople traditionally do goes beyond merely presenting products. They present the company itself in a perspective that makes sense to the buyer. That is the basis for long-term relationships for which the Web is beginning to electronically substitute itself. Explains Whiting:

> In business relationships, very rarely does a customer align perfectly with a vendor suppliers' organizational structure. A customer looks at a supplier only based on products and customer needs. Those typically don't match up. . . . The Internet is a nonlinear methodology of presenting content and navigating through a buying process. Never before have we had a generalized technology with a ubiquitous infrastructure that matches that relationship between a customer and a sales representative.[29]

Even for more routine purchases, the advantages of buying locally diminish with many of the advantages of the Web interface and smart catalogs. Bruce Lowenthal, a vice chair of CommerceNet and Tandem Computer's manager for electronic commerce, noted that traditional local warehouses often had limited availability and little hard information on the differences between parts. Lowenthal, overseeing online purchasing for building Tandem's computers, envisioned a growing electronic ratings service need for a "Siskel and Ebert function" that would provide companies with simple comparisons between parts purchased from the range of global vendors. The need for local supply sources would disappear. Lowenthal foresaw in the mid-1990s that "manufacturing is leaving" the region, a comment that reflected Tandem's own merger into Texas-based Compaq soon after.[30]

The result of this disdain for local supply networks was that the first major appearance of electronic cataloging appeared not in Silicon Valley but most prominently in such places as Harrisburg, Pennsylvania, where AMP Incorporated, the largest industry supplier of electrical and electronic connectors and interconnection systems, launched an electronic catalog system called AMP Connect in January 1996. AMP had traditionally spent $8 million to $10 million per year printing its four hundred separate paper catalogs of its products, most of which quickly went out-of-date and had to be redone every twenty-four months. With its new Web service, it soon had tens of thousands of customers, two-thirds of them engineers, buying its ninety thousand products online from more than one hundred countries.

However, Silicon Valley's CommerceNet partners were central to these new cataloging projects. They had a prime role in the success of AMP Connect and other electronic catalogs. AMP's catalog system was built on a Sun Microsystems SPARC 1000 server using Oracle's Oracle 7 databases and specialized catalog software developed by Sunnyvale-based Saqqara Systems Incorporated, another member of CommerceNet. The director of AMP's online division, Jim Kessler, was just following the experience of CommerceNet partners when he promoted the company's electronic catalog by arguing, "To serve the business-to-business market, companies need to create database-driven dynamic content that can be searched by requirements."[31] By supporting the expansion of Internet commerce, Silicon Valley firms rode an exploding market for their services, even as traditional local production was undermined by the technology they promoted. General Electric by 1997 was doing more than $1 billion worth of business over the Web with fourteen hundred of its suppliers, an amount greater than that of all consumer-based Web purchases in the same year. By 1998, American companies had $43 billion in sales to one another over the Internet, five times the consumer retail total, most of them using software and services provided by Silicon Valley firms.[32]

With estimates of trillions of dollars in supplier-contractor business over the Net within a few years, a new market for software was exploding like a bomb.[33] An expanding pie for business Internet products meant that any diminishment in local production-based commerce in Silicon Valley was being far overshadowed by the new global Internet market's contributions to local employment in the region.

Standards and the Postindustrial Politics of Regions

The example of electronic catalogs is just one part of the exponential growth of electronic commerce promoted out of the furious networking surrounding

Smart Valley, CommerceNet, BAMTA, and all their associated projects. What tied them all together was the creation of the new marketplace of cyberspace and the support for a variety of public goods necessary to create the trust in that new area of commerce: from the development of successful technologies to facilitate commerce such as electronic catalogs to the technology standards that create an open, innovative environment for new companies to enter the electronic marketplace.

The erosion of local production relationships occurring at the exact same time as the appearance of new business associations like CommerceNet is not a contradiction, but actually highlights crucial features of the new postindustrial politics of regions in the information age. It is precisely because the marketplace *does not* naturally produce open standards and public goods in a globalizing economy that new political structures between businesses are necessary. While such political structures do not necessarily have to be based in physical regions, they are facilitated by their having a core of business leaders who share such regional interests and who have the backing of governments that operate politically within fixed geographic space.

Business politics is much like ordinary politics—it is a way to pool resources for endeavors that would be too expensive for individual businesses to attempt or that would likely generate too little monetary returns if any individual business had to make its own investments. What CommerceNet delivered to its members was a test bed for the new electronic marketplace in which, as Mack Hicks argued, "if every firm was to figure out how to provide service . . . by themselves it would cost a lot of money and they'd make the same mistakes over and over." By working together, companies could experiment to find out what tools and practices worked, and what standards would really facilitate a strong marketplace. The inclusion of national and multinational firms facilitates the creation of practices that are applicable globally (which is important if companies eventually want to sell to that global market). At its core, though, CommerceNet was built on the relationships fostered in the environment of the Bay Area and backed by government participation, which helped legitimize CommerceNet's results and practices.[34]

The reason that politics and regions matter is that market forces tend toward proprietary systems. "A lot of these technologies depend on widespread acceptance of standards and open non-proprietary systems," noted Smart Valley's Michael McRay. "Left to themselves, an IBM and others will offer a closed system. These consortia will offer open systems. In a nutshell, they provide a system of collaboration for competing firms to work to their mutual advantage." McRay cited the example of the budding conflict around security for financial transactions over the Internet that evolved between Netscape and EIT. When it looked like two different systems of security were evolving that would have required two different kinds of browsers, the CommerceNet consortium helped them hammer out a joint deal to create a common standard.[35] Similarly, when Visa and MasterCard disagreed on how to protect credit card numbers online, CommerceNet persuaded RSA Data Security to give away its software for encryption that could then be incorporated as a standard into Web browsers.[36]

For companies outside the Bay Area, Smart Valley's projects offered a structured way to create new strategic relationships and participate in the standards creation process. In the BAMTA project, Eastman Kodak invested in a shared technology-testing site (the Collaboratory, mentioned earlier) where combinations of technology could be tested for interoperability and performance. For smaller companies, a public space such as that offered by BAMTA gave start-ups a chance to, in the words of Smart Valley's Karen Greenwood, "partner with these big companies without giving their companies away."[37]

With manufacturing and production relationships increasingly going international and multinationals joining in regional relationships in organizations like CommerceNet and BAMTA, Bennett Harrison is correct in describing Silicon Valley and other industrial districts as part of "a global system populated by big companies perpetually on the prowl for new profitable opportunities [where] the very success of a district can itself bring about changes that give rise to its opposite" in global hierarchies.[38]

However, at the same time, the region itself becomes an arena for negotiations about public goods, including standards, those public standards in turn creating restraints on proprietary competition between multinationals. Even as multinationals seek to harvest the benefits of regional collaboration, their investment in that collaboration in turn shapes a new politics between business that constrains some of their opportunistic activity.

It is this tension to balance competition and opportunism that makes consortia like CommerceNet more and more ubiquitous in the new information economy. Why regions matter for these standards and for other public investment has to do with the nature of innovation; electronic communication facilitates greater and greater dispersion of production from corporate headquarters, leading to a variety of strategic production relationships, but higher-level innovation often depends to some degree on physical proximity. Not only does this reinforce the traditional concentration of engineers and innovators in places such as Silicon Valley, it also makes such locations a natural point for negotiations over standards and the development of shared practices to facilitate new marketplaces. As Tandem's Bruce Lowenthal argued, "The fact is that people have and will work together better if they do personal interactions. . . . I've seen a lot of people move away and try to keep up with what's happening in Silicon Valley. But their productivity declines after half a year. They try but they usually fail. It's an observation."[39] If anything, the electronic mode of communication supplements and enhances other forms of local interaction by allowing those casual contacts to be followed up more methodically through e-mail, which can be dealt with at a more leisurely pace.

Yet CommerceNet's evolution showed both the gains from cooperation and the ultimate temptations of opportunism in a privatized-standards world. As electronic commerce exploded, the focus on the Web turned to updating the basic language of the World Wide Web, HTML, to expand its capabilities. Competition over design of its successor, called XML (for eXtensible Markup Language), would continue to threaten the creation of new proprietary divisions as companies from Sun to Microsoft sought to control their implementation. A strong open XML was crucial for strengthening the ability of business groups—whether stockbrokers, travel agents, or pornographers—to create agreed-upon codes embedded in Web pages to ease buying, selling, and communicating over the Net.

The federal government stepped in with a $5-million grant from its Advanced Technology Program to encourage CommerceNet to form a consortium with a number of other businesses to promote its standard.[40] Working with firms ranging from Sun to Microsoft, CommerceNet developed an XML-based electronic commerce standard, called eCo, to help prevent the splintering of the Web into mutually incompatible commerce systems and codes. Controversially, CommerceNet decided it would be easier to develop the system commercially by creating a for-profit spin-off, Veo, which received half of the federal grant for eCo's development. Veo was sold in 1999 for $60 million to a new B2B company, CommerceOne.[41]

There is little question that CommerceNet helped speed support in the industry for the move to XML, and CommerceOne would introduce its version in high-profile projects such as the new revamped auto-company purchasing Web system, Covisint, involving Ford, General Motors, DaimlerChrysler, Renault, and Nissan and which potentially will involve $800 billion in transactions.[42]

But CommerceNet's economic profit from its standards promotion, including an eventual payout of tens of millions of dollars in stock options that went to the non-profit's top executives, seemed to end the trust that had allowed the consortium to act as an honest broker in the industry. Many felt that

CommerceNet had lost its focus and had walked away from creating a universal standard. Microsoft was soon competing with other groups of companies to push proprietary versions of XML. "They dropped the ball," said Murray Maloney, the former project manager for eCo. Founders of CommerceNet, including Sun and Ariba, soon let their memberships in the consortium lapse.[43] Many felt the consortium had degenerated into little more than a chamber of commerce for the industry.[44]

Other consortia stepped into CommerceNet's role in seeking consensus on XML standards, but the rise and fall of CommerceNet's role in the industry is emblematic of the tensions in the role of consortia and that of regions themselves as forums for standards negotiation. As such consortia become ever more prevalent on the economic landscape, how government should deal with them becomes ever more crucial.

The Role of Consortia in the New Economy

Consortia are relatively new kinds of organizations in the United States and their existence seems to have grown with the rise of the information economy. To understand the role of CommerceNet and its associated organizations, it is worth considering their predecessors in the electronics field. While Japan has officially encouraged the creation of consortia since 1960, most business consortia in the United States have appeared in the 1980s and 1990s. An overwhelmingly large number of these have been in the electronics industry, including the Microelectronics Center of North Carolina, founded in 1980 by the state of North Carolina and seventy company members; the Semiconductor Research Corporation (SRC), founded in 1982 by eleven semiconductor companies (with eventual participation by government agencies and 13 other companies); the Austin-based Microelectronics and Computer Technology Corporation (MCC), launched in 1982; and Sematech, launched in 1987.[45]

Much of this growth was facilitated by the passage in 1984 of federal legislation to regularize the antitrust rules governing shared research among companies (largely because of questions arising out of the establishment of MCC in Austin). By 1994, more than 350 U.S.-based R&D consortia would be registered in the country. These consortia have been a mix of national industry-based endeavors combined with region-specific efforts often driven by local government needs. And like Smart Valley, most experienced the relationship between industry and region as a complicated one, given the need for local collaboration. What is clear is that the growth of such new forms of business association has become a fixture of the new economy.

Austin's MCC illustrates many of the lessons, positive and negative, that inspired the creation of Joint Venture, Smart Valley and the whole array of new regional business-to-business initiatives that emerged in the early 1990s. The success of Austin in attracting MCC and the Sematech consortia and using them to boost technology development in that region was cited in the original 1992 Joint Venture publication *An Economy at Risk* as illustrating the political infrastructure that was lacking in the Bay Area region.[46]

In turn, MCC had largely been inspired by the success of the Japanese VLSI (Very Large Scale Integration) Project consortia from 1976 to 1979, which had allowed a group of Japanese companies to produce the manufacturing technology to challenge U.S. computer and semiconductor manufacturers. Ten of those U.S. electronics firms, including Silicon Valley firms such as AMD and National Semiconductor, would launch the MCC consortia in 1982, eventually recruiting twenty-two participants by 1993. The companies involved were spread across the country and saw the need to focus cooperative research on developing new technologies to compete, both internationally and with IBM.

Where to locate the consortia was the largest issue facing MCC's founders. Texas was selected not only because of its large economic subsidies of the project (more than $70 million in leased property and low-cost loans) but also because of Texas's massive upgrading of the University of Texas at Austin's microelectronics research program. This included tripling the program with the establishment of thirty new endowed professorships in electrical engineering and computer science. The MCC founders had wanted to avoid the top electronics areas, where their project would get little attention, yet were able to work with a newly world-class electronics faculty—the best of both worlds in the view of the MCC companies.

With investments of $500 million over ten years of operation, the consortia would file for 117 patents, license 182 technologies, and publish more than twenty-four hundred technical reports. The Austin region gained new recognition for its already established base of companies and company branches (Tracor, IBM, Texas Instruments, Motorola, AMD) and twenty major companies would locate sites in the city over the following decade. And by the 1990s, new firms were spinning off MCC Ventures Incorporated (a commercializing subsidiary of MCC), and by 1994 a new Austin Technology Incubator (funded by the city government and local chamber of commerce) had helped create forty additional new start-ups. By the late 1990s, Austin and its suburbs were home to more than nine hundred software companies and the largest concentration of microchip plants outside Silicon Valley.

Lesson 1: Regional Spin-Offs Trump Corporate Ties

Yet the companies who created MCC were not happy with the results. Their goal had been to create cutting-edge technology breakthroughs for their own companies, not boost Austin's economy. Technology transfer back to corporate headquarters and manufacturing facilities outside Austin largely failed. The creation of MCC Ventures to commercialize MCC's technology was actually an admission of that failure and an attempt to recoup some of the funds invested in the consortium. What this highlighted was that local collaboration on research and local conversion of that research into commercial products and companies still overwhelmed internal corporate communication about technology mediated over geographic distance. In a sense, this just replicated the lesson of Xerox's failure to capitalize on PARC's innovations in creating the personal computer, which instead were capitalized on by Apple and other firms in the Silicon Valley region.[47]

When Sematech was formed in Austin in 1987–88 by the federal government and top semiconductor firms, almost half of them from Silicon Valley, it applied those lessons. It concentrated on a clear development program with some funds (nearly $200 million per year in the early 1990s) being directed both to outside research centers beyond Austin and to research being done in collaboration with company sites located in Austin in order to gain the benefits of the applied research program. One prime example was Silicon Valley–based Applied Materials Incorporated, the top American semiconductor equipment supplier, which created a design site in Austin specifically to be near Sematech. Other companies followed suit in expanding office sites in the area to take advantage of Sematech's results.

Many of these firms became backbones of the Joint Venture consortium; Applied Materials executive Tom Hayes in fact became Joint Venture's first chairperson, and all the companies involved would take advantage of the lessons of MCC and Sematech in highlighting the gains to local regions of the existence of consortia established in a region. And the lessons of consortia like MCC seemed to show that far from there being a threat that the inclusion of national and even international participation would hurt regional actors, the inclusion of far-flung participants was more likely to support the local region than the other way around. This was one reason why the leaders of Smart Valley initiatives could so confidently welcome national and international participants, knowing that despite electronic communication, technology transfer would still be biased toward those closest geographically to the consortium work.

Lesson 2: The Need for Social Leveraging of Research into Commerce

For Austin and other regional economies, another lesson had been learned. Political coordination of social institutions in a region was crucial for attracting technology resources, and in turn, ongoing social coordination was crucial to making sure that those technological resources would be turned into businesses that would feed employment in the region. Despite the presence of MCC in Austin from 1982 onward, almost no business spin-offs were created from its work until institutional support was built in the form of MCC Ventures Incorporated and the Austin Technology Incubator. Turning technology into regional economic growth followed a step-by-step institutional process that helped create the institutional ties, the "social capital," that could translate into technology exchange and economic development.

When Joint Venture was formed, many business leaders felt that the drop-off in technology leadership by the region in the late 1980s had strong institutional sources. In evaluating both the failures of the 1980s and the recent successes in Silicon Valley of the mid-1990s, Joint Venture in its publication *The Joint Venture Way: Lessons for Regional Rejuvenation* would argue, "Social capital is the glue that holds together a successful economic community. Before Joint Venture, the stock of social capital across the public and private sectors in the region was low. . . . Joint Venture has repeatedly demonstrated that collaborative processes can create social capital."[48]

Silicon Valley's success over the years can be tied directly to an ongoing rejuvenation of its social institutions and relationships, which in turn could attract investments in technology and then, in turn, commercialize those investments into economic development in the region. Its origin came largely from the Southern Pacific fortune of Leland Stanford, with that fortune itself the result of the close cooperation of the "Big Four" Southern Pacific partners—Collis Huntington, Leland Stanford, Charles Crocker, and Mark Hopkins—in mobilizing every civic connection they could (and capitalizing of course on Stanford's position as governor). They would parlay a combined investment of less than seven thousand dollars between them into a fortune of hundreds of millions of dollars, largely generated through federal government funds. While larcenous in its looting of the federal treasury, it was marked by a level of ongoing cooperation among the partners and civic institutions quite in contrast with what characterized most of the cutthroat railroad operations across the country, and that civic participation played a key part in their success.[49]

As noted in the preceding chapter, Stanford University itself was oriented by Leland Stanford toward engagement with the surrounding region's economic life. Frederick Terman turned Stanford into an institutional cornerstone through his leveraging of federal funds in support of new businesses in the region. The threat of the end of some of those federal contracts and the need to mobilize to keep them coming to the West in fact led to the creation of one of the first formal West Coast electronics business institutions, the West Coast Electronics Manufacturing Association (WEMA), later to become the American Electronics Association (AEA).[50]

The second phase of regional institutional organization related to technology was ongoing support for building of the social institutions, formal and informal, that assist commercialization. That took the form of Terman's backhanded industrial policy via the Stanford Industrial Park, the support of local community colleges that were training talent in the region, the more informal networking facilitated by Fairchild's splintering and the Homebrew Computer Club's bringing together of hippie entrepreneurs with established research scientists in spawning the personal computer revolution. When the AEA's offices were moved to Silicon Valley in the 1960s, the organization ignored lobbying in favor of fostering the development of new companies and promoting networking among businesses. Other more specialized associations, among them the Semiconductor Equipment and Materials International (SEMI) and the Software Entrepreneur's Forum (SEF), would promote similar shared information in their industries.[51] The Santa Clara County Manufacturing Group would work closely with the public sector in the 1970s in order to correct the region's transportation, housing, and environmental problems. Anna Lee Saxenian emphasized how the civic engagement of these Silicon Valley business associations contrasted sharply with the aloofness, even antagonism, of Massachusetts's Route 128 high-tech firms to working with the public sector (and often one another). Route 128 firms preferred to see themselves as global companies and rarely as local firms with an interest in the region; all this contributed to the slowing of ongoing commercialization of technology in the Route 128 region.

Smart Valley, CommerceNet, and BAMTA became new iterations on this tradition of cooperation that would inevitably benefit the region. Despite their fierce competition in the microcomputer market, Intel and Advanced Micro Devices (AMD) along with Motorola would team up in 1997 with three Bay

Area Department of Energy research labs (Lawrence Livermore, Sandia, and E. O. Lawrence National Laboratories) in a partnership to work on dramatically shrinking microprocessor circuits—a $250-million investment by the companies that highlighted the public-private cooperation in the region.[52]

Lesson 3: Regional and Global Networking Is More Important Than National

As Silicon Valley firms went from organizing to attract federal funds to consolidating production support in the region, the late 1970s and 1980s would see much more emphasis by firms on organizing to defend their market against the Japanese. In 1975, the CEO's of National Semiconductor, Intel, and AMD met over a meal and founded the Semiconductor Industry Association (SIA) which would spend the next twenty years promoting intervention by the federal government against the Japanese. The SIA would, along with an AEA newly oriented to national lobbying, lead lobbying efforts to demand that the United State government develop tough trade agreements with the Japanese government to protect market share for U.S. companies and open up the Japanese market more to U.S. firms. Most of the electronics consortia of the 1980s, from North Carolina's Semiconductor Research Corporation (SRD) to MCC to Sematech would be justified based on the Japanese threat. It was in this period that most of the original Silicon Valley semiconductor firms started defining their home "region" as the United States rather than Northern California, and local business institutions suffered from the neglect.

Ironically, it was the lessons of Austin that would reinforce the advantages of regional location for innovation and show the limits of national efforts by MCC and Sematech. While both these organizations did assist U.S. firms in regaining competitiveness in real ways, the limits of that help to national firms situated far from Austin, along with the much more dramatic gains for those locating in Austin, delivered a solid message to the Silicon Valley firms of the forgotten advantages of regional proximity. Whatever the national participation in these consortia, it was those firms that strengthened their regional presence in the Austin area that reaped the greatest benefits from the consortium's technology spin-offs.

At the same time, the nationalist thrust of the 1980s was being tempered by the reality of an increasingly global production system in electronics. Many companies based in Silicon Valley would get as much as 60 percent of their components from Japanese firms, often using electronic communication to wire designs to manufacturing facilities in the Far East.[53] Firms like Sun Microsystems were forming tight relationships with suppliers like Fujitsu. Even the decidedly nationalist Sematech, with an explicit buy-only-American policy, began seeing itself less as an institution engaged in a head-to-head struggle to build the strength of U.S. firms versus those of Japan. It now aimed to "help U.S. companies gain sufficient technical muscle to join international partnerships . . . from a position of strength" in the words of MCC and Sematech chroniclers David Gibson and Everett Rogers.[54]

With the establishment of CommerceNet and BAMTA, there seems to be have been barely a thought of stopping at national borders once nonregional members were accepted. BAMTA project coordinator Karen Greenwood noted the "tension between regional and global [and] I say global rather than national," since she saw that as the real tension driving the economic focus of firms in the electronics industry in the 1990s. Like most people who were involved, Greenwood saw the international participation as merely validating lessons that had been learned, since this participation would inevitably do more to help the overwhelming number of Bay Area companies than hinder them.[55]

Lesson 4: Standards, Not Research or Production Relationships, Are the Largest Fruit of Consortia

Despite the best intentions of their founders, most private-sector initiatives cannot deliver the basic research needed for breakthrough innovation or the just-to-market techniques that companies want for immediate production. After Craig Fields, head of DARPA, was fired from his job in 1989 for promoting ties between government and business as the key to moving the innovation of basic research to market production, he was hired as head of MCC, in 1990. After talking with consortia members and examining the tensions in the project, he noted that the four goals of the organization included (1) giving competitive advantage to member companies; (2) helping the United States; (3) advancing science; and (4) advancing industry. Field concluded that these goals were mutually exclusive:

> Twenty years ago . . . these four criterion were reasonably aligned. The way the world has gone, that correlation is much, much lower. Almost all large companies that are part of consortia are multinational, so helping the company doesn't necessarily create high-value jobs in the United States. For many companies, business success is only modestly related to technology advancement; it's more closely related to issues of good marketing, closeness to the customer, and continuous small

incremental gains and improvements rather than some breakthrough technology.[56]

His conclusion was that basic research was useless in private consortia and should be left to the public sector. Private consortia should concentrate on the public goods it could deliver for its members and the public, namely, the standards and agreed-upon protocols that would make collaborative production easier in commercializing the basic research supplied from the public sector.

Partly this was based on a revised view of what had allowed the Japanese to supplant U.S. semiconductor firms in the late 1970s through their VLSI consortium. While the research from that collaboration of Japanese firms was seen as fruitful, including its creation of the then-breakthrough 64K random-access memory (RAM) chip, what became most critical for the commercial success of Japanese firms was a new shared standard among Japanese firms for production. This gave a boost to local Japanese contractors and equipment makers, especially when compared with U.S. equipment suppliers, which had a fragmented set of standards for production. With those new standards of production, the dependence of the Japanese companies on foreign equipment suppliers dropped in three years (from 1977 to 1980) from 75 percent foreign dependence to less than 50 percent.[57] In turn, one of the most successful initiatives of MCC has been the creation of the ATLAS Standards Laboratory as an independent subsidiary to speed the creation and adoption of electronic information standards in a variety of MCC-related research areas.

CommerceNet and other Smart Valley initiatives took these lessons and worked to produce consortium relationships that would create these less tangible public goods. Venture capitalist Bill Davidow used the lessons he learned as an executive at Intel to help launch Smart Valley and says:

> I see these things, whether Sematech or Smart Valley, or Joint Venture, as attempts to address the issues of an economy and a government in transition. . . . So what's going on is that we are going to an information-based economy where what is important in society are the intangibles and not the tangibles. It's not that food isn't going to be important, or a roof isn't going to be important, but a lot of wealth will come from an intangible base. What I see Smart Valley doing is building up the intangible asset base in Silicon Valley so it can be competitive in the world economy.[58]

No basic research was even attempted at the Smart Valley initiatives, and since they could draw on the participation of engineers in companies already located in the region, the consortia could maximize use of the money that they had (orders of magnitude less than Sematech, for example). They concentrated on the networks and standards promotion that would build that asset base and expand the electronic marketplace of public goods. Whether developing the best system of online cataloging or brokering agreements on standards for secure financial transfers on the Internet, new consortia like CommerceNet drew on those already involved in the industry to push forward.

Government funding both facilitated cooperation for the new electronic-commerce standards and gave the results of that cooperation more legitimacy. "There is a movement towards these processes," notes Smart Valley's Michael McRay. "By admitting government funds, it makes it happen faster and gives presence to the consortia. Market forces will eventually come to bear and produce similar results, but many years down the road. The funding is a catalyst." Many industry leaders have worried that recent cuts in the technology programs that fund such standards consortia at the federal level endanger the sort of cooperative endeavors represented by Smart Valley and have allowed competition over proprietary standards to impede the development of the technology.[59]

Lesson 5: Watch for the Private Benefits of Those Participating in Consortia

While consortia members promote the advantages to the public of more stable standards and a quicker development of an electronic marketplace, the flip side is that participants understand how the consortia members can benefit as well if the standards match their competitive advantages. This creates a danger of opportunism, which often leads to stalemate in the creation of agreed standards.

At the least proprietary level, by helping companies get in early and define the shape of the electronic marketplace, companies involved get a strong leg up on later competition. "It's pretty clear from how this is going that early entrants to this area have a significant competitive advantage, and if you wait a year, you may be out of business," argued CommerceNet vice chair Bruce Lowenthal. By working together, consortia members structure a marketplace that reinforces the technical and social advantages they already have in their firms. On the broadest front, the goal of the firms involved in CommerceNet, for example, was to make sure the Internet became the dominant networking standard. This helped a range of firms, mostly centered in the Bay Area, whose existence had rapidly become tied to the Internet, versus companies promoting proprietary private networks that were based mostly outside the region.[60]

From a more specific company position, Lowenthal noted that his company, Tandem Computers, was in a better position to influence not only the technological standards adopted for electronic exchange but also what corporate allies benefited: "If you're in the pilot projects, you help develop the standards that are being used, so it is oriented to the strengths you have, technical or personnel." Large banks had been important customers of Tandem for years, so the company worked through CommerceNet with bank company members to ensure that any financial security transfer systems would favor banks over upstart financial competitors. Since some industry group would have to issue digital "keys" to authenticate financial transactions, Tandem wanted technology that would favor its banking allies. At the time, Tandem supplied the mainframe computers that processed more than two-thirds of all credit card purchases and 80 percent of all ATM transactions, so Tandem had an interest in making sure that any standards would be easily integrated with those banking assets. "It's more in our interest to have banks do the certificate issuing. Banks know us, they like us, they get meals together with us, so it helps us if they create the standards. [By participating in CommerceNet], we not only know the technology, but our friends are doing this," so it cements corporate alliances around the technology.[61]

The rise of companies trying to manipulate standards for their own ends is generally credited with undermining the effectiveness of the old ARPA Internet standards bodies such as the IETF. By 1995, in the words of one participant, the system had broken down, and "what I hear at IETF meetings is people worrying about patents and lawsuits and all sorts of crap we never used to have to deal with."[62] The issue of patents has become one of the hardest issues for standards bodies to deal with, since companies often promote standards tied to "submarine" patents that they hold and that are not yet public.[63] Working out mandatory licensing agreements for such patents became a critical role for many consortia. Tim Berners-Lee, the designer of HTML, has, with some success, promoted the industry-dominated W3C, which he helped set up, to fulfill this role. This was assisted by the Federal Trade Commission's bringing a 1995 case, *FTC v. Dell*. The case targeted Dell Computer Corporation, which had worked on setting a standard for "VL-bus" technology without revealing plans to assert its patents. Once the industry began using the new standard in desktop computers, Dell tried to collect licensing fees. After the FTC filed suit,

Austin-based Dell ended up relinquishing its patents.[64] But this reflects the fact that absent vigorous government involvement, opportunism will often trump successful cooperation on standards.

The Role of Standards in the New Economy

So what, ultimately, is the function of these standards, these "intangible assets," in the economy?

In the production of already established goods, standards push the economy to regularize production and decrease the dependence of smaller suppliers on proprietary standards that leave them hostage to the strategic decisions of large manufacturers. These open standards allow fixed hierarchical relationships between firms (and within firms) to be replaced by a much broader array of market relations in goods. Smaller firms can have less fixed investments in specific relationships (in other words, goods and training associated with proprietary standards) and are therefore freer to deal with a multiplicity of firms in the market. By reducing what economists such as Oliver Williamson label the "transaction costs" of dealing with different firms, market relationships can supplant hierarchical ties between firms.[65] In this limited sense, electronic commerce based on standardized Internet protocols promises a more market-ized set of relationships between buyers and suppliers in a range of industries. In a related manner, electronic commerce through such vehicles as smart catalogs address another traditional reason for tight buyer-supplier hierarchies, namely, limited information about the availability of alternatives.

However, as the evolution of CommerceNet and Smart Valley showed, there is a deeper effect of standards in the new information economy, namely, the creation of an entirely new marketplace for a range of information goods. Instead of merely changing buyer-seller relationships in an established market, the creation of the Internet has brought into being a marketplace of information goods that had not existed previously. And in every step of the creation of that Internet-based marketplace, trust in standards has had to be built so that new buyers would participate in this new marketplace.

CommerceNet's Randy Whiting posed the distinction as one between improving the functioning of individual markets, driven by buyers who might want more efficiency from their contractors, and creating a marketplace, wherein a collection of sellers want to *create* new markets in which to sell their goods:

When you're on the Internet, there is a huge cost of infrastructure that will not come from the [individual customer's] side, so there is a large push for the seller to create the infrastructure. "Buyers make a market, sellers make a marketplace.". . . The founders [of CommerceNet] said they had to get major corporations in, and the corporations, being very entrepreneurial, saw an opportunity to sell stuff; they wanted a marketplace to pull their buyers in. . . . If we build these structures that enable the marketplaces, it allows these markets to spring up.[66]

In the information age, much of the infrastructure is composed of the intangible assets of standards, which convince customers that goods they purchase will be compatible with other information goods—the Holy Grail of interoperability—and that they are not buying into a dead-end technology. Standards also build confidence on the part of developers of supporting technology or "content" that they are not wasting their time on a here-today, gone-tomorrow proprietary system that can be changed at the whim of the system owner (as happened to the developers of Microsoft's original proprietary Microsoft Network). The process of creating an open standard is tied to building trust between a whole range of companies and the buying public. In the case of the Internet and its related standards, this encompasses an enormous range of industries from computer hardware makers to telecommunications companies to banks to software developers, making the trust issues all the more difficult.

The regional nature of the standards process comes from the nature of trust itself. There is nothing that bars organizations based in different physical locations from building the trust needed to make the standards agreements that could facilitate such new marketplaces, but organizations are run by people and it usually takes day-to-day interactions and embeddedness in a range of social spaces that transcend specific business relationship to create that trust. And in pathbreaking new markets, it usually takes the assistance of government, which overwhelmingly operates within regional spaces, to help referee the relationships that allow trust to develop and help legitimize the standards that come out of such processes. The original Internet, developed under the supervision of the government, which was operating overwhelmingly in three or four geographic regions, and the federal government's drawing back from supervision of the process undermined the ease of building trust in far-flung geographic spaces even further.

The role of Smart Valley plus the range of regional companies' deals connected to building the Internet-based electronic commerce standards reflected the need for some regional focus for even such a global networking project as the Internet. Regional participants in the Bay Area welcomed national and international companies, since their ideas could only strengthen the acceptance and trust in whatever standards were supported, but the day-to-day work of the network would depend on personal ties from within the Silicon Valley region. Over and over again, CommerceNet members would emphasize how personalistic the organization was in its operation and how dependent success was on the individual ties people could build through the process, far beyond the formal business-to-business relationships they brought to the organization.

And with 50 percent of the organizational members located within the Bay Area and more than 75 percent of day-to-day participation coming from within the region (along with the highest concentration of Internet domains in the world), the region was bound to have a disproportionate influence and derive disproportionate benefits from the evolution of an Internet moving forward in the direction desired by them. Even Microsoft found that it needed to continually expand its workforce located in Northern California in order to maintain closer contact with software developers and technology innovators at other companies needed to keep Microsoft products integrated with the rest of the technology world.[67]

All technology-based economic changes have had similar regional concentrations that have helped to facilitate the creation of new markets, either through monopolistic proprietary agreements or in the more diffuse models of production districts.

A perfect negative example of the hazards of standards forged outside such regional negotiations was the long, drawn-out battle over the digital video disc (DVD), a critical storage medium for movies and computers. DVDs are an enormous advance from compact discs. On a single DVD, up to seven hours of music or a full-length movie can be recorded; and massive amounts of computer data can be stored and accessed from one disc. A consortium of ten global technology and media companies, including Time Warner, Sony, JVC, Phillips Electronics, and Toshiba initially agreed on standards for the devices, but the result was anything but an open or stable standard. The interests of Hollywood and the recording industry led to tight controls on how easily information on a disc could be shared with computers or other machines—the antithesis of the ease of information exchange on the Internet. Worse, as new machines were introduced in 1997, the consortium partners rapidly broke down into warring standards camps as each sought proprietary advantage over its global rivals; by 1998, there were no fewer than six mutually incompatible DVD standards in development or on the market, which only slowly converged. The result: instead of the 3 million DVD players expected to be sold in 1997, only 500,000 machines were sold that year as consumers sat out the standards war. This war had slowed the expansion of the market for years.[68] And even by 2001, there was still no agreement on a standard for a recordable version of DVDs as a computer storage device.[69]

In the absence of governmental involvement or any degree of regional coherence, competitors often find no way to maintain agreement on standards, thereby hurting not only their rivals but also their own share of smaller pie. A similar story unfolded as American cell phone manufacturers largely lost out to European competitors, largely because Europe's governments agreed on a continentwide digital standard early in 1991, while U.S. manufacturers were left behind as American telecom companies took years to agree on new digital standards.[70]

As the decade of the 1990s came to an end, many technology companies began to welcome more involvement by government to help, especially as regional models like CommerceNet stumbled. In 1998, a consortium called OASIS (Organization for the Advancement of Structured Information Standards), led by the usual-suspect Bay Area firms, among them Sun and Hewlett-Packard, began to promote broader XML standards for business-to-business commerce. Uniting the global with the local, the United Nations trade policy and e-business arm, known as UN/CEFACT, chose to work with OASIS and held a 1999 conference with two hundred industry representatives in San Francisco that would establish a new global standard called EBXML for electronic business on the Web. By 2001, Microsoft was the only major company promoting a serious alternative standard and even they were feeling pressure to support this new standard sponsored by governments around the world.[71]

Companies in Silicon Valley were increasingly recognizing that with weakening organic connections at the regional level, failures on standards agreements could only lead to broader economic failure.

Silicon Valley's Regional Model Versus Its Rivals

The problem for Silicon Valley firms was that even geographic proximity did not automatically create the standards that propel economic growth, especially in the absence of a firm alliance with government. The drive to create CommerceNet and the host of other assorted consortia had been an acknowledgment of the need to structure those regional advantages in a way that built confidence that all participants would gain from the expanded growth of the Internet.

The reality was that despite the Internet's success, many of the firms in the region continually struggled with the danger that proprietary technologies would upset the trust needed to sustain the collaborative model that had fueled the growth of firms in the region. At the top of the list of dangers was of, course, Microsoft. Microsoft had used a combination of its early alliance with IBM and hardball tactics to build its proprietary operating system monopoly on the desktop. From that base, Microsoft would extend its proprietary standards in the 1990s increasingly into the market for large-scale business computing, formerly the province of mainframes or UNIX-based network servers. While the Internet at first appeared as a danger to Microsoft, even a dagger at its throat, Microsoft also saw that success in molding those standards in a proprietary direction could extend the company's control throughout the world of corporate computing.

Through in-house software applications and developer tools optimized for its proprietary standards, Microsoft created an all-pervasive computing environment that promised any corporation that its needs would be met. The Microsoft solution might be less innovative than any particular competitor, but Microsoft's very completeness and pervasiveness across all sectors of computing would make up for its rigidity.

In fact, Microsoft's rigidity could be an advantage when compared with the weakness of the standards that pervaded the UNIX corporate environment in the 1990s. After the heyday of the 1980s when government purchasing requirements had enforced a broad UNIX standard on the industry, the industry had divided into warring UNIX camps and left customers uncertain that their needs would be met in that fragmented environment. By 1997 Microsoft NT computer servers were outselling UNIX servers.[72] Despite the odes by Saxenian and others to the flexible design relationships in Silicon Valley, it was clear that in the absence of government support for strong standards, proprietary models could have a decided advantage in yielding the market stability and the associated monopoly rents that a company like Microsoft could reap.

The geography of this showdown was clear. With Microsoft dictating standards from its Redmond headquarters based on chip designs from its ally Intel, friendly companies or even grudging competitors writing for Microsoft's operating system could locate anywhere that they could attract a good workforce, from Utah to Texas to South Dakota. When Bill Gates was called before the Senate Judiciary Committee in March 1998 to discuss its monopoly practices, Microsoft's defenders on the panel included Michael Dell of Austin-based Dell Computers and Douglas Burgman of North Dakota's Great Plains Software, while Gates's critics were notably all from Silicon Valley.

The UNIX wars made clear that strong open standards were essential if the region were to compete economically against such proprietary steamrollers as Microsoft. CommerceNet and other consortia built around Internet standards were the first step in the process, but companies such as Sun saw the need for broader solutions that would expand open standards from the operating system to the tools used by programmers. The Java language, with its promise that any program would be able to run on any computer, no matter its hardware or operating system, became one key part of that strategy. As described in Chapter 3, Sun worked assiduously to build the consortia links that would turn Java into a broad standard. And the explosion of Java-based firms in the Silicon Valley reinforced the idea that regional firms would reap the greatest advantage from collaboration on such open standards.

Still, Microsoft responded by seeking to co-opt Java and other Internet standards into its proprietary approach by creating versions that would run only under Windows environments, thereby violating the point of these standards. And its continued success in linking contracts for its Windows operating system to installation of its browser increased worries in Silicon Valley that Microsoft could engineer a proprietary approach to the Internet.

With the need to generate stronger global support for its standards, Netscape took the unprecedented, and largely last-ditch, step in March 1998 of publicly revealing its browser source code—the usually top-secret guts of any program. Netscape, soon to be acquired by America Online, invited anybody who wished to do so to modify the code and even resell their own version as long as any modifications to the code were published on a special Web site. Faced with the onslaught of Microsoft's proprietary approach, Netscape decided that the regional commercial commitment to developing standards was insufficient. It needed to marshal the resources of the global programming community and it needed to open its code to gain the kind of trust needed to ensure their support.

Harking back to the original ARPANET vision, the idea was to invite the participation of the Net community in developing the tools and standards embedded in the browser software. "It's no longer Netscape alone, pushing the client software forward, but now it's really the whole Net," said Bob Lisbonne, Netscape's senior vice president for client products at the time. "For Netscape, this gives us a way to engage the creative, innovative abilities of literally orders of magnitude more people than we could ever—really any commercial software company could ever—afford to just put on their payroll."[73] With hacker enthusiasts lauding the decision, thousands of developers would download the source code within the first day, and major modifications of Navigator were released onto the Net by independent developers from all over the country within weeks. While the business strategy did not in fact survive the company's absorption into America Online, the software code that became known as Mozilla did—proving one of the points of "open source" software whereby software survival goes beyond the life or death of particular companies.

Netscape's action highlighted the continued importance of this kind of public-interest software development, which had survived much of the privatization of the Internet. Most dramatically, despite the focus on the Microsoft-Netscape rivalry, the most popular Web server on the Internet was neither company's but rather a free system called Apache, an offspring of the original NCSA Mosaic server software, which the NSCA as part of the government privatization had largely abandoned. Instead, a geographically dispersed group of software programmers, some at universities and some in private business, began collaborating in 1995 to update the NCSA server to increase its power and manageability. Most of the programmers participated out of altruism and the long-term hacker ethic of promoting free software. The result was a Web server that in 1997 was used in 44 percent of Internet sites, compared with just 16 percent using Microsoft and 12 percent using Netscape's software. And that list of sites included those of McDonalds, UUNET Technologies, HotWired, Yahoo Incorporated, CBS, the FBI, and IBM, which passed over its own Lotus Domino server in favor of Apache when it put its Big Blue–Gary Kasparov chess match on the Internet.[74] Similarly, a favorite Web programming language is a free language named Perl that has similarly been modified and improved through a global network of collaborators.

But of even more dramatic importance was the exploding strength of the Linux operating system, a free version of UNIX that in the late 1990s was the fastest expanding operating system in the world, especially in the "server market" for computers undergirding the Internet. A early as 1997, *Wired* magazine described Linux as "[Window] NT's most serious competitor, the only viable alternative to the Microsoft monoculture."[75] And by 2001, while Microsoft's share itself was still growing to a total of 41 percent of the market for servers, the freely available Linux had grabbed 27 percent, supplanting the various versions of UNIX as the alternative operating system for most commercial servers.[76]

Remarkably, Linux was started in 1991 by a University of Helsinki student whose first name, Linus, led to the naming of the language. At that point, a series of free UNIX tools had been developed by programmers connected to

GNU (a self-referencing acronym for GNU is Not Unix) foundation, itself founded by an early MIT hacker named Richard Stallman who objected to the increasing commercialization of university research. What this network of free-software hackers lacked was the core of the operating system, called a "kernel," which would tie all these tools together into an alternative to the commercial UNIX. Linus Torvalds wrote that kernel and from his university post would use the Internet to coordinate improvements in this new operating system with help from hundreds of enthusiasts around the globe.

Based on what GNU called "copyleft" principles—the license Netscape adopted when it released its source code—the Linux operating system could be distributed freely or packaged with documentation and sold for modest amounts (but without copyright restrictions) backed by technical support from Caldera, Cygnus Solutions, and other companies. Extremely popular in developing nations, among them South Africa, Cuba, India, and the Philippines, Linux also began to eclipse other forms of UNIX in the United States partly because of its price but also because many people considered it technically the best operating system in existence, if lacking some of the ease-of-use features to readily challenge Microsoft on the desktop itself. Linux was the first operating system to include Java capability, so every increase in Java programs adds to its functionality.[77]

Contributing to Linux's success were the emergence of new partners for the operating system as other open source software began to emerge. Corel—maker of WordPerfect—announced that it would release a full suite of office applications for Linux. IBM began building Web tools to support the Apache Web server and join with other "Apache Group" developers to help improve the Apache code. Hewlett-Packard, Compaq, IBM, and Silicon Graphics have all agreed to install and support Linux for their hardware customers. New independent companies such as Red Hat and VA Linux, sellers of Linux software and services, were able to conduct billion-dollar IPOs (admittedly sinking to merely multi-hundred-million-dollar capitalizations after the NASDAQ meltdown) based on open source software.

This emergence of open source software throws into relief the different geographies and economics of trust involved in proprietary standards, commercial open standards, and freeware standards. With proprietary standards like Microsoft's, everyone trusts (or fears) that Microsoft will enforce whatever standards it dictates from its company-specific development, so hardware and software partners on such proprietary products have a stable enough environment to locate almost anywhere they think appropriate. However, with commercial open standards, collaborators need the repeated interactions and day-to-day commercial interactions of shared geography such as that of Silicon Valley to generate financial gain while assuring that multiple collaborators all profit from innovation. The alternative to both approaches is the collaboration on open source software systems that uses the Internet itself to build trust based on altruism and the hacker ethic of achievement without the need to share any geography—the extreme example being Linus Torvalds's management of the evolution of Linux from his lonely station in Finland. Without expectation of financially capturing the social benefits of their creation, collaborators are free to innovate without restriction. As Microsoft's proprietary approaches gained ground, many Silicon Valley actors reluctantly saw an alliance with the global freeware model as necessary for survival—abandoning some control and profits in order to maintain the priority on innovation that gave them a regional advantage in the remaining commercial aspects of technology development.

Silicon Valley as Corporate Center for High Technology

However, even as the leaders of Silicon Valley made alliances with the free software radicals of cyberspace, it was in its other international alliances that the most radical changes in the cooperative shape of the region were taking place in the second half of the decade. Given the scale of the Internet, there are vast implications for the changing place of the Bay Area in the global economy and the role of less skilled workers within the region's politics. The physical location of design and production centers has become more and more disengaged from any objective measures of infrastructure (traditional or research assets) or inherent production links and instead increasingly reflects the placement of firms within an international distribution system. Multinational firms see regional locations as places to strategically distribute complementary products for maximum corporate coordination.

Agreement on Internet standards shadows the integration of new players into dominance of the coordination of key global industries, from media to manufacturing to banking. Just as New York City became the central coordinator of international capital and corporate governance in the nineteenth century, the Bay Area was making its bid through its integration of diverse industries to assume large portions of coordination functions in the coming twenty-first century. Venture capitalist and key Smart Valley founder Bill Davidow noted in 1995, "If you look at what's happening in the economy in the last few months, Silicon Valley has become the headquarters of the network-based industry." He argued that the East Coast was five years behind the Bay Area in implementing and integrating its economy in connection with the new information, economy and in a rapidly changing economy, that would allow Silicon Valley firms to achieve leadership over a spectrum of industries. The Bay Area had a combination of finance, defense, semiconductors, biotech, medical devices, software, fashion, media, and tourism that, combined with the new technology, could create an integrated corporate coordination center.[78]

Aside from pure speculative insanity—no doubt a factor—the multi-trillion-dollar run-up in the tech-laden NASDAQ index can be understood as investors' expectations that the new Silicon Valley dot-coms could gain the financial rents of such corporate power by leveraging the technological rents of innovation. The end result of the financial upheavals of IPOs and takeovers was as many local assets being taken over by outside corporate investors as local companies taking over established players, but there was little question that the region had made a shift from being merely a center for technological cooperation to being a center of global financial activity as well. Whether New York or the Bay Area wins out or what portions of global coordination they end up sharing with other global cities around the world is all part of the economic conflict of the developing technology.

However, comparisons with New York City also bring to mind comparisons to the way that the elite of New York, cultivating international ties around the globe, eventually severed much its own prosperity from the standard of living of those at the bottom of the economic scale. The company leaders of Silicon Valley admit as much in interviews and the founding report of Joint Venture: Silicon Valley sketched out that possibility in rather stark terms as early as 1992. In looking to the future of the region, the report proposed three future regional scenarios: a "High-Tech Manhattan" of corporate headquarters and little else, a "Virtual Valley" of niche companies allied in stable industrial clusters, and an innovative "American Technopolis" of thriving start-ups and new technology creation. This last would deliver the most for the region, while the other scenarios would be either a draw or a net loss for others in the communities of the Bay Area. However, the assumption inherent in the scenarios was that in any of them, the companies would do well. The worst scenario, High-Tech Manhattan, still "results in a win for companies" even as it is "a loss for the regional economy and community."[79]

While there was a preference, both personal and technological, by industry leaders in keeping a strong Silicon Valley economy, the implication was clear that companies in a thoroughly global economy would seek out other regions that would fulfill their needs if the Bay Area failed to do so. Given this reality, any offer of collaboration with local government officials took on the character of an "offer they can't refuse" rather than suggesting a long-term mutually dependent relationship. The reality was that while the economic action of technology innovation might be local, the power of corporations to pick and choose their venues was global and outside the control of the local actors who desperately tried to negotiate with these global partners. Those who had more power globally therefore had a stronger hand for negotiating locally within a region, while those unable to muster such power were out of luck. The end result was increasing inequality within regions.

Inequality and the Contingent Future for Non-elite Workers

The immediate benefits of the Internet boom for many non-elite members of the Bay Area community was substantial during the technology boom, with low unemployment and wages rising at rates as much as five times the national average. By 2001, the average wage in Silicon Valley was $62,600 compared with a national average of $36,200.[80] But such averages disguised an increasing economic divide. And beyond the dollar numbers, what is clear was that few lasting institutions were being created to lock in shared benefits for regional members over the long term.

The midcentury wave of manufacturing growth in the Bay Area region and around the country was created through not only corporate cooperation with government in connection with shared economic growth; it was also the result of new institutions that built in a share of that growth for most workers and community members. While Henry Kaiser and his fellow industrialists were enriched by the industrial take off of World War II and its aftermath, labor unions also grew exponentially in the Bay Area with the expansion of jobs in what was called the region's "Second Gold Rush." New health care institutions such as the pioneering Kaiser Permanente health program would outlast the factories that funded their creation.[81] In what researchers Harvey Molotch and David Logan called the age of "growth coalitions," cross-class cooperation would fuel growth in urban areas throughout the country. These growth coalitions would use the proceeds of growth to invest in a range of public goods that would benefit the broad public and lift the wages of all workers (even if the greatest benefits might remain in the hands of the elite).

While many community members shared in the initial burst of employment and higher wages following the Internet boom, it is unclear how real the shared wealth is for many workers and even less clear whether these benefits will be transitory for all but the corporate elite. While the richest 20 percent of households saw their incomes increase 20 percent between 1993 and 1999, the bottom 20 percent saw incomes actually decline in inflation-adjusted terms by 7 percent. In fact, California as a whole saw real erosion of wages.[82] In 1999, a San Francisco Federal Reserve Bank study reported that median family income in the state had dropped by 4 percent from 1989 to 1999, even though it rose 8 percent for the rest of the nation. "In 1998," the bank reported, "a greater number of Californians lived in poverty, a smaller number were in the middle class and a majority had family incomes below those of comparable families living outside of California." Only the top 30 percent of families out-performed the rest of the country in this more recent period, compared with all levels having done well above average for most of the decades after World War II.[83]

Even in Silicon Valley, where average family incomes increased, the average apartment-rental rates increased far more rapidly. Across the country, 60 per-cent of housing is deemed affordable for households with median incomes in their regions, yet in Silicon Valley by the year 2000, only 16 percent of houses were affordable. Partly this was the function of cramped geography and rapid growth as Silicon Valley added hundreds of thousands of jobs but far fewer new housing units.[84] But the low unemployment rate in the region partly re-flected the fact that people were forced to leave the area when they could not find a job. "With the high cost of living, if you don't have a job, you can't afford to live here," publicly argued Lois Koenig, assistant to the director of the San Mateo County Human Services Agency, in noting the pressure on low-income families in the region.[85]

Even for those remaining in the region, the reality of the Internet "boom" was one of rising inequality and greater hardship for those at the bottom of the economic ladder. Twenty thousand residents of Silicon Valley's Santa Clara County found themselves homeless at some point during 1997.[86] In the 1990s, poverty among children in the Silicon Valley region increased from 10 percent of children in 1989 up to 13 percent in the middle of the Internet boom of the mid-1990s, with three-quarters of those children concentrated in only 30 per-cent of the census tracts.[87] When President Clinton's advisory board on race came to San Jose for a hearing, they walked into a firestorm of anger as many of the seven hundred attendees, including San Jose school board member Jorge Gonzales, denounced the 400 percent rise in CEO salaries during the decade, even as the average wages of Latinos in Santa Clara County had dropped 6 percent.[88]

Beyond the increased costs of living and persistent pockets of poverty, there is a sense of the ephemerality of the boom for many workers in the region. If the World War II manufacturing boom pioneered new forms of job guarantees and widened benefits for all workers, skilled and unskilled, the recent Internet boom seems to be pioneering precisely the reverse. Including all categories of contingent workers—temporary, part-time, self-employed, and contract workers—the best estimate of the size of the contingent workforce is 27 percent to 40 percent of workers in Santa Clara County, the heart of Silicon Valley. This sector of the workforce grew two to four times as fast as overall employment and nearly all net job growth in the county was largely accounted for by the growth of contingent employment.[89] Contingent employment has accelerated along with the Internet boom; in the first nine months of 1995 alone, employment in temporary agencies in the immediate Santa Clara area grew by 41 percent. Since the early 1980s, employment in temporary agencies has grown by 150 percent, a rate more than fifteen times the overall employment growth in the region. There are more than 250 offices of temporary agencies operating in Silicon Valley, and Manpower Temporary Services, the largest national temporary agency, operates fifteen offices in Silicon Valley alone, placing more than five thousand people a week by the mid-1990s. Much of this contingent work was facilitated and accelerated by the Internet, through which jobs can be posted, reviewed, accepted, or rejected with greater ease.[90]

A report by Working Partnerships, a nonprofit backed by the South Bay Central Labor Council which covers most of Silicon Valley, points out that what all these contingent workers share is that "their terms of employment stand outside the standard employment relationship on which the framework of employment and labor law was built. The fact that all contingent employees are outside the standard employment relationship means that they are vulnerable to rapid economic change and have difficulty being represented."[91] When the tech slump hit in 2000, many temporary jobs were the first to be cut, often not even being mentioned in public announcements of smaller job cuts of permanent employees.[92]

Even before the tech slump hit, large areas of Internet support and routine software programming were being outsourced from the region. "Safeway [supermarkets] just announced that they are moving their data center to Arizona," noted Bank of America's Mack Hicks as early as 1995, "taking their best-paying white-collar jobs there, leaving the CEOs here to enjoy the Bay Area."[93] In a period of high innovation growing from the Internet, technology companies rapidly expanded local employment in the first burst of standards creation as immersion in regional dialogue had been at a premium. But as development of technology related to the Internet becomes more routine, more of that employment will follow Safeway and other companies out of the region. Overseas software programming is becoming a real threat to the more routinized programming jobs. India's sixteen hundred engineering colleges and technical schools now graduate fifty-five thousand students a year, contributing to a 45 percent annual growth rate in India's software industry.[94] American corporations increasingly subcontract out work to India ranging from Web design to software development to customer service.[95] And beckoning on the horizon is even lower wage labor in China, where the Shanghai-Suzhou region is becoming a new high-tech attractor of overseas investments and corporate contracts. China's universities are annually turning out 145,000 computer engineers who already perform technology work for Taiwanese companies[96]—and Silicon Valley firms are unlikely to be far behind.

This subcontracted work largely focuses less on new products than on support for more standardized software projects, but the very success of Internet standards will create more such projects as the initial burst of innovation ebbs and Internet communication continues to facilitate global coordination of such routine projects. The Internet is a natural means by which to direct customers, suppliers, and internal company communication to various components of a corporation. With the Web, one click of the button can connect customers to online service, sales, or technical support—each of which can be in a different city and even a different continent. All this bodes ill for the more contingent and less elite members of the Bay Area technological workforce.

And beyond the likely geographic redistribution of work, there is the reality that new technology is being used not to just relocate, but to replace, workers. At the lower end of skill, online ordering forms replace many data entry clerks as customers essentially do their own data entry work as they electronically register their order. New software increasingly replaces midrange technical skills, creating an endless upgrading of skills that lasts only until new software explodes in the workplace and employers demand a constant influx of new skills. "I worry about a potential financial bust on the Internet," argued Stephen Roach, chief economist for Morgan Stanley, as early as 1996, "or a culmination of a software product cycle." Roach was highlighting the way technology has been a key tool for corporate downsizing to increase profits and he worried presciently that the current technology-induced cycle of business expansion would falter with the peaking of the Internet wave of technology innovation.[97]

Even without the Internet bust at the end of the 1990s, it was clear that the constant skill upgrading demanded by the technology was itself an instrument that divides the educated elites from other workers. Smart Valley's Bill Davidow acknowledged the inequality being driven by the new technological cycles:

> We have relatively short electronic cycles, but if you look at what has happened culturally to people, it is ridiculous to talk about a profession going away in five years, but that is what you are facing. Two hundred years ago, if you learned how to make shoes at twenty, you could expect to be doing it at sixty. But you have people who learned what they thought were valuable professions, people who thought nuclear energy was the future or who got caught in the mechanical engineering profession when people found no one needed them for years . . . it's nice to talk about constantly retraining people and making them more adaptable, but the rate at which you can do that and the rate at which people will accept that is limited.[98]

Employment is increasingly being divided into one set of jobs where a desperate upgrading of skills is required and another set of increasingly low-wage service jobs that are growing even faster.

Aside from global production, the other key element of the global economy, immigration, accentuates this divide. Rather than training less skilled U.S.-born workers for new technical jobs, high-technology companies, especially in Silicon Valley, have increasingly hired foreign-born immigrants from places like India, Taiwan and other overseas Chinese enclaves. One in four people with science degrees in the United States was born in another country. Leaders of Silicon Valley firms have become strong public advocates for the rights of (skilled) immigrants to come to the United States and lobbied to increase the number of immigrants on H1-B visas, given to those with high skills. Conversely, Silicon Valley engineers are increasingly served by the influx of unskilled Latino immigrants (one ethnicity with markedly little representation in the professional strata of the region) in a range of menial tasks from janitorial work to fast-food service to child care.

In this way, the global economy—through outsourcing, immigration, and the breakneck pace of skill obsolescence in the high-tech field (which only a minority can keep pace with)—are working to tear apart any career ladder or even mutual work dependence between these two classes of workers in the region. The public policy promoted by industry in the region increasingly serves only the upper core of professionals who are part of their regional organization of skills and standards. The rest are left behind as even the physical geography of the region is reshaped in a new politics of abandonment.

Regional Polarization and the Politics of Abandonment

If elite collaboration in the region increasingly concentrates on building the intangible assets of standards and trust, there has been a corresponding deemphasis on the more tangible public works assets that once tied regions together and, moreover, benefited the broader community. It is actually hard to overstate the complete collapse of investments in broadly shared infrastructure in the state. The Public Policy Institute of California estimates (using constant dollar measurements) that capital outlays by the state government has declined from highs of nearly $180 per person per year in the state during the 1950s to less than $20 per person during the years of the 1990s.[99]

What electronic communication threatens to do is reconfigure the economic geography of regions and allow elite information workers to partially bypass the social problems of this declining urban infrastructure. Telecommuting, a prime project of Smart Valley, has emerged as one component in that transformation. Telecommuting over the Internet has opened the possibility of some of the largest reconfigurations of work and home relationships since the rise of suburbia and suggests the possibilities of even larger changes in the future.

For decades, geology had locked Silicon Valley into the rather narrow corridor of the San Francisco peninsula and only slowly had it grown shoots over the mountains into Santa Cruz County and into the southern parts of Alameda County. However, under the pressure of limited housing, a new explosion of development had occurred along Route 280, an interstate highway inland over the hills to the east of the urban centers of Oakland and Richmond (although assiduously avoiding those largely minority cities themselves). The sprawl even reached out to more distant areas of the San Joaquin Valley. The number of people commuting into Santa Clara County from other areas of the region increased from 68,000 in 1980 to 113,000 in 1990, according to the U.S. census, and continued to explode throughout the 1990s.[100] Connecting Silicon Valley to both the residents and the new satellite offices of many technology firms along the 280 corridor became a greater and greater priority for Smart Valley firms.

Kathy Blankenship oversaw the telecommuting project for Smart Valley. She emphasized that telecommuting could expand the radius of effective collaboration in the region. "With traffic congested so badly in the region and housing so expensive, the region is threatened with losing a lot of skilled workers. . . . You could have a greater number of people with job satisfaction in the Bay Area with telecommuting. You'll have less people saying maybe I should get out of here and go raise ostriches in Montana."[101]

Smart Valley launched its "Telecommuting pilot project" in February 1995 and continued to publicize its results to companies seeking the best way to implement it in their workplaces. "We decided to focus on telecommuting in its own right. There is a confluence of service, productivity, economic, and environmental benefits," said Eric Benhamou, a director of Smart Valley and then president and CEO of 3Com. "Telecommuting doesn't depend on massive technology investments."[102]

Blankenship noted that people are much more willing to commute over larger distances if they can telecommute two or three days per week and come to the office the other two or three days for the face-to-face interaction needed to do their jobs most effectively. 3Com's Eric Benhamou estimates that with 20 percent of employees telecommuting part of the time, businesses could get 80 percent of the potential benefits of productivity gains and expanding the regional space. The key is that by eliminating two or three days of commuting per week, telecommuting can make working in Silicon Valley bearable for the professional workers living in the new suburbs of the 280 corridor or San Joaquin Valley.

This raises the issue of what this new telecommuting geography means for workers who have to use the regular concrete highways and public transit each day. It is a faint hope that the telecommuters will ease the traffic snarls and gridlock that are day-to-day realities in the Bay Area. CalTrans and other transit agencies have looked at telecommuting as an alternative to roads, but the result of taking some people off congested roads is that others see that the roads are clearer and decide to drive. With congestion at the levels that they are around the Bay Area, as you take people off the road, new people just get on, so telecommuting's relief for technological workers means little for the rest of the region's commuters.[103]

Overall urban concentrations of business are not disappearing in the age of cyberspace, as academic Manuel Castells has emphasized in noting that the costs of telecommunications infrastructure often reinforces top metropolitan areas where the volume of information traffic makes technical investments most profitable. Throughout the past few decades, larger metropolitan areas have ended up with a disproportionate share of information processing jobs. This regional reinforcement of central cities defies predications of spatial decentralization, but Castells does note the regional polarization between central cities and their suburban "urban villages," which combine the advantages of access to cosmopolitan centers without the costs of the central city.[104]

In the context of Joint Venture's pressure to relax government regulation and keep taxes down, its push for telecommuting begins to emerge less as a source of public-interested policy than as developing the technology for elite engineers "coping" with the hellish driving situation in the Bay Area. With a crumbling road infrastructure and a declining tax base, the political support for infrastructure improvements that benefit the average commuter may completely disappear as higher-income telecommuters find their own strategies to cope with gridlock strategies that are independent of the needs of many nonprofessionals. In fact, one of the main accomplishments Joint Venture claimed for itself in its first few years is helping to pass an exemption from sales tax for most manufacturing equipment[105]—a billion-dollar hole in the state budget that came as savage budgetary battles were being fought in the effort to fund earthquake retrofitting for existing roads and bridges.

In the context of public policy for local transit, high-technology executives have followed the same principles of supporting resources for their core of elite engineers living in or commuting to Silicon Valley while working to deliberately undermine revenue sources and transit that serve non-elite workers. In the core of Santa Clara Valley, millions of dollars were poured into a new light rail transit system carrying just twenty-one thousand passengers a day to stops at IBM, Adobe Systems, Intel, and Cisco Systems. Other major destinations included the San Jose Convention Center and San Jose Airport, with connections to a new $244 million Tasman Rail Project that is designed to bring commuters back and forth to the city of Mountain View with stops at major employers like Hewlett-Packard, Lockheed-Martin, and Netscape.[106] With the rail system receiving an 85 percent tax subsidy for operating costs from local, state, and federal funds, residents of poorer minority neighborhoods such as East San Jose complained at public forums that the bus system serving their areas continued to be underfunded and had few connections to the rail system, leaving them trapped with very limited transportation options.[107]

The technology firms of the Bay Area have been as targeted in pulling in funds for longer-range commuter trains, while undermining the public funds in other areas that would serve the transit needs of poorer residents. Even as Silicon Valley leaders pushed through funding for the new Altamont Pass passenger rail line for their high-tech employees making the long commute from the San Joaquin Valley, they had Alameda County (serving the city of Oakland) contribute an initial $2.9 million.[108] This they achieved in a period when the local bus system had been forced to implement an 11 percent reduction in services used by 225,000 mostly poorer, minority residents of the East Bay. Bus riders in the system formed a Bus Riders Union to protest the billions of dollars going to suburban rail lines even as services were cut for poorer residents riding the bus every day. They patterned themselves after the Los

Angeles Bus Riders Union, which in 1995, in conjunction with the NAACP, had won a lawsuit forcing that county to allocate transit funds in a less discriminatory way.[109]

The *Economist,* hardly a critic of wealth, has described the emerging Silicon Valley as "a grander version of one of California's less attractive creations, the gated community: rich, elitist and insular."[110] The civic-mindedness of its business organizations translated into an extremely targeted use of public funds for the benefit of its core of innovative workers, while deliberately undermining the social infrastructure that does not serve its direct regional needs.

Nowhere was this reflected more than in the 1996 statewide elections, largely hailed as the debut of Silicon Valley firms as an active political force in the state. Locally in Santa Clara, the high-tech business organizations strongly backed a measure to raise sales taxes, falling most heavily on poorer consumers, for a nine-year $1.1-billion fund to pay for local rail and highway improvements to decrease local road congestion and support the light rail construction (with almost no funds earmarked for buses, notably).

At the state level, the main event for technology executives was a $38-million campaign to defeat Proposition 211, an initiative backed by senior groups and consumer lawyers that would have made it easier to sue companies for defrauding investors, especially in the case of pension funds. However, as it became clear that Prop 211 would go down to defeat, Silicon Valley executives leading the campaign backed moving $1.75 million to defeat a statewide tax initiative, Proposition 217, that would have helped local governments across the state to invest in schools and other local funding needs by blocking planned reductions in income taxes for those making more than $110,000 per year—thereby restoring $700 million for local government that the state had taken from those local authorities during the state budget crisis in the early 1990s. The $1.75 million spent by the high-technology political action committee (PAC) against Prop 217 was in addition to large individual contributions from Silicon Valley executives such as Tom Proulx, the co-founder of the financial software company Intuit, who gave $410,000 to defeat Prop 217. Yes votes for Prop 217 had been leading before this last-minute infusion of Silicon Valley spending but finally was defeated by a margin of 49–51 percent. The end result of that autumn political action was a commitment of new sales taxes for improvements in local Santa Clara transit, while the executives were able to block funding for less wealthy counties throughout the state, counties that were desperate for funds to serve their needs.[111] The irony was that the tech executives were promoting local sales taxes as the tax source of choice, even as they began lobbying at the federal level against any attempt to allow local governments to collect sales taxes from sales on the Internet—an issue we will return to in Chapter 6.

Many Silicon Valley labor unions, such as those of the building trades, and local Democratic officials had, in defiance of statewide labor unions and the Democratic Party, supported the technology companies in their anti–Prop 211 campaign, but they were outraged when that campaign morphed into an anti-217 campaign vehicle in the last weeks of the campaign. But this merely reflects how disposable technology executives feel cross-class alliances have become, a marked contrast to the midcentury days when business and labor had created strong alliances in supporting funding for education and infrastructure state-wide.

As the following chapter will document, even traditional industries such as banking and utilities, which had previously been natural allies in support of public-interested investments in broad-based regional growth, have been affected by the new technology in ways that have increasingly cut their self-interested links with poorer communities. All this has contributed to the paradox of a vibrant regional politics that matters most to multinational companies at the same time as communities are increasingly disempowered by the technology at the local level.

5

When Joint Venture was launched to revitalize
the Silicon Valley economy, the founders turned
to Bank of America to supply an initial $250,000
to get the effort off the ground, a crucial investment that was a catalyst for the effort. Similarly,
one of the largest ongoing supporters of Smart
Valley was Pacific Gas and Electric (PG&E),
which contributed more than $100,000 to the
effort, far more than any of the computer companies involved. Pacific Bell, along with millions
spent on its own CalREN networking project,
contributed almost $500,000 to Joint Venture's
efforts to wire Silicon Valley schools, also a
larger contribution than any of the computer
companies involved.

Why so much money from these particular
companies for a "common good" when so
many computer companies might have been
more obvious angels in supporting these regional efforts?

What all three of these companies shared is a
traditional commitment to the region, a commitment that has spanned almost one hundred

years in all three cases, one that is tied to the fact that each is a classic regional network conveying financial information, power and communication respectively. Banks, power companies, and telephone companies are the classic components of what David Molotch, John Logan, and David Harvey have called fixed capital, the bedrock of regional coalitions to promote growth and economic development.[1] Add in Southern Pacific Railroad in earlier years, and these same companies were the key components of regional growth coalitions half a century ago and even seventy-five years ago.

Across the country, local community banks historically played a vital role in recirculating local savings into economic development—backed by government policies that kept banks focused on serving those local customers. Growth in consumer credit would keep local commerce expanding, which would in turn feed a cycle of expanded local business and stronger banks, which could extend more credit in a virtuous cycle. In California, that expansion was concentrated in the form of Bank of America, which became the largest bank in the world following World War II, thriving on exactly that cycle of serving local depositors and regional businesses. Its founder, A. P. Giannini, would make it a creed in the bank that the bank's success was inevitably tied to the economic success of working-class depositors and the prosperity of the region in which those workers lived.

Similarly, like other utilities across the country, PG&E came into being as a merger of the numerous city and regional power companies found new economies of scale in integrating regional power production and distribution systems. By its nature early on, generation and distribution of electricity had to be local, so power companies gained a keen interest in supporting the local expansion of industry and commerce that in turn would spur new energy sales. Hydroelectric and other power projects became key capital investments in regions, spurring employment and a range of subsidiary commerce. Of especial importance in the West was the fact that these integrated regional utilities would deliver subsidized rates to rural agricultural areas, thereby expanding economic development and access to the electricity grid to a range of new customers. In a similar way, hundreds of local phone companies would be consolidated under AT&T, which would be a holding company for the separate state phone companies. Each state company would have its own budget and strong local public participation on its board of directors and would be governed by broad public interest guidelines set by local state utilities commissions.

All these companies would becomes fixtures in the political life of their regions, often acting as a political bridge between working-class organizations and the rest of the corporate elite. Their constancy of involvement in the civic economic life of the region stemmed from a simple congruence of self-interest and public interest for these companies whose customer bases and financial investments were so thoroughly tied to a single region. This is something that most local industrial companies can rarely say, with global markets and (in the back of their minds at all points) global production facilities beckoning them in the future. The fixed circulation of capital between such local banks and utilities and their consumers created the basis for the long-term cross-class cooperation that built the civic economic life of regions across the country. And, however imperfectly, it was those cross-class collaborations that brought many poor and working-class families into economic life, through commitments to broad banking services and subsidies for universal service by local utilities. It was these subsidies in the postwar era that continued to expand the base of consumers for these basic services and gave those families a connection to the broader economic life of the expanding regional economies.

However, just as the new pressures of industrial competition have led to the downgrading of private investment in basic research by many private firms, there has been an accelerating trend of technological change, competition, and deregionalization of these traditional regional anchors of banking, power companies, and telephone service. Subsidies that had flowed from big business and wealthier customers to smaller users of these services have begun to be reversed as changed regulations and new technology, particularly the Internet in recent years, have allowed sophisticated industrial customers to cream off the benefits of new technological innovation for themselves, even as they abandon regionally based markets that had delivered universal service. And as those subsidies disappear and as previously local banks and utilities become global players through both expansion and merger, it is less and less clear from where any private source of public goods not exclusively focused on business itself can and will originate. In that sense, the participation of the traditional anchors of the regional economy in the Joint Venture Internet projects may represent less a last gasp of that tradition and more their emergence as key players in the new "gated community" elite politics of the region.

In the end, technological standards and the investment that make them possible are tied as much to pricing and legal arrangements as they are to specific feats of engineering. As will be explored in this chapter, earlier banking, telecommunication, and utility models of accumulation and network building were based on pricing models that allowed those who were committed to expanding the network the opportunity to recover the fixed costs of their investments. In turn, that commitment to fixed investments in the community would be tied legally and financially to subsidies that would expand the number of people brought into the network, thereby increasing the value of the network for all. The bottom line for such regional actors was that, given the limited geographic scope of their markets and the ability to recover almost the full value of their investments with various government guarantees, it was in their interest to maximize growth in every sector, rich and poor, in that region.

As those geographic restraints have been lifted both politically and technologically, new companies have begun to operate through use of new technologies, including the Internet, in multiple geographic markets, "skimming the cream" of high-profit customers by taking advantage of already established public investments. These new companies essentially profit from the long-term inherent value of the network as a whole while only having to expend their own funds on servicing the most valuable members of those networks. New technology has gone hand in hand with legal changes in the geographic boundaries of these markets, legal changes that would alter in pricing models, moving the focus from fixed cost recovery to marginal costs regardless of the total public and private investment needed to maintain financial and utility networks.

Ironically, since political concerns about new network construction, network interconnection, and regional investments have not disappeared, "deregulation" has actually ushered in an era of almost constant tinkering by political leaders and regulators seeking to achieve the regional, economic, and social goals that were implicit in the more laissez-faire days of delegating day-to-day management of the regional service networks. Instead of having designated regional actors who had a private interest intertwined with the public interest and could be left more to themselves to implement directives that could be periodically reviewed, political leaders are now having to debate and design the "ideal" market needed for each service, police against predatory rent-seeking, and compensate for loss of regional subsidies with a range of jerry-rigged measures to ensure some degree of universal service.

As analysts Steven Vogel and Jill Hills have argued, the move to "deregulation" in no way decreased the level of government involvement in industries such as financial services and telecommunications.[2] In fact, much of this political movement, which Vogel labels "reregulation," was in many cases an attempt by specific regulatory agencies to rationalize and strengthen their control of the industry from previously independent monopolies or from other government bodies. In the case of the United States, the creation of new markets in the banking, energy, and telecommunication industries has amounted to a massive transfer of political control from local government authorities (which traditionally governed these regional actors) to federal regulatory agencies such as the Federal Communications Commission. Despite the odes to government decentralization, most of "deregulation" has been the emasculation of the power of local government in favor of national government regulators. This change in political venues has translated into an erosion of the cross-class regional collaboration that had been a hallmark of the earlier era of local governance in favor of a stance that promoted global competitiveness of these industries in relation to other countries. Noting that the ideological promotion of telecommunications started in the United States, Hills argues persuasively that much of "deregulation" globally can be understood as part of a U.S. strategy of asserting information technology dominance in the world by opening new markets for its corporations.

This chapter will outline how the technology of the Bay Area region would interact both with these elements of regional fixed-capital industries and with the global economic alterations initiated by these technological changes. National investments in the Bay Area would contribute to the networking technology that in turn, with specific political arrangements enacted to support this new technological regime, would help launch financial revolutions in the banking industry. These would undermine traditional regional banking systems around the country, including in Northern California. In turn, the region would face new challenges for local banking that would push Bay Area banks to use the newest Internet technology not only to compete in the new global banking system but also to seek to dominate it. In the area of power companies, computer technology and the Internet would make possible new national systems of power distribution and coordination that are now beginning to undermine regional utilities. And the phone system would go through regulatory earthquakes as telecom technology would both initiate and be affected by the swirl of new computer technology. In each of these changes, large business customers would overwhelmingly benefit from the new global systems even as local customers would end up shouldering the burden both of network maintenance and the economic losses of the transition. These losses would range from bailing out savings and loan associations to blackouts and rising energy costs to slowness in upgrading local telephone service for high-speed Internet access. Technology was again leaving those trapped in local geography with the fewest options and least power in the new era.

How Microchips Lead to Megabanks

The film *It's a Wonderful Life* pictured the local banker as a fixture of the community—whether as evil oppressor Mr. Potter or local savior Jimmy Stewart's George Bailey. Either way, it was hard to think of the community without seeing it as a reflection of the local banks.

These local banks, however, are increasingly a thing of the past, done in by regulatory changes and technology that has encouraged bigger national and even international banks. Electronic banking has emerged as the central strategy for banks in going global and cutting costs, while the disappearance of local bank branches is just the most obvious sign of the deep cost-cutting involved. Automatic teller machines constituted the first wave of electronic banking to begin to replace the local branch, with Internet banking now coming on fast. With alternatives to the traditional bank branch costing as little as 15 percent of comparable branch transactions (and with costs dropping quickly), banks are seeking every opportunity to push their way into electronic banking.[3] The two largest California-based banks, Bank of America and Wells Fargo (which both went through massive mergers in the early and mid-1990s), have closed thousands of traditional branches—in the case of Wells, well over half its branches have been closed since 1980.[4]

In 1996 Internet banking would become widely available at banks across the country; only 7 percent of banks were offering Internet banking at the beginning of that year, but by the following year, 59 percent of banks were planning to do so.[5] If any entrant into Internet banking showed the promise (or the threat) of banking without geography, it was Security First Network Bank (SFNB), appearing in October 1995 as the first independent bank with no physical branches and carrying out all business electronically. The creation of an otherwise obscure savings and loan from Kentucky, SFNB built on a collaboration with Hewlett-Packard to launch the bank and a software subsidiary, Five Paces, to sell Internet-banking software to other banks.[6] Within months, SFNB had a few hundred customers, which grew to seven thousand customers and $201 million in deposits by the third quarter of 1996. SFNB never became a big direct player, yet its success was a warning sign to more established banks. More important was its relationship with Hewlett-Packard, through which it sold its Internet software to banks across the country.[7] By 1998, banks totaling $800 billion in assets were using SFNB's Internet software.[8]

Nowhere was that threat felt more than in Northern California, where a majority of SFNB's initial customers lived, clearly revealing that while the Internet was global, the Bay Area was the focus of both the production of the new technology and the effects of its global economic impact.[9] Bank of America and Wells Fargo, the largest banks in the region, realized that their traditional territory was under threat from banks that could reach their customers from anywhere in the world over the Internet. Bank of America's Mack

Hicks would argue that "in the information age there is no boundary for the delivery of services so there is no reason for regions."[10] Bank of America moved quickly to establish home banking services on the Internet, but watched their smaller rival Wells Fargo move even more aggressively to use the Internet to leverage itself into being a major international player. By April 1997, Wells Fargo was adding ten thousand new online customers every week as it expanded far beyond its traditional California roots at the expense of many local banks in other regions.[11]

Even in the core area of loans, smaller banks across the country are losing out to bigger banks like Wells Fargo as the new technology cuts out the use of local intermediaries to evaluate loans—a radical change even for larger banks such as Bank of America in which local branch managers once had the central role in evaluations. Home mortgages are increasingly based on national lending criteria determined and managed by central banking institutions such as Fannie Mae and Freddie Mac. Using an infrastructure of fast computers, database software and high-speed communication networks, new "data mining" techniques allow lenders to identify patterns and relationships in customer data to determine whether loans should be granted. All these technologies require technical and statistical sophistication, which small banks cannot afford.[12]

Home mortgages have been moving toward centralization for the past two decades, but even more significant for regional economies is the recent emergence of a national loan market for small-business loans—once the quintessential part of local economic-development decision making. Until only a few years ago, large institutions saw little advantage in loaning to small business; each loan was evaluated based on business plans, balance sheets, cash flow, and profits. There were few economies of scale. Because of the uncertainty of small-business survival, the assumption was that only local banks had any chance of betting on the right loans, so local bankers with knowledge of an area had free rein in the area of small-business lending.

Computer technology has changed all that as small-business loans have begun to be evaluated by means of the same statistical methods as those for individual consumer loans. Using a system known as credit scoring, lenders no longer perform detailed financial reviews of each borrower. Instead, they use a few essential pieces of information to predict the probability that a borrower will repay, then offer the loan to those deemed worth taking a risk on. The cost of processing each small-business loan thereby drops from thousands of dollars to a mere few hundred, opening the way to apply mass-market techniques to these loans.

Big banks are displacing local banks in this game for two reasons. First, the costs of the computers, software, and databases involved in this new system are far beyond the resources of most small banks. "The capital cost is significant, and if you are not doing volume, you can't recover it," noted Bank of America small-business lending chief Janet Garufis. In that sense, the balance of power is tilting toward large regional or national lenders. "This hits at the fundamental strategy of community banks in the small business market," noted Vikram Capoor, a consultant with the Advisory Board Company, based in Washington, D.C. "It shifts success to the lowest cost provider."[13] Most banks involved in small-business lending are now using off-the-shelf scoring software such as that by Morris Associates and Fair Isaac, which was developed from data pooled from a consortium of seventeen large regional banks.

However, Wells Fargo was one bank that pioneered the development of an in-house system to give it a dominant edge over rivals. It began the process in the late 1980s as it was trying to escape a dangerous concentration of loans in commercial real estate and leveraged buyouts. Turning to its weak small-business lending department, the bank decided that with the right concentration on information technology, it could dramatically increase its presence nationally in the small-business market. Using information collected on its own customers as a guide to evaluating data to predict losses, it began adding to its database as it expanded small-business loans in California, beginning in 1991. This was combined with an increase in technology in back-office operations, to reduce paper and drive costs down. "We engineered out the number of times a human hand touches a loan," said executive vice president Lucy Reid, who heads the bank's national small-business direct-marketing program. Within California, the results were spectacular. In 1989, Wells Fargo had only 1 percent of all small business loans held by California banks. By 1995, it had more than 16 percent of all small-business loans.[14]

Wells Fargo was ready to go national, using direct-marketing techniques to attract customers outside California. In 1995, five million small-business loan solicitations were mailed nationally. Outside California, Wells concentrated solely on loans less than one hundred thousand dollars, using credit scoring to sort potential customers into eight risk categories with different loan rates offered to each group. By June of that year, total commercial loans of less than $1 million at Wells Fargo had jumped to 117,392; that year, the bank earned a stellar 32 percent return on equity in its small-business credit portfolio. By 2000, Wells Fargo had become the largest small-business lender with $9.88 billion in loans.

Smaller banks are being squeezed not only because they cannot afford to invest in the computer technology and in-house databases, but also because they just are not big enough to treat loans as large as one hundred thousand dollars as statistics that can repay or default based on probability tables.[15] Local knowledge of the economy no longer provides the competitive advantage it once did for local banks as low-cost, mass-market loans become possible as a result of the new computer technology. Big banks end up with the advantage as they can average out good and bad returns over a much broader capital base.

Direct mail, telemarketing, and now Internet marketing of such loans nationwide to small business is replacing the local community bank as the source of small capital. And as banks grow more confident in using these computerized techniques, they are moving toward targeting larger business loans, further threatening community banks. All these changes are undermining traditions that linked local businesses and local capital in a shared fate connected to the growth of their region. As capital moves to the national level, local businesses are increasingly able to find loans regardless of the health of local banks, while local banks no longer have to depend on growth in their local region to find places to invest their capital.

Technological Regionalism and Global Banking

There is an irony in the fact that Wells Fargo, once the epitome of banking for the business elite, is emerging as a dominant small-business lender. Of course, this version of small-business lending is in many ways merely a high-tech continuation of earlier elite strategies that have simply used computer technology to move the same centralized approach downscale. While its lending practices are increasingly nonregional, Wells Fargo has used tight regional relationships with Bay Area technology firms employing Internet standards to leverage itself into an emerging dominant position in the new banking order.

The early 1990s would see a number of proprietary attempts to establish electronic bank payment standards, chiefly as Intuit (maker of Quicken financial software) tried to create its own network of banks beginning in 1990. However, by 1995, less than 2 percent of Quicken users were paying their bills using its proprietary network. "It's hard to get a whole culture to accept a new payment scheme," Intuit chairman Scott Cook would admit. Also, Microsoft had envisioned using its original proprietary Microsoft Network as a mechanism for dominating online banking. A proposed merger with Intuit was blocked because of antitrust concerns, after which Microsoft abandoned its whole proprietary approach[16] in favor of connecting its Money software customers directly to the Internet and offering banks themselves the server software needed to integrate their banking operations on the World Wide Web.[17]

A variety of competing standards for electronic payment over the Internet would vie for attention in the following few years. The Secure Electronic Transaction (SET) protocol emerged out of a CommerceNet initiative backed by Visa, MasterCard, a range of banks, Netscape, IBM, and Hewlett-Packard and a number of other technology companies, including Microsoft. Wells Fargo would become the first bank to issue "digital certificates"—a means of authenticating the identity of customers—to businesses wanting to collect payments over the Net. Whenever a sale was made, a merchant using Wells Fargo digital certificates would relay information on the sale to the bank, which would then verify the purchaser's credit card over private networks, then relay that information back to the business instantaneously. Wells Fargo worked closely with Verifone Incorporated—a Silicon Valley firm best known for producing the devices used by merchants to "swipe" credit cards—to develop the tools to link merchants to Wells Fargo. Not coincidentally, the marketing director for Verifone's efforts was Cathy Medich, a former executive director of CommerceNet, who was instrumental in pushing forward agreement between various firms in regard to electronic payments. And in April 1997, Hewlett-Packard announced its buyout of Verifone for $1.15 billion, permitting Hewlett-Packard to further integrate its networking technology into financial systems and strengthen its alliances with banking and other financial sectors.[18] In a move that analysts saw as paralleling Hewlett-Packard's involvement in CommerceNet, Hewlett-Packard launched a new consortium called First Global Commerce with credit card companies, EDS, and other partners to promote the range of online technologies needed for rapidly expanded e-commerce that could directly link its customers' homes.[19]

This elite-based approach to dominating global banking is in sharp contrast to the historical legacy of locally oriented banks, especially the regional traditions of Wells Fargo's archrival Bank of America, whose role in the earlier age as a facilitator of the civic organization of economic life was unmatched probably anywhere in the country. In the section that follows I will explore the contours of that experience and the political context that shaped it, then turn to the way in which that technology pioneered by Bank of America itself began to undermine the very structure of regional development it had pioneered in California.

Bank of America and the Birth of the American Dream

A. P. Giannini, the founder of what would become Bank of America, achieved legendary status in his own lifetime, incongruously for a banker, as a populist hero pitted against other bankers even as he would turn Bank of America into the largest bank in the world. Born in 1870, this son of Italian immigrants nurtured his ambition to start a bank out of his experience working in his stepfather's produce business as he watched banks, even those of fellow Italians, refuse to lend to poor, newer immigrants who needed help more than did the established businesses on which banks focused their efforts. After a brief stint on the board of one small bank in San Francisco, he quit in 1904 to establish his own, the Bank of Italy, using the proceeds of his stepfather's business as initial capital. His first customers were many of his old customers in the fruit and vegetable trade, people who had never been near a bank before—workingmen, fish dealers, grocers, bakers, plumbers, and barbers. He pioneered small loans, some as low as twenty-five dollars, something no other bank did. His explicit goals were to build both the local economy and loyalty to his bank and to share in the growing prosperity of his clients as their expanding deposits circulated through his bank to new businesses.[20]

The first test of his vision came with the 1906 earthquake that leveled the city of San Francisco. As other bankers held back loans for six months to evaluate what was left worth investing in, Giannini famously hauled ten thousand dollars in cash from the vault in the rubble of his bank, set up a plank on some barrels on the city's wharf, and reopened his bank immediately for business in the open air. He lent money to shippers to bring back lumber from the Northwest to rebuild homes and urged others to do the same, making him a local hero. The combination of his open loan policy and skillful self-promotion in newspaper advertisements would bring new attention and growth to his enterprise. He sought to quickly expand deposits, opening at night and on Sundays to accommodate the schedules of working families.

His next innovation was the creation of branch banking, inspired in the United States by its success in Canada. The idea was to use local talent focused on local needs, while making available the capital reserves of a larger banking network—a radical departure from practices of the big banks of the East Coast that lent money to far-flung corporate interests only from central offices. His first branch was in his hometown of San Jose, his second in San Mateo. He would also acquire banks in Southern California and the San Joaquin Valley. By 1928 he presided over 138 branches with $358 million in assets, but in each branch he retained advisors and employees from the area and drew upon local visibility to run the day-to-day lending of the branches.

Giannini's expansion could simply be explained by the fact that California legalized branch banking, whereas other states, notably New York, Texas, and Illinois, severely restricted the practice, fearful of big banks. Countering this explanation is the fact that similar fears in California could have and almost did lead to similar legislative restrictions on Giannini's expansion in the 1920s. However, even the harshest critics of branch banking expansion, such as state superintendent of banking Charles Stern, would grudgingly allow Giannini "the unquestioned benefit your branch offices have brought to many specific localities."[21] Combined with nativist attacks on Giannini's immigrant and Catholic background by nonbranch bankers, the same attitudes as in other states could easily have led to the same restrictions on expansion as in those other large states. Yet it was precisely because Giannini was not J. P. Morgan but had built his network around local investment in the bank itself and local economic development that he could marshal the political support to continue expanding. What was clear was that the policy dynamics in every state was one of attempting to focus banks on local economic development, and Giannini's expansion was based on innovating a system that accommodated that public policy regime. In opposition to those running most large banks, Giannini would strongly support Roosevelt's New Deal, including the Banking Act of 1935. Despite that support for Roosevelt, the same New Deal administration would frustrate Giannini's desires to directly integrate scattered holdings of banks in other western states directly into Bank of America—reinforcing the focus of the bank on regional development of California.

It was in that regional development role that Bank of America would shine and prosper. The bank would become a key source of start-up capital for the largely immigrant-run movie studios of Southern California, lending to an industry largely shunned by other banks, and would use its branches around the state to track the needs of different agricultural sectors to supply capital as required for regional investments and to stave off various agricultural disasters. The supremacy of local branches would keep the focus on small stores, home buyers, and the growing consumer market for loans—all tying together Giannini's ideas about integrated economic development. By the end of the 1930s, Bank of America was handling more than six hundred thousand small loans each year and would extend start-up loans to industries ranging from jewelry to oil wells to furniture to lumber to textiles and clothing. Giannini had almost single-handedly nursed local wineries through Prohibition and would continue to support the expansion of the wine industry throughout the post-Prohibition period. When World War II came, the bank would be a major force in supporting wartime conversion and organizing local civic economic planning. Branches of the bank would organize local pools of manufacturers to apply for wartime contracts and federally guaranteed loans. On the broader regional level, Bank of America strongly supported industrialists such as Henry Kaiser in building the new industrial base of the region.

With the end of the war, Bank of America was at the forefront of the new consumer-led boom in credit and spending. This boom was backed by the federal government through a combination of subsidies and a regulatory regime that favored bottom-up consumer-led growth. The subsidies came overwhelmingly in the form of Federal Housing Administration loans that were supplemented after the war by GI loans to veterans. Bank of America alone would extend more than $600 million in GI loans and would lead the new mass market in housing and the financing of homes.

More important, New Deal regulation would focus banks away from the speculative investment that had led to massive bank failures in the Great Depression and toward focusing on bank services to the broader public. Tied to separations between different financial institutions in serving different financial functions and a ban on interstate banking, the heart of the regulations was Regulation Q, which limited the interest rates paid by banks to depositors. This was linked to antiusury laws at the federal and state level that in turn limited the interest rates that banks could charge to loan borrowers. This set of regulations was passed to prevent banks from overextending themselves in competing for deposits, a problem that had contributed heavily to the collapse of banks at the beginning of the Depression. Since most competition for deposits would have been focused on upper-income depositors, set rates of interest leveled the playing field between richer and poorer bank customers and kept basic bank fees relatively low. This system fit with Giannini's view of a world of small depositors, and his bank would thrive on small depositors in the postwar period as the largest bank in the world. Fitting with his vision of sharing wealth, Giannini died with an estate worth less than $1 million, most of which he left to a charitable foundation—a minuscule estate from a man who had created a bank with more than $6 billion in assets that he had personally built from scratch.

While Giannini's legacy would live on for a number of decades, new technology combined with a political assault on the New Deal regulatory regime would undermine this whole system of subsidies and level bank rates that had benefited both regional development and low-income bank customers. A new generation of Bank of America managers would embark on a global banking vision that would abandon much of the community-oriented banking that had been its lifeblood, even as its regional rivals would join them in both the technological and economic leadership of this emerging new era.

How Bank of America Created the Credit Card—and Saw It Destroy Regional Consumer Credit

In the postwar period, Bank of America had embraced installment credit as a way for consumers to finance a range of new purchases. Auto loans were a notable example, and Bank of America would make 85 percent of such loans in the state at its peak. Partly to ease the burden of endlessly evaluating myriad individual small loans, Bank of America in the 1950s began exploring the possibility of an all-purpose credit card. Individual companies from gas stations to Sears Roebuck had introduced company-specific charge cards and a few small banks had tried experiments, but Bank of America would be the first to build a multivendor card in wide use. It would build its BankAmericard credit card system (which would eventually become the Visa card we know today) and set the standard for all other credit cards in use.

In launching the card, the bank faced the dilemma that they needed to encourage the participation of store owners as well as customers, with the catch-22 that customers would only want the card if merchants accepted it, and vice versa. To create this integrated credit system, Bank of America solved the problem in 1958 by mailing out sixty thousand credit cards, one for every resident of Fresno who did business with Bank of America—the city chosen because of its high concentration of customers. (This strategy was an anticipation of the Web browser/server division in networking that Netscape would solve years later through similar Internet "drops" of browser software in order to sell server software to businesses.) With so many cards suddenly in the hands of residents, most merchants, especially the smaller ones, signed up, since participation would relieve them of the burden of tracking the small credit accounts they ran for customers all the time. The credit card system was now possible because of the new computers used by Bank of America that could alert the bank when credit card customers were spending more than their limit allowed. It was therefore natural for the credit card to emerge in the technology center of Northern California. Essentially, the credit card transferred the work of small-store bookkeeping into the offices of Bank of America and its computers. In 1956, Bank of America had introduced the first electronic check-handling system, MICR, developed by the bank and the Stanford Research Institute. The first fully automated checking account system would follow in 1959, the same year that, flush with its Fresno success, Bank of America would make similar credit card "drops" to two million Californians around the state, using its regional concentrations to leverage the adoption of the credit system city by city. New technology was launching a revolution in how information about financial transactions would be tracked from this point onward.[22]

Heavy initial financial losses in some areas (especially Los Angeles) as a result of careless accounting and loose card distribution actually helped the bank by scaring off the competition from trying to enter the credit card field;

by the early 1960s, Bank of America was making large profits in a credit card market in California that it had to itself until 1966. With other companies readying to enter the field both in California and across the country, Bank of America began franchising the BankAmericard to other banks in other states; each could sign up local merchants and reap the profits regionally for a nominal fee to Bank of America. The bank was happy with this franchise approach; in its view, the more universal the card, the more likely it was that California residents would want to have one for travel, or that tourists from out of state would make BankAmericard purchases from California merchants signed up with Bank of America. In this way, each bank could expand its own regional market while taking advantage of a universal credit card system.

In 1967, four of Bank of America's California rivals would form Master-Charge (which would merge with credit card systems of other national banks to become the MasterCard system) and the following three years would see an explosion of credit cards across the country as more than 100 million credit cards were mailed to customers nationwide. Abuses and fraud that arose in the confusion led in 1970 to a federal ban on the practice of mailing physical cards without customers' permission, but by that time credit cards had become a permanent national fixture.

The system used by BankAmericard banks to approve purchases from one another's cards had become cumbersome by 1970. Few banks were happy that Bank of America, a potential competitor even at its most well behaved, was running the information exchange system. For these reasons, Bank of America turned over control of the credit card approval system in 1970 to a new organization, called the National BankAmericard Incorporated (NBI), which was eventually renamed Visa in 1976. NBI would produce a nonproprietary exchange system that would be trusted by all involved, a key step in creating a universal financial network.

What was clear was that computer technology was needed as the backbone of the new credit card system. Technology could render a complex set of financial transactions between banks simple and invisible to the average user, creating the trust needed for wide use of the credit card. Dee Hock, who left Bank of America to run the Visa system, set up two monster mainframes in San Mateo, California, to run the national network of computers in the NBI system, managing 200 million transactions in 1974 and nearly 6 billion by 1992. Replacing a cumbersome system of phone calls, NBI drew on technological advice in the Bay Area to create a system costing only $3 million that would save member banks $30 million in its first year of operation. This first broad-based computer network of banks would soon extend to computer devices at individual merchants—the largest computer network in the United States until the advent of the Internet.

With a technological jump on rival MasterCard's Interbank system, the BankAmericard system would catch up quickly on the larger number of banks in the MasterCard system. But the very success of the BankAmericard system would open opportunities soon to undermine the regional credit systems of all the banks involved. In the 1970s, Citibank had emerged as the main rival to Bank of America in size; Citibank had even set up an R&D facility in California to draw on the same technological resources available to its rival in innovating new consumer uses of the technology. Citibank had been a long-time member of the MasterCard credit card system, but when lawsuits forced the NBI system to allow banks to issue both BankAmericard and MasterCard cards in 1975, Citibank soon signed up as a member of NBI. And when Bank-Americard changed its name to Visa in 1976, Citibank took advantage of the confusion to mail a staggering 26 million pieces of direct mail telling people that their "new" Visa card was waiting for them, never mentioning the word *Citibank* as the company grabbed 3 million Visa card customers from regional banks. This one campaign transformed Citibank into the second largest Visa card issuer after Bank of America and, adding in its MasterCard accounts, made Citibank the largest issuer of credit cards in the world.

In one fell swoop, Citibank had left its New York base and created a national system of consumer lending—using the technological system that Bank of America had built through the support of various regional banks. It had steamrollered over the local systems of lending that each of the regional banks had constructed over decades. These actions by Citibank would, in the words of financial writer Joseph Nocera, make it among other bankers "the world's most hated bank" and cause bankers "[to spew] unbankerly venom at an institution they had all come to despise."[23] Much as Netscape (and then Microsoft) had used technological networks built up by others to steal control of the Web, Citibank had used the Visa technological network to steal customers from the regional banks around the country, banks that had laboriously built the social capital of trust in each regional component of the broader Visa network.

Citibank had been able to bypass investing in building the local lending networks itself in favor of wholesale marketing to already existing customers of the credit card network; by the mid-1980s Citibank would emerge as the largest bank in the nation, with one out of five American families having some kind of banking relationship with Citibank. However, because of Citibank's deception and fears that something similar could happen with any new network, when automatic teller machines (ATMs) were introduced, Visa banks refused to integrate them into the Visa system. They preferred proprietary ATM systems, which would give them greater control over who could use their ATMs and ensure that each bank could collect its fees. After a decade in which regional banks built trust in open, nonproprietary systems of financial information exchange, Citibank had undermined trust in the network and halted the movement toward an integrated financial information system until the Internet appeared two decades later.

How Silicon Valley Technology and National Political Muscle Undermined Regional Banking

Citibank was hardly alone in using the new technology to grab customers from traditional regional banking. A combination of Silicon Valley technological knowledge and political lobbying for banking regulatory changes (which Citibank was one of the few banks to support) was eroding the traditional banking system in the United States. In the 1970s, when inflation was reducing the value of personal savings, the attraction of technology for upper-income investors looking for a better deal for their money became a political steamroller that would wipe away decades of bank regulations. The result would be devastating for lower-income savers, in the form of the savings and loan disaster, higher banking fees for average depositors, and the closing of branch offices across the country.

On the West Coast, one of the first money market funds was created by a San Jose analyst named James Benham, a Merrill Lynch broker who quit that company to create a fund investing in nothing but U.S. Treasury bills as a way to help consumers obtain a better interest rate than at their local bank. Joined by Henry Brown and Bruce Bent of the small Manhattan investment firm Brown and Bent, Benham in 1972 would use deceptive filings with the Securities and Exchange Commission to get approval for its new unregulated money market fund. Their fund could now compete directly with the banks—something that violated the decades-old intention of the Regulation Q restrictions on interest rates. Computer software would be used to make these new bond funds mimic the actions of a bank account, further assisting in the blurring of the difference between money market funds and savings accounts. Such funds would explode from $1.7 billion in assets in 1974 to $200 billion by 1982.

It was in this period that SRI, now independent of Stanford University and looking for corporate clients, took on a $1.5-million study of the financial industry, a study funded by more than fifty corporate sponsors. The most surprising finding was the conservative attitude of families toward their savings: the typical "affluent" family's savings account had more than forty thousand dollars in it, while average earners (those making fifteen thousand dollars a year) had roughly ten thousand dollars in savings. Savings and checking accounts amounted to 16 percent of total household assets, just behind real estate. Money that was then routinely recycled through banks, mostly in local regional investments, were seen by the SRI study as prime targets for companies investing in global capital markets.[24] One of SRI's first major clients was Merrill Lynch, for whom in 1975–76 they helped create the cash management account, which would tie money market accounts to stocks, bond funds, checking accounts, and credit cards. The system was immensely complicated technologically and required that millions of dollars be invested in new computer systems, but the result was a simple instrument that could attract the savings of the most affluent bank customers. Cash management accounts at Merrill Lynch would eventually swell to $250 billion in assets.

One other main Bay Area contributor to the emerging deregionalization of financial assets was the San Francisco–based discount brokerage firm of Charles Schwab, which would computerize its operations years before other Wall Street firms. By electronically updating each customer's trades, Schwab could offer deep discounts to a new class of upper-income families entering the stock market. And Schwab's success in the Bay Area would result from not only the technological know-how he could tap but also the new class of "computer jocks" from Silicon Valley who were being paid in stock options. Members of this new breed of equity owner would need to trade their shares over time, creating the critical mass of a market that could launch Schwab's type of firm.

As much as the regional banking system was undermined by nonbank institutions using the new technology, the regional banks themselves contributed to its decline through their abandonment of their local markets and branches. By the 1980s Bank of America would become a symbol of this decline as it teetered at the edge of bankruptcy and jettisoned almost every remnant of its Giannini tradition. A new generation of corporate-oriented bankers had entered its career ladder in the late 1940s, following Giannini's death, and by the end of the 1960s, Giannini's prized bank branches were losing out to the ambitions of the new bank leaders. Individual branches no longer handled corporate accounts, weakening the ability of the bank to work closely with rising start-ups and new industries as the bank had for decades before. Aside from a few investment forays, such as an ill-fated alliance with Memorex, Bank of

America would miss most of the opportunities for investment in Silicon Valley in the 1960s and 1970s as its new corporate leaders sought to compete with nonbank financial institutions in global markets. Instead of recycling regional capital between home mortgages and local business start-ups as it had in the past, Bank of America would spend much of the 1970s trying to recycle Arabian petrodollars into third-world loans.[25]

Bank of America was becoming even less interested in its traditional mortgage business and was persuaded in 1977 by traders at Soloman Brothers (which would pioneer mortgage trading on Wall Street) to package its home loans in the form of bonds, to be sold by the firm. This was the first private issue of mortgage securities in the country, a practice that by the 1980s would help lead to the wholesale conversion of local mortgages into more than $150 billion in bonds by 1986. It was largely this speculative trading that would help bring on the savings and loan debacle by the end of the decade, a debacle that would hit California especially hard as ground zero for a disproportionate share of savings and loan failures.[26]

One other mark of Bank of America's disengagement from its own regional market by the 1970s was its antiquated banking computer systems, now a haphazard combination of near-obsolete IBM mainframes connected by a Babel of software programs unable to talk to one another. Unbelievably, it was the London office that had to pioneer the bank's computerized system of information for its global investments. When Bank of America finally updated its computer systems in the early 1980s, it was in the context of seeking to leverage the daily $1 trillion worth of transactions moving through its electronic networks into (an ultimately unsuccessful) proprietary fee-for-service electronic network to service other banks. Gone was the tradition that had built the open BankAmericard electronic standard on the assumption that rising use would strengthen the California economy (and thereby Bank of America); now, the explicit view of the bank leadership was that direct loans were losers and the real money was to be made in fees from elite customers, especially other banks. This was also the period when Bank of America, previously a laggard in deployment of ATMs, would expand its ATM network into the largest in the state even as local branches began to close, 134 closing in 1984 alone. In 1985, Claire Hoffman, the last remaining member of the Giannini family still on the Bank of America board of directors, resigned her position and publicly denounced the bank's abandonment of its role as a "corporate trustee of great public purpose" and its forsaking its relationship with customers in favor of technology.[27] It would be left to the following decade for Bank of America and, even more enthusiastically, its old elite rival Wells Fargo to reconstitute regional civic economic coalitions employing the new model of using region-based electronic standards to leverage global economic dominance.

In this context, the growth of the venture capital system in the Silicon Valley region appears less as a bold innovation than as a mutation of that older Giannini entrepreneurial investment tradition—albeit a mutation that no longer cycled capital between rich and working families in the state but rather moved superprofits between the hands of elite investors and soon-to-be millionaire computer jocks who would invest their proceeds in new brokerage firms such as Charles Schwab. This evolution split off the process of business creation into an elite financial world separate from other regional economic development.

There was some hope that at least a broader range of investors nationally might be benefiting from the venture capital nexus as the NASDAQ Internet boom took off in the late 1990s. But it turned out to be a vicious financial mirage that only reinforced the economic inequalities between insiders and the broader public. Between March 2000 when the tech-heavy NASDAQ reached a staggering 5,048 and the following year when it dropped below the 2,000 line, more than $4 trillion in value was wiped out. Some of that lost value was in the form of purely paper losses by stock option insiders who had never paid full price, but large chunks of that lost value were in losses suffered by individual investors who had bought into the Ponzi scheme that became the Internet stock bubble. Financially interested analysts mindlessly promoted buy recommendations to their customers, even as insiders who had the best information sold off their shares. In fact, even where the law required insiders to hold onto stock, investment banks devised a new financial service through which they would promise to buy a venture capitalist's or tech executive's "locked" stock later at the stock's high early issue price while selling sophisticated computer-calculated financial instruments to "short" the stock as a way to insure against any loss for themselves. So insiders were guaranteed profits at the expense of the broad range of investors lured into the market by the seeming commitment of those insiders to their shares. Michael Perkins, a founding editor of the Internet business magazine *Red Herring,* flatly declared that the "high-tech financial bubble represented the greatest-ever legal transfer of wealth—from retail investors to insiders."[28] Small investors had traded in investments in their community for the ephemeral promise of this new global Internet El Dorado.

What is worth emphasizing in all these financial changes, especially in regard to the floating of bank interest rates and the breaking of regional mortgage markets, is that while new Silicon Valley technology made all these changes possible, it was political changes that legitimated them and even pushed them forward. The SRI study had shown that most people were conservative about and loyal to their bank accounts, so it was not overwhelming demand that pushed the new financial systems forward. But once money market funds slipped into the marketplace, the federal government left them unregulated, giving them a decisive advantage over traditional banking, which was restricted in what interest rates it could offer. Broker firms were encouraged to engage in new cutthroat competition through the breaking of agreements regulating brokerage fees.

State governments attempted to apply state regulations to restrict the new "casino economy" among small savers, but were met with a tough political response by the new investment firms, which could mobilize their elite base of customers to attack the Regulation Q regime of equal interest rates for all consumers. The state of Utah would become the test case when it sought to apply its banking regulations to money market funds in 1980; Merrill Lynch and its fellow offerers of money market funds would launch a $1-million lobbying campaign to defeat the attempt and scare off other state governments from attempting similar legislation. And when New York State refused to lift its usury laws, which limited the interest rates that could be charged to consumers, Citicorp moved its credit card operation to South Dakota after that state wrote tailor-made legislation for the company that abolished usury laws in the state. With Supreme Court decisions making the home state of a banking operation the relevant legal standard for customers from other states, this move by Citicorp effectively abolished usury laws across the country as Maryland and Delaware followed South Dakota's example in abolishing usury laws in their states. Federal legislation in late 1980, which deregulated bank interest rates and freed savings and loan associations from a range of regulations, would only push the ongoing financial revolution forward. With Merrill Lynch's CEO, Don Regan, assuming the position of Treasury secretary in the incoming Reagan administration, the elimination of the regional banking regime of the New Deal would be completed.

The conversion of local family mortgages into speculative bonds was not a natural event, but *was* based on legislative changes at the federal level. That first sale of Bank of America mortgage bonds would become a precedent for the wholesale conversion of the sleepy regional mortgage business—once the glue of "fixed capital" in the shared self-interest of regional growth coalitions—into a key part of global speculation. The sale of such bonds was legal in only three states before federal legislation allowed it everywhere. With interest rates uncapped by 1980 legislation and local banks hemorrhaging because of old loans, a new tax break in 1981 would pay savings and loan thrifts to sell their bonds on this new global market. As Michael Lewis, a Soloman trader in the 1980s, would write in his book *Liar's Poker:* "On September 30, 1981, Congress passed a nifty tax break for its beloved thrift industry. . . . Wall Street hadn't suggested the tax breaks, and indeed, [Soloman's] traders hadn't known about the legislation until after it happened. Still, it amounted to a massive subsidy to Wall Street from Congress. . . . The market [in mortgages] took off because of a simple tax break."[29]

The Collapse of the Regional Subsidy System for Poor and Working Families

While some transformation of the financial system was inevitable given the power of the new technology, the abandonment of regional banking systems was a product of political rather than technological sanction. The success of the early electronics-based money market funds were based on the high margins that were possible in dealing almost entirely with high-income depositors, while leaving the higher-cost low-income depositors to traditional banks. Such cream-skimming operations had specifically been blocked in the past by Regulation Q, which mandated equal interest rates for all depositors, thereby blocking competition on rates that would overwhelmingly benefit upper-income savers. It was the relaxation, then outright elimination, of those regulations that refocused banks not on investment of regional deposits—which had encouraged the continual expansion and deepening of those regional markets—but on a model centered on fee-based transactions in the global financial market, a model that emphasized upper-income individuals, for whom the margins on fees are the highest.

For individuals, the new technological model of banking has the potential for banks to tailor services to customers' needs (especially in the case of upper-income savers) as well as the more sinister potential for banks to tailor fees and loan offers so that the most money possible is extracted out of each transaction. By using the massive computer databases detailing customer characteristics and purchasing information about how the customer responds to other forms of telemarketing or electronic solicitation, banks can use repeated offers of different services to set a price for each individual for different kinds of transactions. Using this information, banks have a leg up in knowing when, in the words of one bank analyst, a bank's "prospect is likely to need a particular financial product and [when] to present an offer for that product at the right time."[30] For society as a whole, this presents the danger that bank personnel and others with whom people conduct day-to-day financial transactions can distort financial markets through privileged access to customer information. With direct knowledge of customer buying patterns, they will be able to undercut other competitors on the basis not necessarily of lower prices but of knowing when to deliver a bid to a customer and knowing when they are likely to accept a higher-priced offer. Communications professor Oscar Gandy Jr. observes that banks are approaching the position of "perfect discrimination where the organization captures not one penny less than what a customer might be willing to pay. The consequence is a market that becomes increasingly inefficient as monopolization expands."[31] The banks that, like Wells Fargo, are most committed to electronic commerce and smart cards (cards used in place of cash) see little threat in the loss of their revenues from traditional deposit fees, since they can make up the difference in increased monitoring of consumer buying habits as these consumers substitute the use of smart cards for the use of coins and bills. Wells Fargo's executives argued that banks now had opportunities either to sell this new massive amount of data on customers to other corporations or to use it for targeting their own in-house products to smart card users.[32]

The deeper fear from the new technology is that banks will have even more detailed information allowing them to choose which customers to avoid altogether. Banks have been closing physical branches at locations where they think customers are too expensive to service and using higher fees to push such customers to use ATMs and other alternative services. The next step has been to evaluate not just groups of customers but each individual customer to determine that person's overall profitability to the bank. Only with recent technology and the internal networking of different departments within a branch have banks been able to tally whether the net gain in fees, interest payments, and use of customer's deposits pays for the costs of servicing that particular customer. Most analysts for the banks estimate that only about 20 percent of customers actually make money for a bank, while three out of five customers cost more to service than they generate in revenue. Many banks are hiking fees across the board in anticipation that many customers, especially low-income and less profitable customers, will withdraw their accounts altogether. If a more desirable customer complains, however, that individual's fees are waived automatically by tellers who have been alerted to which customers to nurture.[33] The result is that the same services are priced differently for different customers, with higher fees being charged in many cases to the poorest customers in order to drive them out of the banking system altogether. The new Internet-only banks were able to offer higher interest rates partly because of their lower costs, but also partly, as they acknowledged, because the Internet itself limited their customer base to mostly well-off consumers. "We feel we are getting the cream of the crop from the Internet," stated one Internet bank executive in 1998.[34]

As banks market to new customers, especially in the areas of loans and credit, the implications of using information technology become even clearer. With the right databases, banks will make sure they do not solicit business from lower-income and less profitable customers. As an analyst wrote in *Bank Marketing*, "At this point, the concept that information is power becomes very clear. Competitors that gain access to valuable consumer information aren't going to go after your unprofitable customers—they have enough of their own."[35] This will expand redlining in minority and poor communities even more broadly, albeit with a bit more selectivity to avoid the most blatant racist patterns. But as Gandy argued, even if discrimination resulting from failing to offer services or from offering higher prices for the same service is focused on groups identified through "multivariate clustering techniques," it is discrimination nonetheless. In fact, Gandy sees this form of the discrimination as more invidious than others, since those being discriminated against lack the information to know how they are grouped or to easily gain the collective consciousness to act collectively to struggle against this "categorical vulnerability."

The irony of "deregulation" is that it has opened an era of constant tinkering with bank laws and the jurisdictions of various financial entities. The federal government has even occasionally made half-hearted attempts to pursue the same regional financial goals of equity and regional civic economic cooperation that "deregulation" itself has undermined. Even as the government changed the laws that had supported regionally focused banks, it enacted new ones to deal with the consequences of failed savings and loans, fiddled with the tax code to spur venture capital to replace lost local business loans, and struggled with capital flight from crumbling inner cities. The greatest irony was that just as federal government banking rules and technology were undermining regions as relevant lending units by banks, the Congress in 1977 passed the Community Reinvestment Act (CRA) to scrutinize banks even more closely on the fairness of their lending practices within regions. But if the general banking rules were not naturally channeling deposits back into the region where the deposits were made, the CRA legislation's attack on capital flight from the inner city became as natural a development of the new era as attempts to promote venture capital through capital gains tax cuts. Each was a jerry-rigged attempt to support regional capital accumulation in a new era of increasingly nonregional banking.

CRA regulations would never require that as much be invested in inner-city minority communities as was deposited by their residents. However, it would create an increasing paperwork surveillance of bank actions that would, if banks were found deficient in their attention to community needs, slow approval of bank expansions, leading some banks to consider dropping their bank charters purely to escape CRA regulations. No such expansions were ever actually blocked under the act, but at least two banks—Atlanta's Decatur Federal Savings and Loan Association in 1994 and Chevy Chase Federal Savings Bank, in Washington, D.C., in 1995—were forced to sign consent decrees committing funds to minority areas following evidence of discrimination. Those decrees targeted the banks because of historical traditions of where they made loans, making such a regional history a liability for banks.[36]

When Bank of America merged with Security Pacific Corporation in 1991, it was forced to agree to a goal of $12 billion in loans targeted to communities covered by the CRA, a total loan volume of 9 percent of expected new consumer loans (excluding significantly credit card loans). Similarly, when Wells Fargo merged with First Interstate, it pledged $45 billion in CRA loans over a ten-year period.[37] However, with much larger regions across which to spread those loans and with banks using the new technology to market to specific, multivariate categorical groups (to use Oscar Gandy's phrase), the traditional CRA tools used to analyze compliance using census tracts may fail in the face of the statistical power used to create targeted marketing and services. At the deepest level, though, the rise of the national marketing of banking outreach over the Internet is undermining the traditional basis for even holding any specific bank responsible for reinvestment in any specific community. However hard it is to evaluate discrimination in the limited regional reach of community banks, it becomes nearly impossible to analyze national marketing and loan approval processes given the sophistication of statistical grouping by banks. When a bank markets over the Internet, what region gets defined as "their" region? When bank executives proclaim the end of regions, they have a strong self-interest in doing so, since the disappearance of regions in essence erases most of the political regulations tied to regions, regulations that empower communities to demand resources from those banks. In the end, these billion-dollar CRA commitments by Bank of America and Wells Fargo ended up being paid with little regret by the banks as the price for once-and-for-all disentangling themselves from regional responsibilities as they merged their way into global markets.

When Bank of America was bought out by NationsBank in 1998, most of the bank executive functions would be transferred to NationsBank's headquarters in North Carolina, but the corporate lending offices of the combined banks would remain in San Francisco. The intimate connection to the California consumer that had been Giannini's creed was now disposable, but the connection of the bank to the region's financial and technological elites was seen as indispensable—a symbol of the collapse of cross-class regional collaboration in favor of the new community of elite business relationships in the region.

Electrifying the Internet: The Marketization of Electric Utility Networks

If bank capital had been the traditional circulatory system fueling regional economic growth, power utilities had supplied the energy at the base of industrial expansion just as that industrial expansion in turn had fueled growth and jobs in regional utilities. While there has been some erosion in the correlation between energy use and economic growth in recent decades as less energy intensive information-based industries have become more important, the critical role of energy supplies remains unabated in the growth of regions.[38] With $200 billion in retail sales of electricity alone to business and residential customers, energy still plays a massive role in the economy.[39]

However, what is changing, and at a rapid pace, is the way a market in energy transactions is replacing traditions of public-service utilities that were once the "growth statesmen," in the words of John Logan and Harvey Moltoch, of local regional growth and universal access.[40] Instead, the 1990s saw the accelerating trend in the "deregulation" of electric utilities—the quotes around *deregulation* reflecting the complicated increases in regulation needed to replace the broad functions of the regionally based utility system with a new, much more complex market system of national energy exchange. In this process, dynamics paralleling the change to a national banking system are emerging: a dependence on new technology to facilitate the change, the replacement of local regulation by national regulation, particular regional dynamics of innovation fueling new national initiatives, and the rise in inequality in a new national market. Universal energy rates are increasingly abandoned in favor of deals that benefit larger, more profitable customers who are cheaper to service in a period when marginal profit rates replace the fixed rate returns of an earlier era.

A key part of new national regulation that facilitated these changes was tied to the new information technology, particularly the Internet itself. In fact, even as the media was hyping the initial hundreds of millions of dollars of commerce being conducted over the Internet by the end of 1996, the federal government and the utilities were quietly launching $50 billion of wholesale electricity sales onto the World Wide Web. Mandated by the Federal Energy Regulatory Commission (FERC) as a component of the industry's deregulation, the new Open Access Same-Time Information System (OASIS) would become the marketplace for utilities to reserve energy from producers across the country. The OASIS Internet system would guarantee open access to information about transmission capacity, prices, and any other data critical to making energy purchases or sales.[41]

Through a structure compared to an airline reservation system, each utility (or in the new welter of jargon, transmission provider) is required to continually update the total transmission capacity of its individual area while listing the available transmission capacity at any moment. Any producer of energy may request a "seat" on a utility's electricity grid from one point to the next, possibly across as many as a dozen grids to a final destination. This request is affected by thousands of other energy producers attempting to place similar transmission reservations, all asking for a similar "seat" to get the best price at the right "departure" time. The continuous nature of power distribution makes the real-time aspects of Internet information exchange critical to the whole system as power producers seek to sell their energy to utilities, which in turn will retail it to their consumers (or will have multiple "power marketers" competing to offer it to consumers). In order to enforce fair competition, utilities are being forced to separate their functions into three kinds of divisions: retail marketing, power production, and transmission provider functions.[42]

The motivation for using the Internet, embodied in what is known as FERC Order 889, was to create standardized access to information with no time-based advantages for any competitor. All utilities with transmission capacity would now be required under the rule to post a common set of data about that capacity on the Net in consistent data formats with common transmission protocols.[43] To accomplish these goals, national regulation of electricity has increased as utilities have been forced for the first time to make firm calculations on a continuous basis of what their transmission capacity is at any time, calculations that were unnecessary in the past when most of that capacity was tied to long-term internal power production. The "unbundling" of different utility divisions has required greater and greater scrutiny by the FERC to determine exactly what information needs to be available to competitors to ensure equal access. In a sense, each decision by the FERC is a decision on what an ideal marketplace for energy should look like. Each regulation is a choice from a range of possibilities to create the information standards that will yield the "ideal" market result, an ideal that has been and continues to be sharply debated as lobbyists from all parts of the industry line up over each phase of the FERC's mandates that relate to creating an energy marketplace.

The creation of a national market regulation involving energy utilities dates from 1978, when Congress enacted the Public Utility Regulatory Policies Act (PURPA). PURPA slowly expanded wholesale competition in the sale of electricity by requiring the interconnection into the utility grid of small energy producers, especially "cogeneration" plants (producing energy in the course of other industrial activity). More recently, Congress enacted the Energy Policy Act of 1992, which enhanced FERC's authority to order wholesale "wheeling," the industry term for equal access by all comers to each utility.[44]

Also in 1978, natural gas prices were deregulated and a gradual marketization of the gas pipeline system was instituted. It was that earlier experience with natural gas contracts that deeply influenced the FERC's movement to the Internet. After years of customers simply purchasing natural gas from the nearest pipeline, the FERC had in a 1993 order mandated the unbundling of the marketing and the distribution of natural gas, thereby encouraging the explosion of new agreement systems, including electronic bulletin boards that allowed online contracting, making purchases of natural gas from distant sources much easier.[45] Those with excess capacity from long-term contracts were required to post that capacity on the electronic bulletin boards for it to be resold to the highest bidder. However, while there was some standardization mandated by FERC related to information exchange standards, proprietary systems created by different companies had balkanized the industry into conflicting approaches that made information exchange nearly impossible (and extremely expensive) for numerous different participants. It was estimated that the industry spent more than 750,000 hours of labor each month on information transaction bids because of the repetition of data entry needed to deal with different systems. By 1996, the FERC began detailing proposals on how gas companies would have to standardize business transaction information even as many industry analysts began eyeing the Internet as a better alternative.[46]

These later proposals explicitly cited the electricity industry—more recently moving toward wholesale market competition—which after a December 1995 FERC decision was mandated to move information exchange to the Internet. FERC had been urged to move in this direction by the Western States Power Pool (WSPP), a group of utilities, federal power marketing agencies, and rural electric cooperatives in twenty-two western states and British Columbia.

WSPP had begun using a computer bulletin board, employing standardized software, for selling power between the utilities and saw the Internet as a good next step. The WSPP saw mandated use of the Internet as the best way to continue successful energy pooling with an expanded set of participants. FERC had established technical task forces on how to implement information exchange, and the Palo Alto–based Electric Power Research Institute (EPRI), the electric utilities' research consortium, proposed using the Internet with communication standards that EPRI had developed and that were tied to the Internet's TCP/IP standard.[47] By mandating Internet information standards for electricity before competition entered regional electricity markets, the FERC would avoid the problems of the natural gas industry, where already existing competition had undermined free information exchange.

While the technology centers of the West Coast played a strong role in contributing both support and expertise to the movement of the electricity market onto the Internet, it was the traditional energy centers in Texas that would leverage that tradition into a central role in the new electricity market. Companies including Enron, Reliant, El Paso Gas, and Dynegy would become the new names of electricity production in many areas. Houston-based BSG, a computer system integrator for the energy industry, was able to convince utilities representing half the transmission capacity of the country to hire the company to establish a Web site that could integrate their Internet energy transactions. The collection of mostly East Coast and Midwest utilities, called the Joint Transmission Services Information Network (JTSIN), specifically hired BSG because of its experience with information systems involving the natural gas industry and knowing what to avoid on proprietary systems. "The industry in aggregate," argued JTSIN co-chair Jeff Geltz, "would have spent a ton of money if they had gone off in their own directions and tried to develop their own OASIS programs. Not only that, they would have ended up with the Tower of Babel syndrome they have in the natural gas industry." Using security software from Tradewave Incorporated, an Austin-based security software company, the JTSIN system reflected how high-tech investments in Austin helped to engineer an regional concentration of energy-related technology companies in Texas as energy sales have gone high-tech and an estimated $25 billion a year of energy commerce flows through BSG's JTSIN Web nodes.[48] Similarly, Enron's fifty-story headquarters in Houston, employing one thousand traders before its economic meltdown, became a new center for buying and selling electricity as a commodity market, especially as retail electricity markets exploded with deregulation in states such as California.[49]

Establishing clear Internet information standards was seen as even more crucial as many states moved beyond the federal creation of competition in the sale of energy between utilities themselves (a $50-billion-a-year enterprise) toward allowing competition in the direct sale of power to retail customers (a $200-billion potential marketplace). The unanimous passage in August 1996 of California's AB 1890, a law that opened up competition in the state's $23-billion retail electric utility industry, would largely follow prescriptions laid out by the California Public Utilities Commission (CPUC). The law created a nonprofit Independent System Operator (ISO) to manage the physical trans-mission grid and a separate Power Exchange (PX) to manage the state's inter-nal market for buying and selling power on the grid. The PX and ISO themselves became "nodes" in the national OASIS system of power purchas-ing and transmission as they became operational in January 1998.[50] Based on the idea of assuring competition in the production of electricity in the state, PG&E was forced by the CPUC to put four generating plants, with a total capacity of 3,059 megawatts, on the auction block. In the same period, PG&E paid $1.59 billion to purchase eighteen power plants in New England in order to position itself as a player in supplying energy to consumers across the coun-try. Symbolically, it separated itself into two parts: a subdivision dealing only with retail California sales and a much larger (and more profitable) national corporation buying and trading in the new national electricity market. Similar to previously regional banks, PG&E decided that its self-interest was no longer going to be tied to a specific region but rather to marketing to specific slices of customers in a national market.

The Politics of "Deregulation": Blackouts and the Collapse of Energy Price Equality for Working Families

Changes in technology have accommodated the changes to a market-driven electricity system, but it is worth emphasizing that technology is not destiny. Although there was a logic to centralized integrated utilities in the earlier part of the century, the existence and success of federally owned power systems, rural cooperatives, and municipally owned utilities show that any technology can inspire multiple economic formations. The balance is determined more by politics than by engineering. And while the logic of the new Internet-related technology opened up new possibilities for the federal government to encour-age the wholesaling of power between various generators and power agencies, the rapid movement to retail markets in electricity followed little logic other than that of the political demands of big business at the economic expense of residential customers. The most striking result was the dissolution of regional energy planning and shared energy rates between large industry and working families, with the blackouts and surging prices that began to sweep California in 2000 being the most visible sign of the collapse of rationality in the system.

After California passed its early legislation opening up retail sales of electricity to competition, one state utilities commissioner reported to the utility industry's publication *Public Utilities Fortnightly*: "What defines our whole exercise, our whole restructuring, is our high rates and the strong cry by the large users for customer choice. Those are the two driving forces." In the same article, another state regulator noted that industrial pressure was driving retail competition across the country: "If the 100 largest industrial customers in every state want to have some kind of restructuring, there's going to be tremendous pressure put on the commission. And I think if the commission doesn't respond in some form, there will be pressure put on the legislature and I'm sure the governor's office in every state. . . . Those are the realities. I don't think it's the utilities; it's the large industrial customers driving this."[51]

Industry Week, a manufacturing trade magazine, was explicit in seeing a payoff for big industries of $80 billion to $100 billion a year in lower electricity costs throughout industry with 20 to 50 percent savings at individual industrial plants. The threats driving states to open up retail competition was clear and written about approvingly by the magazine: "Some large companies are threatening to close or move facilities unless they get rate reductions. Such threats are drawing responses from states. Michigan, for example, is so fearful of losing industry that it now allows new industrial customers to shop for power."[52]

If big business was to save 20 to 50 percent on its energy bills—far more than any projected general technological savings—where were their savings going to come from? Part of the answer was that the large companies would likely lock up long-term contracts before competition was fully opened to small business and residential customers. "The more jaded observer will note," wrote Pennsylvania State University professor Frank Clemente early in 1996, responding to the market competition hype in *Public Utilities Fortnightly*, "that this 'women and children' last approach ensures that large users get a big piece of a pie that may not be further divided."[53] *Inc.* magazine noted that small business organizations were attacking the new changes. "What you see is small business getting screwed," said Julie Scofield, executive director of the Smaller Business Association of New England. *Inc.* condemned the "fine print" that gave large customers freedom to negotiate rates while leaving smaller customers still tied to electric utilities with fixed rates.[54] As for residential customers, it was even less clear how they would fit into the new market system.[55]

Another factor was the way in which market competition dealt with what was euphemistically referred to as "stranded assets"—the less profitable power plants owned by utilities that would become nearly worthless once new, lower-cost energy sources were allowed to cannibalize the electricity markets that the older plants were built to serve. Built in the 1970s when energy demand was expected to increase more rapidly than it actually did, these stranded assets included many nuclear power plants costing $90 billion to $150 billion. Early on *Fortune* magazine stressed that it would be only after "the costs [of the stranded assets] are paid down *over the next decade or so* [that] households should see prices fall" (italics added).[56]

The deadly part of this was that market competition created the opportunity for large industry to find its own deals, releasing it from the political imperative of the past, which had forced it to work with the broader community in energy planning. California sold deregulation to the public with a rather bizarre jerry-rigged system that awarded public subsidies to the utilities to help them pay for their $20 billion to $30 billion in stranded assets. While the state legislature decided that there would have to be an official rate cut of 10 percent for consumers in order for the package to be sold politically, the state authorized the utilities to issue up to $10 billion in revenue bonds, paid for with a "competition transition charge" of 2 to 4 cents per kilowatt-hour on electricity with taxpayer-funded interest subsidies.[57]

Early on, a number of regulatory commissioners across the country raised the issue of "information inequality" in an arena where complex issues may leave the elderly and the less educated vulnerable to deals that look good but deliver less than promised. (In a sense, the description applied to the whole California retail electricity competition law, which cynically sold a rate hike to voters that the voters themselves would have to pay for over the following decade through the subsidized bonds sold by the utilities.)

But as the electricity crisis erupted, the issue of information inequality became the driving force of the blackouts and price gouging in 2000 and 2001 that threatened power supplies not only in California but across the western states. The state set up its own retail electricity Internet auction sites, which gave producers a real-time view of electricity demand and the "choke points" for distribution, creating the possibility of opportunistic manipulation of the system by the new national power producers. With emergency power supplies subject to last-minute auctions, strategic withholding of power could drive prices into the stratosphere. State power grid officials began to observe that power producers would triple their prices within minutes of a system bottleneck. As attorney Mike Aguirre argued, representing the city of San Francisco in the suit against the power producers, the Internet had the power producers "marching lockstep and communicating in the way a cartel communicates, only they are actually much more sophisticated—you aren't going to catch them together in some smoky room."[58] One of the first solutions implemented by state officials in the wake of the blackouts and power surges was to shut down a number of the auction sites with the view that less information for the producers would curb their ability to manipulate prices. Stanford University economist Frank Wolak, heading a team of economists surveilling the California electricity market, estimated that in 2000 alone, $8 billion out of the $27 billion that Californians paid for electricity resulted from such manipulated overcharges.[59]

Instead of the information revolution delivering cut-rate price competition as originally promised, it had delivered a new tool for corporate collusion on price manipulation. Estimates of the costs of the crisis, from overcharges to losses to the economy from blackouts, have been estimated as high as $100 billion, a figure that places it in the same league as the savings and loan bailout and the NASDAQ collapse, both of which left average voters and consumers holding the bag for the failures of market "deregulation."[60]

The Loss of Regional Power Planning and Research

The headlines about price gouging merely mask deeper losses arising from the end of regional power planning. In many ways, the high costs of the stranded assets held by utilities resulted from the very successes of regional regulators in encouraging conservation and cutting energy demand far beyond expectations. If energy demand had risen at the rates projected in the 1970s when many of those plants were built, those white elephants would instead have been seen as profitable smart investments.

And the fact that energy demand did not rise at the rate expected can largely be attributed to regional power planning that encouraged energy conservation across the country. But instead of the fruits of that energy conservation and lower prices being shared by everyone, the breaking of universal rates in regions was a political method through which the rewards of those years of regional planning accrued to the narrower benefit of big industry.

Analysts have noted with irony that since the early 1970s, courts have required thorough socioeconomic-impact statements on all power plants before they are built, to assess their effects on jobs, on energy rates, and on the environment.[61] Yet the entire system of power management was thrown out the window with virtually no hard numbers produced on how the changes would effect the regional economies of communities across the nation.

In California, as one example, it was the CPUC that inaugurated many of the most innovative energy conservation programs in the country. Jerry Brown had taken office as governor in 1974, and his appointments would promote a new focus on energy conservation. The CPUC eliminated the volume discounts that encouraged high energy consumption by industry and structured energy prices for residential customers in "block rate" systems that discouraged consumption. In addition, penalties were exacted for failure to pursue conservation. The result, in combination with federal laws and subsidies for conservation such as housing weatherization, were programs supported by utility advice and low-interest loans to homeowners that led to a broad decline in energy demand compared to the projections of the mid-1970s.

At the same time, the CPUC promoted unregulated "cogeneration" at industrial plants to expand the domestic energy supply. Along with government legislation requiring utilities to buy power from nonutility energy producers and encouraging the R&D into new technologies, the regulators created the expanded energy sources that have now lowered the prices of electricity production. Paradoxically, that very success of the regulators undermined the system of universal rates and fixed cost utility returns that allowed such regional planning. Barbara Barkovich, the policy director of the CPUC for the last five years of the Brown administration, argues that a focus on only the lowest marginal rates on energy leaves regions without the ability to create incentives and rate structures to deliver conservation and other long-term goals. "Regulators no longer have the flexibility," argues Barkovich, "to make tradeoffs between the present and the future or among customer groups. . . . Rather than having a fixed number of customers and amount of sales over which to spread the utility's revenue requirement, they must face the possibility that sales will disappear as customers leave the utility system."[62]

It then becomes impossible for utilities to expend resources for non-revenue-producing goals such as conservation or new technology and expect to make it up in general rates. Conservation programs becomes nearly impossible where consumers, especially large industrial users, can abandon the utility system for low-cost producers who are concentrating only on day-to-day production and are unburdened by other costs associated with longer-term management of energy consumption. In fact, such new producers often undermine that longer-term management by bringing back the volume discounts and pro-consumption incentives that the regional utility system worked to abolish. Similarly, many of the incentives that were created regionally to discourage use of environmentally destructive production processes increasingly fall victim to national competition that imports high pollution energy to regions that once would have prohibited its production.

The irony is that market competition in electricity was only possible because of technology created because of research investments made possible by the regional power system, investments that are being slashed in the new era of competition. The traditional consolidated regional utility structure had a technological logic that had shaped the geography of electric power in the United States. Supplemented by the long lines tied to the alternating-current technology (the technology that had defeated the smaller-scale generation designs pioneered by Thomas Edison), consolidated regional utilities became the norm across the country. Each was built around large generators that required heavy financial investment backed by the guaranteed rates of a fixed market. These integrated utilities delivered subsidized rates to rural agricultural areas, thereby expanding economic development and access to the electricity grid.[63]

What has changed in recent decades is that new technology has made possible power generators that can be scaled in size to meet marginal demand in a more flexible way than what was possible in the past, even as new information-based technology has dramatically increased the efficiency of energy generation and transmission. Large-scale coal-fired and nuclear plants are increasingly giving way to a new generation of natural gas–fired plants with much higher efficiencies than in the past. Using turbine technology partly borrowed from the aircraft industry, these new gas turbine plants combine fast construction, lower capital costs, and the ability to be scaled in size from relatively small to quite large to meet energy demand. The new plants allow much more flexibility than did the older economies of scale, which almost automatically demanded mammoth plants with guaranteed markets. Especially with the low cost and plentiful supply of natural gas, the new technology has helped to fundamentally realign the comparative costs of traditional large-scale plants versus newer "modular" gas turbine designs.

With the addition of advances in "renewables" such as wind power, biomass, and solar technologies, a new range of small-scale power has been added to the generator mix. Fundamentally small-scale and, given their intermittent production of electricity, needing to be supplemented with additional power generation, they have helped to challenge the paradigm of large-scale, centralized, utility-based energy production.[64]

Many of these innovations along with critical advances in the use of information technology would be directed from the Bay Area through the offices of the Electric Power Research Institute (EPRI). EPRI had been founded in the early 1970s, when the real cost of electricity was rising for the first time in a century and the traditional producers of power research, large contractors such as General Electric and Westinghouse, no longer seemed able to capitalize the research needed to deal with new energy challenges and rising environmental mandates. Faced with Congressional bills proposing the creation of a government-run power research agency funded by a tax on electricity, the industry reluctantly agreed to create a private consortium that could share the costs of research to supplement the disparate energy programs of the federal government. Jesse Hobson, a successor to Fred Terman as head of SRI, had pushed the idea of such a consortium in the 1950s. Coincidentally, Palo Alto was picked as the site for EPRI headquarters, in 1973, based on many of the same criterion that led Xerox and so many other engineering-based companies to site research centers in the area. The idea also was that being far from Washington, D.C., would help keep its research agenda out of the political fray. The oil shocks of the 1970s would only sharpen the focus on alternatives to traditional fossil fuel plants and on increasing efficiency in all aspects of the industry.[65]

The results, after twenty years of research at EPRI, federal research facilities, and technology being borrowed from aerospace, would lead to dramatic changes in the cost-effectiveness of energy technologies and help to drive down the costs of electricity once again. Along with assisting new technologies, the infusion of microelectronics and "smart software" into older plants increased their efficiency dramatically, and planning software had made efficient management of a mix of technologies much easier. But some of the most dramatic changes in technology came not in the production of electricity but in its transmission—a key innovation as bulk power transfers between utilities increased fourfold between 1986 and 1996. With 40 percent of the electricity that was generated in the United States being sold on the wholesale market by 1996, the need for efficient transmission management of those power transfers became critical. The answer would be a new class of solid-state controllers called thyristors, a development sometimes labeled the "second silicon revolution." Able to control electricity in the grid with the same speed and precision as a microprocessor but with power levels a billion times higher, thyristors allow utilities to safely operate much closer to a system's thermal limits and expand the transmission capacity by 20–25 percent. By making transmission more economical, this new system, called the Flexible AC Transmission System (or FACTS) would open transmission capacity to new producers and make it more economical for them to extend the geographic reach of their marketing. For those seeking to marketize electricity transfers, EPRI's director,

Richard Balzhiser, has argued that with these new controllers, "the electronic controllers of the new FACTS system are indeed setting the stage for widespread competition."[66]

However, utilities are already responding to the collapse of the fixed rate system by slashing their funding of R&D. Under the fixed-rate system, the utilities could collaboratively invest in basic research with the knowledge that the results would benefit each utility in their regional area. As collaboration between utilities becomes competition in the new system and low-cost producers enter the scene using the fruits of research without having had to invest in it themselves, the utilities can no longer expect to fully capture the value of their investments in research by passing it on through their rate base. The result has been that the EPRI has been reorganized to emphasize specific proprietary projects rather than basic research. Utilities now fund only the products from which they expect immediate returns, undermining the long-term research that delivered the technological gains fueling the low-cost power that inspired competition in the first place.[67] Government spending on energy research has been slashed even more dramatically than that by utilities, falling 80 percent between the 1970s and the late 1990s; research dollars dropped by half between 1992 and 1997 alone.

In practice, large industry is enjoying the fruits of that research and the research gains derived from the regional system of rates of the past while residential customers seem likely to bear the disproportionate costs of competition in paying for stranded assets. And if the cutback in R&D leads to little new technological improvement over the coming few decades, then the immediate cost savings that large industry is enjoying by breaking off from regional universal rates may be the only savings to be had as energy demand expands.

Rethinking Regulation in the Era of National Electricity Competition

Even if the initial gains from market competition go to big business, proponents have argued, the society as a whole will gain as supply and demand replaces the regulatory intervention of bureaucrats and as business savvy replaces the dead hand of political interests. The problem with that argument is that as competition has increased, so has regulation and political lobbying by all energy interests involved. The real difference is that the ability to regulate has been stripped from regional authorities, such as state utility commissions, and has been transferred to national authorities, such as the FERC, and directly to the U.S. Congress as it writes laws to design the new system. The early part of this chapter detailed the extensive regulatory interventions of the

FERC to shape and encourage marketplaces for energy on the OASIS Internet system, but those are just the beginning.

As long as all retail energy sales were carried out by local utilities, traditional federalism had left regulation to the states. However, as energy was increasingly sold across state lines directly to customers, the Interstate Commerce Clause of the Constitution increasingly paralyzed the power of local regulators in favor of the federal government. FERC officially stated in 1996 that "to the extent that retail [competition] involves transmission in interstate commerce by public utilities, the rates, terms and conditions of such service of subject to the exclusive jurisdiction of the Commission [FERC]."[68] So in an age of rhetorical decentralization of power to the states, "deregulation" is, ironically, largely a nationalization of power over electricity whereby national politics, rather than regional imperatives, will shape who wins and loses in the fight for the $200-billion electricity market. As the California crisis developed, it became clear how powerless local regulators were in the face of national price competition, even as the FERC, in its belated move to impose price caps in the late spring of 2001, revealed its clout as it helped to end the crisis that had been fueled by its initial hands-off approach.

With so-called deregulation in place, lobbyists swarmed the Capitol and hired lobbyists by the bushel. The top two dozen electricity combatants spent more than $32 million in 1996 on lobbying, along with millions more on political research, polling, television advertisements, and organizing grassroots support for their interests. Enron, one of the largest power companies in the world, launched on its own a political advertising campaign in 1996, costing upward of $23 million, to push for opening up more markets for direct retail sales by the company. The Edison Electric Institute, the main lobbying arm of the utilities, pulled in key Democrats and Republicans onto its payroll, including former representative Vin Weber, a close ally of Newt Gingrich, and Haley Barbour, the former chair of the Republican Party. This spending was quickly matched by that of a coalition of independent power producers called Americans for Affordable Electricity.[69]

The reason for the lobbying is that God, or rather profits, is in the details of how retail competition is structured. At issue is exactly how independent power companies get access to the electricity grid and to the customers at the retail level. As systems develop, each utility's transmission grid will have to calculate, according to FERC's directives, its available transmission capacity, a figure that it had never measured before; moreover, there is no unified approach on how to calculate this figure. How it is defined by FERC has a major impact on how much is charged power providers and how much the utilities make in transmission. These new requirements that power companies provide precise information on a wide range of data—in contrast with the older fixed-rate system in which unified management made precise measurement unnecessary—is just one example of the perversity that "deregulation" is increasing bookkeeping for a range of business and increasing the complexity of business transactions to the point of near chaos. Market "deregulation" has created a blizzard of rules by the FERC on exactly what functions and data sets will have to be specified in each bid in each round of transactions from production to transmission to retail sale.

These details and regulations, multiplied tenfold on the national level, structure winners and losers in each stage of transactions from generation to distribution to retail sale. In light of how often competition advocates proclaim the end of the "natural" monopoly for utilities that integrated generation, distribution, and the retail sale of energy, it is worth considering the proliferation and expansion of regulation needed to separate those functions to create the "natural" market system. What is clear is that, in a complicated system such as energy delivery, markets are as much a construct of government as monopoly utilities ever were. Barbara Barkovich, the former policy head of the CPUC, sees exactly this continuity from her own proconservation interventionism to the market-oriented approach of her successors at the CPUC: "The roots of this new competition-oriented ideology are not inconsistent with those of interventionist regulation. Both market-oriented regulators and interventionist regulators are motivated by doubts about the ability of utilities to manage their businesses to provide the greatest benefit to their customers."[70] In both cases, whereas decision making on the details of how to structure energy delivery had once been left to designated regional utilities, the new interventionist regulations require constant fine-tuning of how to shape that delivery from production to transmission to retail delivery.

Where the old and new regulators differ is in the control mechanisms and technology that they encourage to circumscribe the behavior of the energy companies involved. With the traditional regional system of monopoly utilities, a relatively fixed demand for energy (that could be modified over time by interventionist policy) at a mandated price helped determine which technologies were deployed and how much capacity to build. Once regulations were made to determine what rates to charge customers and what the mandated rates of return for utility stockholders would be, regulators had little to fiddle over in regard to the actual day-to-day details of maintaining network reliability and allocating costs between the internal functions of the utility. Those could largely be left to the companies, since their own long-term financial self-interest required regular maintenance of the transmission grid and long-term investments in upgrading the system as a whole.

Under market competition, power producers treat their own plant capacity as fixed moment to moment (since all prices will be based on marginal costs, not on longer-term rates of return, as with the utilities), while prices will fluctuate across the country as demand adjusts to prices changes. Regulation is required because marginal cost decisions will not include calculations relating to maintaining the system as a whole, forcing the issuing of new regulations at each point in the distribution system in order to bring those market transactions in line with the need for stable service, reliability, access, and long-term investments in the transmission grid. It is the shift from a few key macroregulations to a proliferation of microregulations.

As the California crisis engulfed the country, large numbers of political leaders and prominent analysts began to rethink the empty rhetoric of "deregulation" in favor of a more nuanced understanding that regulation is being transformed, not eliminated, and not for the better. Washington State governor Gary Locke wrote an editorial proudly noting his state's rejection of retail markets for electricity, commenting that he did not want "the legal obligation of regulated utilities to serve" consumers to be replaced by independent producers in a "market structure [providing] inadequate incentives to construct new plants—or, worse, perverse incentives, not to generate electricity."[71] Even the *Wall Street Journal,* in a column provocatively titled "Maybe the Utilities Weren't So Dumb After All," citing conservative economics icon Ronald Coase, noted that the regulatory morass of separating electricity production from distribution had missed the original economics of vertical integration that had made the utility monopolies successful.[72] In solving its crisis, California itself rapidly retreated from full-blown market competition to a system in which the state government will, in some form, play a much more centralized coordinating role. Even large producers have been unwillingly hauled back into a shared power-buying pool, a change they are fighting to reverse but one that in the meantime promises a slightly fairer sharing of the burden of the costs of the market fiasco.[73]

Still, what worries many people is that political pressure in the new national regulatory environment easily creates microregulations that favor short-term profit for producers over the reliability of the system, a dangerous proposition for networks transmitting the lifeblood of commerce across the country. Even before the California crisis in 2000, many critics of the move to competition pointed to the West Coast blackout of four million homes that occurred in 1996 just weeks before final passage of the California market competition legislation. A July 1996 report by the North American Electric Reliability Council, the umbrella for the nine regional utility councils that manage the national electricity grid, warned that with greater and greater national transmission of power, the thermal limits of power lines will be pushed farther on a day-to-day basis than ever before. All this will happen in an environment where short-term profits will encourage stretching the system to the limit, even as utilities that formerly cooperated in management of the grid increasingly become direct competitors.[74]

The hope is that through the careful information mandates involved in the OASIS Internet system, the FERC will maintain real-time management of demand in a way that overcomes those dangers. But if a loose wire can shut down the West Coast in thirty-five seconds at a time of peak demand, it was predictable that swings of market demand would easily cause blackouts in a system in which knowledge of national capacity was no longer coordinated in any central way.[75]

It is ironic that the Internet played such a central role in the debacle of the California crisis, since the Internet is itself a product of supposed decentralization based on centralized federal regulation and has had similar problems of interconnection and service integration. Utilities are beginning to form alliances to allow reciprocal transmission rights for electricity over their lines, much as Internet backbone services agree to exchange information freely, but it is precisely those kinds of decentralized agreements that lead to erratic flows on the Internet, which cause congestion and delays in that network system.

As will be elaborated in the following section, the Internet evolved in exactly the same environment that is pushing forward market competition for energy—new competitors breaking from regional telecommunications systems to find profit through servicing wealthy customers at the expense of working families. Historically, this created a telecom system based more on a politics that serves those elite interests than any abstract idea of "efficiency" and, as part of that policy, promoted a privatization of management of the Internet grid that served those profit-making concerns more than ensuring reliability for customers.

Cream-Skimming the Old Bell System, or How Subsidies and Regulation Made Phone "Deregulation" and the Internet Possible

With the privatization of management of the Internet from 1992 to 1995, the Internet industry quickly began trumpeting its success as proof of how unregulated market competition had led to the explosion in Internet participation. No players were more likely to trumpet the success of this new free market than the independent Internet Service Providers, or ISPs, in the incessant lingo of the industry. From veteran Whole Earth Networks to upstart Netcom to giant America Online, these ISPs not only beat back proprietary networks such as Microsoft, they also delivered to their customers (local businesses and upper-income individuals) an unlimited "all you can eat" flat-rate price for service that made the high prices for long-distance phone service seem laughable in the face of the new technology. The use of Internet phone calls, now made nearly for free over the Net, began to bypass traditional long-distance phone services, and the Net seemed to promise limitless connections at a price that the mastodons of the old regulated phone system could only dream of.

Two events in April and May 1997 would undermine the "free market" bravado of the ISPs as these Internet free-marketers made loud, extremely public appeals for the FCC to protect them from market prices in order to "save the Internet" (and their own profit margins.) The first event was the announcement by UUNET (described in Chapter 2), the national backbone service that was given control by NSFNet of most of the Internet access points used by ISPs, that it was no longer providing free peering to the midsize Internet providers, companies carrying roughly 20 percent of Internet traffic. Peering is the practice, crucial to the formation of the Internet, of automatically transmitting information from point A to point B whether it originated on a particular machine or not. However, the unregulated expansion of traffic on the Net rapidly outstripped the infrastructure for carrying the traffic, leading to congestion, lost messages, and the "World Wide Wait" of connections.

UUNET announced that it would continue the practice of peering with other backbone providers, including MCI and others that invested heavily in infrastructure, but that the midsize ISPs, a large proportion of them located in the Silicon Valley area, would have to start paying for the right to exchange messages over UUNET's backbone and connections. Smaller ISPs had been paying UUNET for the right to directly connect their customers to the Internet, but the charging for peering has emphasized that the traditional cooperative model of infrastructure building on the Net has broken down.

However, the midsize ISPs immediately cried foul, and a number of them tried to turn to regulators to settle the score. David Holub, the president of San Francisco–based Whole Earth Networks, which has about twenty thousand customers in the Bay Area, immediately scheduled a meeting with the state Public Utilities Commission to protest UUNET's decision and indicated that he would turn to the FCC to regulate Internet carriers like phone companies, invoking rules that bar those that route other phone companies' calls over their networks from imposing discriminatory fees on different competitors. That proposed regulation would have prohibited UUNET from peering with MCI while charging a fee to ISPs such as Whole Earth—a change that had thrown the system of Internet pricing into chaos and accentuated how unstable the whole marginal pricing system of Internet competition had become.[76]

At the same time (and more successfully for the Internet providers), the ISPs, along with AT&T, Netscape, and Microsoft as well as Compaq, IBM, and a host of other computer companies were demanding FCC intervention to prevent market pricing on the other end of the information pipeline, namely the local telephone companies used by ISP customers to reach them in the first place. Since the initial breakup of AT&T back in 1983, the FCC had exempted Internet providers from paying the same kind of per-minute access charges to local phone companies that long-distance customers had to pay to connect their customers. This allowed Internet providers to pay the flat business rate to local phone companies that ordinary local business customers paid—which in turn allowed them to offer flat-rate service for the Internet to their customers.[77] What this means is that for a customer being connected to an Internet provider, payments for Pacific Bell were one-twentieth of that for connections to normal long-distance carriers. This was all despite the fact that the costs for handling each kind of call are exactly the same for the local phone company.

Even worse for the local phone companies is the fact that Internet calls are on average much longer than either local or long-distance phone calls. According to a study by Pacific Bell in 1997:

- 30 percent of the total time of customers' use of the phone system generated by dial-up Internet traffic came from calls lasting three hours or more and 7.5 percent came from calls lasting twenty-four hours or more, while the average voice call lasted only four to five minutes.
- With the concentration of ISPs in its Northern California region, Pacific Bell has been affected far more than any other local phone system; of Pacific Bell's 772 switches, approximately one-third in 1996 serviced ISP hubs concentrating Internet traffic. By January 1997, 62 of those switches had already exhibited congestion so that voice calls were being degraded or even blocked by the level of Internet traffic.

Pacific Bell cited one Silicon Valley ISP hub at which traffic levels in late 1996, driven by a single ISP, undermined service in the whole area. The ISP represented only 3.6 percent of total office lines, but accounted for about 30 percent of use during the busiest hours of the day. The result was that one out of six phone calls was being blocked due to the congestion. Pacific Bell spent $3.1 million cost to fix that one hub alone, and it estimated that it would spend $100 million on Internet traffic upgrades in 1997 and would spend $300 million on upgrades by the year 2001. Given the ISPs' exemption from paying long-distance access changes, Pacific Bell maintained that it would earn only $150 million from additional revenues in that period because of ISP traffic.[78]

Worse than the actual costs of the upgrades was the fact that those investments were being made in traditional analog voice phone lines and switches, instead of the ISP phone traffic being moved onto high-speed digital switching systems that would be more efficient in upgrading all data traffic. Most of the Baby Bells began offering such high-speed digital services for ISPs in 1997, but the Internet providers had little incentive to pay for such services as long as they could convince the FCC to allow them to use the local phone lines in the same way as ordinary business users.[79]

And in May 1997, the FCC, under intense lobbying from both computer companies and Internet users, agreed to continue the ISP exemption from access charges, although it ordered essentially minor concessions to the local phone companies by raising all charges on second phone lines, the logic being that these would likely be used for Internet connections. Some of the smaller Internet providers complained that the additional charges on all their incoming phone lines would hurt them, but larger ISPs such as America Online declared victory: "We will see an increase in our charges, but we do see that on balance we need to accept the additional charges because they are flat and they are nominal," said Jill Lesser, America Online's deputy director of law and public policy. "A permanent access charge would have been orders of magnitude worse for AOL. Even at one cent per minute, we would have incurred a charge that would have been in the neighborhood of $100 million and which we would have had to pass on to the customer. So when you look at an increase that is 1/10 of that, that's a fairly modest increase."[80] The broad coalition of computer companies had successfully protected the subsidized status of Internet providers.

By spring 1998, the FCC was in the even more farcical position of refusing to apply access charges to companies using Internet systems for actual telephone calls. A new class of companies, including Qwest Communications and

IDT Corporation, had grown massively by allowing customers to use ordinary phones to access Internet lines and make phone calls more cheaply than with competitors. Since these new companies were using IP packet switching, they could claim to be Internet service providers and thus to be exempt from long-distance access charges. Since those local access charges add up to about four cents a minute (on calls costing ten to fifteen cents), the exemption for IP phone calls gave these new companies a massive window for profit, essentially subsidized by the local phone companies. While IP networks were carrying only 1 percent of phone calls in 1998, such phone traffic was growing rapidly, promising losses of billions per year in access charges for local phone infrastructure.[81]

The irony of these lobbying campaigns by the Internet industry is that its spokespeople had pictured the privatization of the Internet as the "end of government subsidies," creating a gap into which the free market had successfully stepped. The reality, as these FCC decisions highlighted, is that the profits of the private Internet industry had derived substantially from the cannibalization of past and present investments in the local phone infrastructure. Local phone users, mostly of lower income and without a computer in the home, were seeing investments diverted to industry and higher-income Internet users that could have been targeted for upgrading the overall network or delivering new technology for schools, hospitals, or other public places serving the whole public. The specific private subsidies for the Internet industry helped fracture planning for the overall local phone system and blocked overall upgrading of data traffic.

Whereas federal investments once fueled overall economic and technological advancement in regional economies, these new "market competition" policies end up sucking funds from the infrastructure serving low-income and local users to subsidize those using the Internet for national and international purposes. And the forced segmentation of "competition" into their own boxes of long-distance, local service, ISP, and other regulated divisions has so fragmented phone service as to make comprehensive investments for upgrading the overall system nearly impossible.

Now, if this had been a small sin to help the Internet get off the ground, it might be a minor, even admirable hiccup in regulatory history, but this pattern dates back to the first attacks on the integrated AT&T Bell system. And with competition and "deregulation" of telecommunications becoming the metaphor and model for other network-based industries such as electricity, it is important to understand how MCI, Sprint, and other new telecommunications companies were a product of regulatory subsidy and infrastructure cannibalization rather than market efficiency as their mythmakers claim. The last part of this chapter will explore that history, its implications for understanding network economies, and how this legacy has thwarted more comprehensive regional economic approaches to digitally based telecommunications.

Many proponents of competition pooh-pooh concerns over investments in phone infrastructure, noting that in the early decades of this century, full-throated competition led to a massive expansion of phone service across the country—which is absolutely accurate. AT&T had emerged by 1880 with a monopoly on the telephone industry with its control of the Alexander Graham Bell patents and expanded service to 260,000 people by the time its patents expired in 1893. Full-scale warfare broke out between the Bell system and three thousand independent competitors to compete in building infrastructure across the country, with AT&T retaining only half the market of a vastly expanded 6 million phones by 1907.[82]

But it was an infrastructure that frustrated most of the customers, since they could not call friends in the same city if they belonged to a competing network and would be unable to call whole cities if those towns were controlled by networks hostile to the hometown service. The Bell system was the only service that provided anything approaching a comprehensive long-distance phone network. For the rest, competition made most of that expanded infrastructure unavailable across lines of hostile businesses—a state that led to pressures toward consolidation and regulated utilities. Facing this dilemma, in 1907 AT&T board of directors hired as president Theodore Vail, a longtime advocate of universal service within the company, who would largely shape the modern telephone structure and regulatory regime. Vail changed corporate strategy and withdrew from direct competition all over the country, favoring interconnection agreements with small operations in which AT&T was not competing—de facto absorbing them into the Bell system. As AT&T began to also purchase other phone companies, Vail also reversed Bell policy and accepted government regulation of the industry in order to maintain high-quality technology and uniform pricing. A 1913 consent decree with the Justice Department officially put AT&T purchases of other phone companies under the regulation of the government and required non-Bell companies to be connected into AT&T phone lines, all in the context of negotiated agreements that turned AT&T and the independents from competitors into collaborators in maintaining the phone infrastructure. State utilities commissions strongly supported the movement to consolidation and in 1921 federal legislation, the

Willis-Graham Act, placed AT&T under the jurisdiction of the Interstate Commerce Commission and exempted it from antitrust restrictions on purchasing other telephone companies. AT&T would purchase 223 independents in the following thirteen years. The new Bell system was largely decentralized, with AT&T operating the long-distance lines and acting as a holding company for separate state telephone companies which would be regulated by state utilities commissions. Each of these state Bell affiliates would become, in the public-service vision of Theodore Vail, a fixture in the economic development of their regions with strong public participation on their boards of directors. This was the structure that would last largely until the breakup of AT&T in 1983.[83]

Latter-day market advocates argue that all that was needed were regulations to require mandatory interconnection between services, and the country could have preserved the benefits of both competition and interconnection (much as is promised today with phone competition).[84] The problem with this retrospective viewpoint (and present advocacy) is that it ignores the basic economic implications of Metcalfe's Law—the rule of thumb that the value of a network increases not arithmetically but geometrically with the number of participants in that network. This means that the economic value of interconnection for small networks to much larger systems is astronomically high, while the main value of the investments in infrastructure by large networks comes from precisely the fact that they can offer such a large geometric network value whereas smaller networks cannot. Mandate interconnection and much of the value of that larger network's infrastructure (and the incentive to create it in the first place) disappears for its owner. Regulation of customer phone service rates may be eliminated, but government regulation will still be required to establish the rates paid for interconnection, a regulatory intervention that will most likely either be too high to encourage new entrants to the marketplace or—more likely, given larger networks' preference for no interconnection (in other words, an infinite price)—a price set too low for the larger network to maintain the quality and breadth of its infrastructure for all users. In such a situation, the most profitable position is to be a smaller network that services high-income, high-profit individuals or businesses that can as needed reach the low-profit customers of the larger network because of mandatory interconnection regulations.

This is the position of cannibalization in which Internet providers are presently situated, but is also the position that opened up competition to MCI and Sprint. That competition would service large industry at the expense of funds to assure access and investments in the local phone networks serving working families.

Subsidies and Separations: How MCI Cannibalized the Bell System and Sold the Myth of Market Efficiency

There was no technological innovation, no business efficiency, that made MCI into a multi-billion-dollar competitor. That is the major fact to understand about telephone deregulation in the 1970s.

MCI's profits and growth (as well as those of Sprint and other new competitors) came purely from convincing regulators to give them discounted interconnection to the Bell system and allow it to shift resources from ordinary ratepayers into the hands of its business customers in the 1970s. It was that simple. Since AT&T was bound by regulation to continue investing in local phone infrastructure while the new competitors were not, the result was predictable: new competitors would underprice AT&T's rates. But the true success of MCI and other competitors was that they sold this regulatory success as a triumph of the free market rather than skillful political and legal maneuvering. Alan Stone in his book *Wrong Number* has noted the irony that:

> AT&T, the largest firm of all, was committed through goals such as universal service and rate averaging to the interests of the small subscriber, whereas those with whom AT&T came into conflict, such as MCI, were primarily concerned with the interests of large business subscribers. The main political achievement of AT&T's rivals was their collective ability to portray their own interests as the 'public interest,' and in the process to gain important allies at every decision-making level.[85]

The key to MCI's success was the confused and fragmented nature of the U.S. government in dealing with precisely the issue of regional development and investments. Even as official federal policy was to favor investments in local expansion of the phone network, universal access, and low-cost service to rural and poor communities, MCI would use divided regulatory structures and economic confusion to expand its markets at the expense of such investments and universal rates. Backed by an alliance of populist liberals who distrusted AT&T's size and ideological market libertarians, MCI would help push forward the breakup of AT&T in favor of market competition, a policy opposed by large majorities of the public and large sectors of the government.

Beginning in World War II, the FCC (the regulator of AT&T as a result of the Communications Act of 1934) had required the AT&T Long Lines long distance division to pay a portion of every long-distance call to local phone company exchanges based on the portion attributable to using that local infra-structure. This allowed local regulatory boards to keep local rates low and expand service to rural and low-income users. This created an extremely com-plex accounting system for the Bell network (this complexity making it easier for later competitors to sow confusion for later regulators) based on "separa-tions" between these two parts of any call. In the 1950s, the National Associa-tion of Railroad and Utility Commissions, the national lobbying arm of state regulatory agencies, pushed for changes in the separations that would allocate a larger portion to the local phone company divisions. This increase in long-distance rates occurred as new technology was reducing the costs of long dis-tance and rising wages were actually increasing the costs of labor-intensive local services. Essentially, public policy was diverting the technological divi-dends of long distance to the costly investments needed to insure universal service at the local level. The FCC would approve a series of increased separa-tions payments to local service in the coming decades, creating an ever increas-ing gap between the price charged for long distance and the actual costs of delivering the service.[86]

The first challenge to universal phone rates had come in a 1957 FCC decision as private industry, promoted by Motorola, which provided the equipment needed, sought to bypass AT&T using a newly available radio spectrum to construct private phone lines over microwave equipment between different corporate locations. AT&T opposed these private lines, since they would un-dermine universal rates for all customers, contribute nothing to local infra-structure investments, and drain off some of the most profitable customers of the long-distance system. Worse yet, AT&T was barred by the FCC from offer-ing competitive rates to those business customers in order to salvage from them some income for the system. When AT&T tried to drop its rates, it was accused of predatory pricing, while the regulatory rules against its dropping its rates would lead critics to accuse the AT&T system of inefficiency, since it didn't match the rates of new rivals. This "damned if you do, damned if you don't" straitjacket defined the ideological vice that would eventually destroy the integrated phone system over the following quarter century.

This private-line exception did not affectthe Bell system directly, but in 1963, a small, almost bankrupt company called Microwave Communications Incorporated, or MCI, sought to establish a private-line service from Chicago to St. Louis that would resell capacity to individual corporations with offices in both cities. The key to their proposed service, however, was that local cor-porate offices would need to use local Bell telephone services to connect to the MCI microwave transmitter. The transmitter would then relay the call to the other city and would then use local Bell service connections at the other end to connect the call to each company's other corporate office. This was a dramatic advance from the 1957 private-line decision, but in 1969 MCI's application was approved by a close four-to-three vote of the FCC commissioners, although this was treated as an experiment for a single line that would be closely watched. But AT&T officials protested loudly that the "experiment" was an inevitable slippery slope toward full long-distance competition that would undermine the whole system.

The recently hired CEO of MCI, Bill McGowan, would aggressively run with the application and expand MCI with a series of interlocked companies creating private lines city by city. McGowan came to MCI not as a technology person but as a finance and marketing expert, and he knew that managing politics would be the key to MCI's expansion. While McGowan would sell MCI to investors and regulators as a "high-tech" company, the technology it used was simply the microwave radar technology from World War II. The essential factor was politics. In his authorized biography of the company, business journalist Larry Kahaner gave a sympathetic view of McGowan but related that: "McGowan knew that the future of MCI hinged on managing the regulators, the FCC. He had learned . . . that you could build a business by thinking of regulatory bodies such as the FCC as something that had to be managed, the way you would manage any other item that affected your business. Lumber companies managed natural resources like forests; MCI would manage the FCC."[87]

In the case of MCI, the goal was to clear-cut the traditional AT&T phone system with the hesitant support of the FCC. To facilitate continuing intervention with the regulators, MCI hired FCC commissioner Kenneth Cox, who had cast the deciding vote in the 1969 decision—a legal if unseemly offer of employment following a decision in MCI's favor. By mid-1970, the FCC faced more than two thousand applications for private-line services, most of them from MCI companies set up in different towns.

Contributing to the original FCC decision against AT&T were recent phone problems resulting from unexpected surges in demand for phone services, specifically in well publicized difficulties in New York's Wall Street district, where technological changes in the finance sector, detailed earlier in this chapter, were driving the need for multiple phone lines and increased data transfer. New consumer activism at state regulatory agencies blocking local rate increases and a shortage of new capital as a result of high interest rates combined to further burden the local phone companies within the Bell system. The surging demand for telecommunications services increased pressure on the phone

system just as MCI was leading the charge to help the most profitable customers of the system avoid paying the long-distance charges that helped finance
that local infrastructure.

Ironically, MCI could not turn a profit on the private-line business and by
1973 was facing bankruptcy (helped along as AT&T dragged its feet on supplying interconnection in many cities). So MCI created a service called Execunet
in 1974 to allow its customers to call anyone in a city serviced by an MCI
line, essentially providing traditional long-distance phone service directly in
competition with AT&T. Much as the original creators of money market funds
had obtained approval from the Securities and Exchange Commission by ob-
scuring what they were doing, MCI filed the new service not with the FCC
commissioners themselves but with the routine tariff division. When the FCC
commissioners found out what MCI was doing, they were extremely angry
and ordered the company to stop. The commissioners had hit the tolerance
limit for their experiment and decided to draw the line at full competition on
long distance. State regulatory agencies were already up in arms over the FCC's
previous approval of MCI's private-line services and their national lobbying
arm had called in 1973 for the FCC to be stripped of jurisdiction over long-
distance rates and how those rates subsidized local phone operations.

But decisions concerning the phone system were spinning out of the control
of both the state regulators and the FCC. In 1974, crusading liberal anti-big-
business lawyers at the Justice Department filed an antitrust suit against
AT&T. Downplaying the big-business customer base of MCI and the complex
subsidies for local phone companies, the Justice Department liberals saw the
case as one of defending scrappy upstarts such as MCI against the "largest
corporation in the world." Federal courts overruled the FCC's attempts to,
belatedly, rein in MCI, with a major 1977 decision arguing that (as AT&T
initially worried) once they granted competitive access, they could not then
limit the range of services offered by such companies as MCI. The end result
was that MCI and other competitors were now in direct competition with
AT&T for long-distance customers—but with discounted access to the local
phone system, they could undercut AT&T on price while actually spending
more to deliver the service.

Perversely, the FCC in 1970 had dramatically increased the separations payments by AT&T to local phone operations under what was called the Ozark
Plan, even as the FCC was allowing MCI to start cannibalizing the long-distance service that made those payments possible. While long-distance rates
had fallen over the years, the separations payments had channeled most of the
improved efficiency in the system into lower rates and more universal access

in the local operating companies of the Bell system. The cost of monthly residential telephone service would by 1980 be *one-third* in real dollars of what it had been back in 1940, even as the quality increased and the percentage of households with phones had grown from 37 percent in 1956 to more than 93 percent by the 1980.[88]

By the time MCI offered its Execunet service, AT&T separations charges accounted for two-fifths of long-distance costs. MCI and other specialized carriers negotiated a rate of payment for interconnection equal to only half that amount, so their rates could be a mammoth 20 percent lower than AT&T's even if their costs were exactly the same as those of the Bell System. That regulation-imposed difference in costs would allow MCI to reach $1 billion in revenues by 1983 when the Bell system was broken up by the antitrust court decree. And that regulatory difference in costs would attract entry by companies with significantly higher costs than the Bell system had, a complete undermining of old antitrust arguments that competition bred the most efficiency. Yet many of AT&T's opponents would watch the growth of competing firms and argue that their growth showed that the Bell System was inefficient and that competition was desirable. The reality was that productivity gains in the Bell system during the 1970s exceeded all but one of sixty-three industries surveyed by the Department of Labor, including fifty-one industry groups with productivity gains one-half that of Bell.[89]

Technology and Interconnection

While long-distance competition fueled by the separations subsidies was one pressure on the Bell system, the other side of the escalating conflict leading to breakup was the battle over interconnecting new technology to the local phone systems. It was this fight that fundamentally hobbled the ability of the phone system to comprehensively respond to the emerging computer revolution and instead precipitated the proliferation of proprietary networking technology, which the Internet would only overcome two decades later.

Fights over computerization and equipment had actually made up the longest-lasting conflict between regulators and AT&T in the postwar period. Long Lines long-distance service and local operating companies may have been deploying the technology that drove down phone rates and improved efficiency, but it was AT&T's Bell Labs division's R&D and its Western Electric manufacturing unit that were building the technology that drove breakthroughs not only in the telephone industry but, as mentioned earlier, also in the computer industry as a whole. Back in 1949, the Justice Department had challenged

AT&T requirements that customers use equipment from its Western Electric division, and it had sought to break the company up in an antitrust suit. The 1956 consent decree that emerged out of this lawsuit kept AT&T intact, but prohibited Western Electric from producing nontelephone equipment and forced AT&T to license its computer-related patents (thereby spurring the migration of technology development to Silicon Valley). With massive expansion of the phone system in this postwar period, AT&T was willing to concentrate its energy on fulfilling phone service orders, even as Bell Labs continue to churn out innovation after innovation that AT&T could not develop itself.

The conflicts arose as computer technology began to converge with telephones in the late 1960s. Where "equipment" stopped and new additions to the telephone network started became increasingly unclear. AT&T had long restricted what equipment, labeled "foreign attachments" in Bellspeak, could be linked to the phone system in order to maintain consistency in the network (and, of course, maintain their revenues from equipment sales). But in 1968, the FCC ruled that a device called the Carterphone, which relayed sounds from a telephone to a mobile radio, was a legal attachment to the phone system, thereby opening up the market for attached equipment from answering machines to private branch exchanges (PBXs) to modems.

Opposed to the FCC decision were the state regulatory agencies, which feared that lost revenues from equipment sales would drive phone rates higher and that additions to the phone network would create a massive, complicated regulatory mess (a point on which they were absolutely correct). The new equipment opt-outs were allowing companies to create their own internal private phone infrastructure without paying the costs of expanding access to expensive rural customers and lower-income users. While MCI was undermining subsidies from national phone use to local infrastructure, equipment opt-outs were undermining intraregional subsidies.

But beyond the economic issues, equipment opt-outs were undermining technological standards and fragmenting interconnection of data on the network. As computer use of the telephone advanced, AT&T found itself in the perverse position of being a foremost innovator in computer technology while being barred from selling equipment in that area. As other computer companies began processing information using the phone system, they inevitably would perform quasi-telephone-company functions as they switched data between different customers, opening the question of whether those information services should be regulated by the FCC and what their contributions should be to the overall network. Time-sharing computers were at the heart of these early disputes, and, in response to a stock information service called Telequote

IV, the FCC made a ruling in March 1971 called *Computer I,* which removed regulation from most computer interaction with the network. AT&T itself was barred from offering such data-processing services, and the FCC even ruled that the company could not provide data processing to its own divisions (although this was overruled by the courts on appeal). But it was clear that any step by AT&T toward computerization of its networks would be met by FCC challenge, so it could do little to upgrade its own interconnection operations to provide an alternative to the proprietary versions then appearing. It was in this context that AT&T had refused to explore computerized packet switching networks when asked to do so by ARPA in the late 1960s and 1970s.

It was fears of just such fracturing that led the National Academy of Science through its Computer Sciences and Engineering Board to publish a report that opposed uncontrolled connection to the telephone system precisely for fear that they would undermine network performance and standards for compatibility of data exchange.[90] The rise of PBX systems and other proprietary devices would justify those fears in the next decade. AT&T was working to build a digital backbone for the entire phone system and slowly extend it to smaller and smaller units. AT&T projected that in the long run proprietary digital devices tied to an overall analog system would be a waste until they could optimize the entire system. However, customers wanted the newest digital equipment for immediate needs, and they turned to other suppliers. Those short-run decisions would increase long-term costs in connecting disparate parts of the phone network, and this would also retard upgrading the system as a whole to digital standards given the technological lock-in of individual proprietary digital devices using the analog phone system. When a 1980 decision barred AT&T from selling data terminals to customers, the coordination of information standards with the underlying digital architecture of the phone system was undermined further.[91]

The Breakup of AT&T

The Justice Department–led breakup of AT&T in 1983 would put the final nail in the coffin of coordinated data standards promoted through the phone system, even as other parts of the government would strive to build an Internet to put the humpty-dumpty of fragmented networks back together again.

The question of whether to break up AT&T divided the government to the last moment. State regulators, most of the scientific community, and the Defense Department acted as the loudest defenders of the AT&T system, while the hybrid populist liberal and libertarian conservative coalition that led deregulation efforts in a host of industries in the 1970s pushed forward the partitioning of the system. An ironic twist for AT&T, in a battle with endless perverse twists, was that the very connections of the phone system to promoting economic civic life and its engagement with public service would be a decisive, if unintended, factor in its destruction. When the Reagan administration won the White House in 1980, the new attorney general, William French Smith, was dead set against continuing the antitrust suit. However, Smith was a long-time board member of Pacific Telephone from his previous work in Los Angeles and had to recuse himself from any consideration of the case, and his deputy attorney General's law firm had done work for AT&T, so he had to recuse himself as well. This left almost total control of the lawsuit to the new head of Justice's antitrust division, Stanford law professor Bill Baxter, who was a longtime ideological opponent of the AT&T monopoly. The lead opponent of the breakup in the Reagan administration was Bernie Wunder, the head of the National Telecommunications Information Agency—an agency that would play a crucial role throughout the 1980s in building the Internet. Wunder's opposition was joined by the secretary of commerce and secretary of defense; the latter's department would issue a 1981 document that argued: "[T]he Department of Justice does not understand the industry it seeks to restructure. . . . All that divestiture as outlined by Justice could possibly cause is a serious loss of efficiency in the manner the network operates today. . . . We believe that [divestiture] would have a serious short-term effect, and a lethal long-time effect, since *effective* network planning would eventually become virtually non-existent."[92]

But fears that any dismissal of the lawsuit would lead to accusations of conflict of interest on the part of William French Smith left Baxter in control and the final result was a partitioning of AT&T on January 8, 1982, beyond anything Justice had originally envisioned. All local telephone exchanges (two-thirds of the million employees in the system) would be grouped into seven "Baby Bell" companies that would remain as regulated common carriers, while long-distance service would be combined with Bells Lab and Western Electric as a purely profit-oriented company competing with MCI and all the other new telecommunications companies on an equal footing. Partly, this was the result that AT&T wanted by this point, since with the financial noose of regulatory restraints slowly strangling the company, its executives saw fully embracing competition and abandoning its public-service tradition as the only way to survive. Unfortunately for AT&T, lobbying by the American Newspaper Association (which feared a threat to its classified ads) added provisions to the settlement that barred the company from electronic publishing for seven years—a further retardation of innovation in the name of competition. This divorce of the system's innovative heart from the day-to-day maintenance of local phone infrastructure would be the final shredding of the fabric that had funneled the fruits of technological innovation into the needs of average phone consumers.

While access charges on long-distance telephone calls would preserve some of the funds flowing to local service, the aftermath of the consent decree would dramatically increase the price of local phone service, and the costs of installation and repair of equipment would increase tenfold in some regions. As market competition increased, a range of services rose in price for local residents. By 1997, even local pay phone costs were skyrocketing in price as the FCC overruled local regulatory controls on the cost of calls from pay phones—an especial burden on the 20 percent of low-income families without their own phones who depend on pay phones as their only source of phone service.[93]

Before the breakup of AT&T, 80 percent of the public questioned in a *New York Times* poll had expressed satisfaction with their phone service and in 1985, 64 percent of the public declared the breakup a mistake.[94] All that was left of the old system would be the jerry-rigged system of access charges that would funnel a (decreasing) amount of money to the local phone companies for infrastructure but would prevent any fundamental technological upgrading of the system.

Was There an Alternative? Minitel Versus the Internet

Before turning to the other effects of the consent decree on regional telecommunications, it is worth considering what opportunities were lost in the manic push to competition in telephone services, especially when compared to the experience of the French Minitel system—the most developed public information network in the world before the expansion of the Internet. In looking at the French experience, it is clear that the regulatory push in the United States for competition in telephone service undermined the ability to deliver an integrated electronic information network and delayed its availability to the public by a decade or more.

Many Americans do not realize that in 1994, as most people were hearing about the existence of the Internet for the first time and Netscape was just being founded, France had for ten years been running an integrated electronic information network that had pervaded homes and workplaces across that country. With more than 6 million subscribers and twenty thousand commercial services in 1994, the Minitel system in France completely dwarfed the

Internet and private information services such as Prodigy and CompuServe. Almost 50 percent of the French population had access to Minitel terminals at home or at work encompassing a wide diversity of the nation, while 30 percent of businesses (and 95 percent of businesses with more than five hundred employees) were connected to the system. Almost every bank had developed home banking services, allowing customers to check their accounts, pay bills, and trade stocks. Travel agencies, insurance companies, and retailers had all created services on the system, generating $1.4 billion in revenue by 1993.[95]

Minitel consoles were found in bars and hotels, libraries and offices, rural post offices and mountain railway stations, making any debate about information "haves" and "have-nots" almost irrelevant. As one U.S. analyst observed in 1994, "The real difference which I noticed in France is the French system's omnipresence. You see it everywhere; or rather, you know it is there but you don't see it—the system has achieved the technological goal of invisibility currently being set by product developers for the Internet."[96]

What is clear is that it was the integrated nature of the French telecommunications system and a financial structure built around fixed returns on capital rather than marginal market transactions, that allowed this broad information network to emerge so much earlier in France than in the United States. Where business and U.S. policy pushed for clear market segmentation between the telephone and computer industries, the French government early on was committed to integrating them. In 1975, two researchers were commissioned by the French president to develop a strategy for computerizing French society. The report by the researchers, Nora and Alain Minc, became a bestseller and coined a new word, *télématique,* which defines the merger of computers and communication technologies as the key to innovation in government and business. France in the early 1970s had one of the least advanced telecommunications systems in the developed world, with only 7 million telephone lines serving a population of 47 million people. With massive investments needed to finance the expansion of the system to achieve universal access, the promotion of information services became not simply a visionary goal but a financial necessity to subsidize the expansion of traditional phone service.

An initial concept was to have an electronic phone directory, to eliminate the rising costs of printing phone books and staffing directory assistance. With prototype terminals installed in test cities from 1980 to 1983, the system seemed to be a technical and sociological success. A full-scale distribution of free terminals to customers was begun, with more than a million terminals in place by the first year and more than 3 million distributed by 1987. The free distribution of the terminals, initially hotly debated by the government-run phone company, was a vital innovation in attracting the critical mass of users quickly and giving the enterprise the aura of democracy that generated quick acceptance. The other key component was adoption of a common, standard protocol that interfaced the telephone network easily and cheaply with the packet-switching data network. The phone company would remain a common carrier, while a range of other businesses could easily "set up shop" on the system with Minitel collecting fees in a simple and cost-effective way that would show up on customers' phone bills. As would happen in the United States' Internet, electronic mail and chat would unexpectedly lead to an explosion in use of the network, as would controversial "messageries roses" sex-related services. This traffic would overload the system in 1985 and give both fame and notoriety to the new system as it was reengineered for the increasing level of traffic that continued to expand.

Financially, the system could only succeed because France Telecom could subsidize the start-up costs and the free distribution of terminals with the expectation of long-term returns because of its monopoly position. Given the link between the expansion of basic phone service and the Minitel service, it has been broadly disputed exactly how well the Minitel system has done in returns on investment, but audits have shown that the system reached the break-even point in 1989 with Minitel terminals paying for themselves within 5.7 years. By the year 2000, the Minitel system had earned an internal rate of return of 11.3 percent over the 1984 to 2000 period. But only an integrated telecom monopoly generating fixed cost returns could have invested with that long-term a view of financial returns and with the confidence that it could capture the revenues from its investments over that period.[97]

In the United States, the hemorrhaging of fixed cost returns for AT&T as new businesses were allowed to grab high-end telecom markets meant that there was little chance of such comprehensive investments, especially as AT&T faced the increasing reality of its dismemberment. And, even if AT&T had been willing to take the financial risk, the regulatory segmentation of phone and computer markets (in the name of competition) meant that the company was legally barred from even offering, much less subsidizing, Minitel-style terminals to its customers. With no central phone system to enforce standards, the result in the United States was a slew of niche networking markets, each with its own proprietary technology and standards. These proprietary technologies fragmented the electronic information services market, a fragmentation that, in contrast with the situation in France, delayed the mass availability of an electronic information network for more than a decade.

Even as one set of government officials had fragmented the telecommunications marketplace in the name of competition, other government leaders centered at ARPA would create the Internet system of standards and finance the backbone lines that would, belatedly, create an alternative data exchange system as an overlay on the phone system. As detailed back in Chapter 2, the Internet leadership would use government and university facilities to create the critical mass of users to make Internet standards compelling to other private networking companies. As the system expanded, the exemption of Internet providers from paying access fees to local phone companies would create subsidies that would encourage further, rapid growth.

The Internet was a brilliant technological solution to networking fragmentation, but its effects on regional investments in overall telecom infrastructure and on general equity were the reverse of the effects of the Minitel system. France's system subsidized the expansion in that country of traditional phone service for lower-income users, while traditional U.S. phone system users had to subsidize the Internet. And whereas Minitel made information services accessible on a democratic basis through the free availability of terminals, in the United States one had to own a computer to obtain access to the early Internet, so subsidies in the system invariably flowed to upper-income and business users.

If the Internet had a relative strength, it was that in having to overcome the patchwork of proprietary systems within the United States, it became a potent vehicle for overcoming the different electronic standards of various countries and therefore emerged, with the economic and political strength of the U.S. government behind it, as the international networking standard. Even France began promoting use of the Internet by the late 1990s, started introducing dual-use terminals that could access both Minitel and Internet services.[98] But if the push in the United States for networking had been combined with long-term investments that AT&T could have pursued with the technological innovation of Bell Labs behind it, there seems a high likelihood that electronic networking could have expanded earlier and more democratically within the United States.

Expanded Regulation and the Aftermath of the AT&T Breakup

While the integrated AT&T system was obliterated with the consent decree, the need to regulate not only did not diminish, but it expanded, as the FCC and other bodies were forced to microregulate the charges paid at every step in the new market of telecommunications. Moreover, while the subsidies from long distance to local operating companies diminished in the new system, they did not end, and the FCC and local regulatory bodies would have an increasing role in detailing exactly how to allocate those subsidies. This time of intimate involvement by regulators was in contrast to the period of the integrated AT&T system that had left those day-to-day decisions to AT&T management.

The breakup of AT&T would unleash a flood of new and increasing regulations and lobbying as the telecommunication division of the Washington, D.C., Bar Association, previously a nonexistent category, would double each year as the splinters of the telecommunications industry each demanded regulation in its own self-interest.[99] Exactly how much to charge long-distance companies for access to local phone companies networks would remain a contentious issue. Initial FCC proposals to raise local rates eight dollars a year to replace previous payments by AT&T provoked a firestorm of opposition and encouraged 1984 legislation by Congress to maintain at least part of the previous subsidy system funding local infrastructure from long distance. Local rates would rise, but continued regulatory battles would determine by how much.

While the original conception of the local operating companies was that of purely common carriers that would engage in no equipment-related or other competitive markets, the reality was that the Baby Bells were not economically viable with lessened subsidies, so they were given ownership of local Yellow Pages, the right to sell phone equipment (but not to manufacture it), and entry into a host of competitive markets in order to bolster revenues. Of course, with local phone monopolies operating in these local competitive markets, this only reopened all the issues that had led to the regulatory controversy that swirled around AT&T, so a new host of regulations were needed to govern the entry of the Baby Bells into these competitive markets. With cellular phones, the Internet, and a variety of more specialized services demanding interconnection into the phone system, a continual stream of microregulations would be needed to govern fair rates for the mandatory interconnection that local operating companies would have to accept. The FCC would spend years embroiled in endless debates over which economic theory should be consulted in order to set rates for interconnection. In a sense, the question was, How do you simulate the market price for a good (namely, interconnection) that the phone company would refuse to offer if given a free choice in the marketplace—an amusing theological question that would consume endless hours of regulatory time.

As the Telecommunications Act of 1996 opened up competition in local phone service, a blizzard of new regulations at the FCC and at state regulatory bodies began to govern where interconnection would occur, what network elements would be provided to new market competitors in individual "unbundled" markets, and how to arbitrate prices for each element.[100] New regulations would mandate that geographic distance rather than density of traffic or costs of service would govern rates—a practice designed to benefit hard-to-access rural areas but one that created a new round of regulatory contention as wireless phone companies complained that this undercut their potential competitive advantage serving such rural areas.[101] As the ISP access charge exemption showed, regulations on if and how rates are set for interconnection would fundamentally shape which technology was deployed. With each regulatory decision representing a choice between economic competitors pushing their own technology, "deregulation" has made politics rather than cost a more determining factor over technological direction than it had been in the days of AT&T's dominance.

As many of these disputes ended up in federal court, it also became clear that at stake was less whether regulation was expanding (since it obviously was) but who would control that regulation, federal or local authorities. When a federal court ruled in 1997 that the FCC had overstepped its bounds in limiting the power of state regulators, there was the spectacle of one set of companies, largely long-distance carriers, lining up to support federal regulation by the FCC, and another set of businesses, largely the regional Bell companies, lining up to support local regulators.[102] Despite the rhetoric popular in Washington, D.C., of "returning power to the states," the reality was that in telecommunications, as in so many other areas, federal regulators were attempting to strip power from local authorities while radically expanding their own jurisdiction.

Adding to the morass of new regulations was the issue of what should be funded with the access fees dedicated to "universal service" and how to ensure that new entrants to the marketplace were paying their share of the overall infrastructure costs. As competition entered the local market, a host of companies would begin seeing themselves as possible recipients of those funds (and could thereby score an economic advantage if their rates could be supplemented with the universal service fund). Additionally, while the extra costs of subsidizing phone service in rural communities and for the poor had been the traditional focus for such funding in the days of AT&T, the 1996 Telecommunications Act mandated that universal service be defined in terms of "evolving levels of telecommunications services."[103] The FCC implemented this mandate by creating a new assessment on all phone services to give public schools and libraries discounts of up to 90 percent on local phone service. This was combined with mandates from local regulators to offer those institutions high-speed lines for Internet access. Exactly how these requirements would be applied created a whirlwind of lawsuits and ad hoc decisions by regulators.[104] "I don't know any economist who thinks the current system makes sense in any terms other than politics," said Michael Katz, a professor at the Haas School of Business and former chief economist at the FCC, in surveying the 1997 decisions of the FCC.

The End of Regional Phone Companies

One key to the old Bell system was that it combined an integrated phone system with strong decentralized operating companies in each state. Since each operating company had to raise its own debt and operate within its own regulatory environment, local companies became fixtures of those communities in representing the Bell system's commitment to universal service.

However, the postbreakup Baby Bells increasingly abandoned their local focus in favor of new, more profitable ventures serving higher-profit customers. Even before the megamergers between two pairs of the Baby Bells (NYNEX and Bell Atlantic, Southwestern Bell Corporation and Pacific Telesis), a merger and investment frenzy had permeated the operating companies. US West Incorporated invested $2.5 billion to acquire a 25 percent interest in Time Warner Incorporated; NYNEX Corporation put $1 billion into Viacom International in its fight to take over Paramount, receiving preferred stock and two seats on the board. SBC Communications made a $1 billion joint venture with Cox Enterprises. The failed merger of Bell Atlantic and cable giant Tele-Communication Incorporation (TCI) only pushed the merger frenzy further.[105]

This was all quite a change from the desperate financial straits in which the operating companies had found themselves immediately following divestiture. But in many ways it was a logical result of an economic environment that left their core business open to economic cannibalization by other market competitors, thereby making investments in that core business unattractive from a shareholder perspective. The change for California customers was probably the most dramatic, since Pacific Telephone has been one of the prime beneficiaries of the old AT&T system. With its rapidly expanding population and its tight pro-consumer regulatory environment in the 1960s and 1970s, California had received a disproportionate share of the long-distance separation subsidies. (AT&T saw one of the greatest advantages of divestiture as escaping its financial support of California's expanding phone infrastructure.) Postbreakup, Pacific Telesis was left so overloaded with debt from those expansion costs that it remained financially solvent only because Congress erased $1.5 billion of tax liability in 1982 legislation.[106] With its new solvency, it quickly began leveraging its regulated phone infrastructure assets into new ventures in competitive telecommunication markets. Between 1984 and 1993, analysis by Boston-based Economic and Technology Incorporated estimated that 95.7 percent of capital for those new ventures came from local phone assets.[107]

One result of those competitive investments was Pacific Telesis's cellular phone system, which it spun off as a separate company, Airtouch, in 1994. With the aggregate value of the separated companies $6 billion more than before the spin-off, the shareholders of Pacific Telesis had done well off the investments diverted from regulated infrastructure to competitive markets. The utility advocate group TURN demanded that ratepayers receive $1 billion as its share of these returns, but an increasingly conservative CPUC granted only a $50-million payback.[108] What Pacific Telesis also delivered to its customers was lagging investments in upgrading outdated infrastructure needed to accommodate advanced networking technologies. Despite having headquarters in the expanding technology centers of Northern California, Pacific Telesis ranked dead last among the Baby Bells in use of fiber-optic cables, the key to delivering high-bandwidth applications into homes and small businesses across the region. In 1995, only 6 percent of its access lines were fiber-optic cables (compared with 11 percent for leader Bell Atlantic), and it ranked near the bottom of the Bells in installation of digital switches as well.[109] Between mandated subsidies to private Internet providers and its own diversion of resources into market ventures such as the Airtouch spin-off, Pacific Telesis had essentially looted the infrastructure serving average consumers in favor of ventures that met the needs of private telecommunications systems serving high-income customers.

The death knell for regional telecommunications in California would come on March 31, 1997, as the state's utility commission approved the takeover of Pacific Telesis by its fellow Baby Bell Texas-based SBC Communications. This new entity would have a combined market value of $47.9 billion serving seven of the ten largest markets in the country; through SBC, it would have global telecommunications stakes in Mexico, Chile, South Korea, Taiwan, France, South Africa, and Israel. Clearly, this new telecommunications colossus had outgrown the day-to-day regional universal service concerns of its heritage.

There would be one last-ditch attempt to claim some funding for the community before its once proud community institution disappeared into its multinational merger. On the basis of California law requiring that at least half the profits from any utility merger be returned to ratepayers, CPUC's Office of Ratepayer Advocacy and the consumer advocacy group TURN estimated those merger profits to be at least $2 billion, with the public's share thereby being at least $1 billion. The California Telecommunications Policy Forum, an advocacy organization formed largely by communities of color, mobilized a coalition of public interest, community, and labor organizations to demand that the state utilities commission take this as the opportunity to "ensure that all sectors of California's diverse economy and population derive measurable and substantive benefits" from the merger. To this end, in addition to insuring that long-term infrastructure investments were made, the coalition advocated that half of the estimated $1 billion be given back as a rate cut and the other half be dedicated to a "telecommunications trust fund" that would finance community-based technology centers in sixty of the poorest communities in the state; pay for consumer education and advocacy; provide additional funds for infrastructure in low-wealth school and library districts; and fund scholarships for low-income students so they could enter the telecommunications field.[110]

However, ignoring its own staff's recommendations for a $590-million rate-payer refund, the state utilities commission ordered only a $248-million refund and ignored recommendations for the telecommunications trust fund. As Pacific Telesis was swallowed up by SBC, Pacific Bell customers could expect a rebate of 25 cents per month. TURN would declare that the decision "sells out California's autonomy for a very few pieces of gold."[111] Much as Wells Fargo and Bank of America were escaping regional obligations through merger (and as PG&E looked to do in coming years), the regional phone company anchor of civic economic life had ceased to exist in California as global mergers swallowed Pacific Telesis.

The Failure of "Competition": Broadband and the Telecom Meltdown

At the beginning of the 1990s, before most people had even heard the word *Internet,* the vision of the imagined Information Superhighway was one of high-speed interactive multimedia—movies on demand, interactive video, instantaneous downloads of music. Cable and phone companies would connect every home with high-speed "broadband" technology to bring this new world into every living room.

After a decade of deregulatory hype, the results could only be described as dismal. Despite all the promises by cable and phone companies, by 2001 only

6 percent of homes had high-speed Internet access, with phone companies servicing less than a third of those homes.[112]

Yet these dismal results were the product of one of the largest capital investment binges in telecommunications history in a period when the market capitalization of upstart fiber-optic and high-speed telecom firms topped trillions of dollars. Between 1997 and 2001, firms invested more than $90 billion in high-speed optical fiber, laying more than 100 million miles of fiber, or enough to reach the sun. Under the slogan "If you built it, they will come," investors rushed to support an ever expanding spiral of investment in high-speed bandwidth.

Unfortunately, hundreds of millions of miles of fiber were laid everywhere but to the homes of the consumers who were supposed to use the high-speed Internet services. The chronic underinvestment in local infrastructure remained, making this fiber-optic investment essentially useless. By 2001, only 2.6 percent of fiber capacity was actually in use, much of it destined to remain "dark" forever.[113] As this reality of speculative stupidity sunk into the financial markets, telecom firms suffered some of the most severe crashes of the NASDAQ plunge. One of the most high flying of the new upstarts, JDS Uniphase, whose market capitalization had topped $200 billion, in July 2001 reported the largest loss in corporate history, $50.6 billion for its fiscal year, as its market value plunged to $11 billion. Of the high-flying fiber firms, only Qwest survived, because it had used its pumped-up market value to acquire the "old economy" Baby Bell U.S. West which provided the revenues necessary to, if not provide profitability, at least allow Qwest to avoid complete financial hemorrhaging.[114]

There is a historic déjà vu to this process: more than a hundred years ago, speculative private investment in railroads followed a similar pattern. Companies built rail lines between various cities without linking them to downtown rail yards. Presaging the proprietary fight for control of the fiber lines, different companies would use different rail widths, thereby increasing the costs for switching trains between lines. In the 1870s, a massive overbuilding of rail lines led by mid-decade to mass bankruptcies in the rail industry, helping to plunge the whole nation into depression as it recovered from the speculative excesses of the private investors. Unfortunately, the United States seems to have repeated that pattern as the world slipped into recession partly from the glut of speculative technology investments.

The failure of private deployment of broadband services was foretold in the failure of local phone competition. Similar promises after the 1996 Telecommunications Act were made about competition and better local phone services. Billions were invested, yet by 2000, out of 191 million phone lines in the country, only 6.4 percent were served by competitive phone companies. And more than two-thirds of those lines connected big businesses, since the profit margins for average users were seen as too low to justify investments in that sector.[115] While the 1996 law had required the Baby Bells to lease use of their wires to competitors, few competing companies were interested in doing anything but cherry-picking the higher-profit business customers. A few companies even built their own local phone networks as an alternative to the Bell local infrastructure, but again they were serving almost exclusively downtown urban areas and business customers. According to FCC data, whereas 17.5 percent of big business had competitive local phone options, only 3.2 percent of residences and smaller business had any competition.[116]

With high-speed Internet access, the problem was that there was no existing infrastructure to which other companies could even interconnect. While competitors complain that Bell opposition to competition was the source of slow broadband deployment, it seems unlikely that it would forego its share of the much hyped profits from high-speed deployment merely out of spite. The Bells have made a good case that if every time they pay to build new infrastructure, competitors can automatically lease use of them at some regulated "average cost" price, the result is predictable. The competitors will cherry-pick the most profitable customers with special deals and the local phone companies will get stuck with the rest. Looking at that reality, the Bells largely avoided deploying high-speed data lines, especially in poorer or rural areas, since they feared losing the profitable customers that balance out poorer, costlier ones.

Increasingly, even many legislators who had originally chanted the mantra of "competition" as the solution to all problems were losing their fervor. In April 2001, House Commerce Committee chair Billy Tauzin introduced a bill that, without changing the competition rules on voice service competition, would allow the local Bell companies to make high-speed data-line investments without letting competitors have access to those new services. Critics rightfully noted the failures of the Bells in recent years, so a coalition of House members representing inner-city and rural constituencies pushed through an amendment in committee to mandate that the Bells achieve 100 percent rollout of broadband connections to all their central offices within five years.

If that sounds familiar in a nostalgic way, it is what was once common in Ma Bell's universal service rules. The next step would no doubt be consumer-rate regulation to ensure reasonable prices by the Bell monopolies. Given the gridlock of telecom politics and competing bills in both the House and the

Senate, this return to regulated mandates was unlikely to pass right away, and the Bells were hardly thrilled with the ideas, calling the mandates "very onerous." Still, the bill's initial passage was a clear sign that the mania for "competition" as a solution was waning in the face of the failures of the 1990s to deliver on the promises of the high-speed Internet.

Back in 1992, then vice presidential candidate Al Gore had suggested that the government should take an active role in building the infrastructure for the proposed Information Superhighway, a position the Clinton administration, once it took office, soon backed off from under pressure from corporate lobbyists. Summing up the corporate critique, in 1993, then chairman of Ameritech William L. Weiss argued, "The government seldom manages anything as well as a private enterprise does. We think we (in private industry) should build it and manage it."[117] It is hard to imagine how the government could have done worse than wasting more than $90 billion on unused capacity while delivering services to less than 10 percent of the population. Government makes many mistakes, but it has rarely built highways while forgetting to build the off-ramps.

Economic Standards/Technological Standards: The Collapse of Regional Economic Models

What is clear across the three sectors of banking, energy utilities, and telecommunications is how the radical shift in networking technology has been inextricably tied to changing economic models. Technology largely originating in Silicon Valley would cascade across these sectors, facilitating the expansion of models of market competition while undermining the regional reinvestment models that were once an essential part of cross-class agreement, stabilizing income and equality for all regional actors.

Previous technological regimes had been backed by economic pricing models that allowed important regional actors to expand their network of customers and recover the fixed costs of the investments needed for that expansion. Banking would depend on Regulation Q and other restrictions on interest rates to prevent a focus only on high-income borrowers, while both energy utilities and the phone system would mandate universal rates for all customers. With fixed rates of return on capital investment more or less guaranteed under these political regimes, each regional industry's fate would depend on expansion of its customer base and the general prosperity of the region—a critical element in the companies' commitment to the civic economic life of their regions.

New technology combined with aggressive political lobbying would break open older regulatory regimes in favor of new market-oriented regimes of pricing. New technology would flow across the networks: computer-managed money market funds would open the first major breach in universal interest rates; use of finance technology would put new pressures on the phone infrastructure (as in the dramatic failures on Wall Street in 1969), thereby encouraging businesses and high-income customers to bypass the regular phone infrastructure; and the emerging telecommunications networks would allow big business to begin wholesale direct purchases of energy for the first time. These new technological possibilities were seized on by big business and the wealthy to push for regulatory changes that would allow them to escape the cross-class investments that had been at the heart of postwar regional growth coalitions. Usury laws were repealed, new long-distance companies such as MCI were largely exempted from obligations to fund local phone infrastructure, and large industry was increasingly allowed to bypass fixed utility rates, even as these broad macroregulations were replaced by increasingly detailed rules governing the proliferating segments of each marketplace.

In the process, while business absorbed the economic fruits of technological innovation (largely created through the investments promoted under the older fixed-return system), it managed to shift the costs of the transition to market-based regulatory systems onto average working families. Subsidies that had flowed from big business and the wealthy to smaller users began to be reversed as market competition allows sophisticated users to cream the benefits of technology for themselves. This would leave the remaining residents of the region with failing savings and loans, discriminatory lending, the NASDAQ meltdown, stranded assets and blackouts in the power utilities, the telecom meltdown, and higher phone bills for average users, and all this as wealthy customers would be given a new array of options, including an Internet that consumed greater and greater portions of regional resources, largely to the advantage of a narrow segment of society.

The fraud of "deregulation" in each of the industries discussed here was that government intervention, rather than disappearing, would become even more crucial and, in many ways, more intrusive. The fixed-capital model of regional development had largely been one of delegating economic development to designated regional actors, whether in finance, energy, or telecommunications. With their markets broadly defined and restricted, relatively light regulatory goals would ensure that the public interest would be served and that the infrastructure needed to undergird commerce would be sustained and managed well. With the rise of market competition, however, the maintenance of the infrastructure, especially in the area of technology standards, becomes less automatic, and higher levels of government intervention become needed. These government interventions may be supplemented by attempts by private consortia to collectively sustain the infrastructure, as seen in the actions of Silicon Valley firms creating alliances centered around electronic banking standards or energy and technology companies in Texas underpinning energy management structures. However, the underlying rules of commerce, of how to ensure fair market access to each market, and even to define where one market starts and others end—all would require that government intervene with careful regulations to guarantee that the infrastructure of each underlying network of transactions be maintained. And with the traditional alignment of self-interest and public interest largely dissolved under the pressure of market competition, whatever shreds of commitment to regional development and economic equity remain now require even greater regulatory intervention to address those goals.

Many of the strongest proponents of market competition saw the solution to this increasing regulatory morass as an even more decisive move to cut the link between private profit and public goals. End any attempt to achieve equity or access within the industries themselves and just let the government issue cash vouchers to the rural and urban poor. Let the poor buy their goods on the open market without the government trying to engineer indirect subsidies.

While such proposals deliberately ignore the regulatory structures that would still be left in calculating the rates for interconnection to and maintenance of these networks, the real hypocrisy of such proposals is that most of those promoting such direct subsidies are precisely the same political actors who have supported the elimination of other direct subsidies to the poor, from school lunches to heating oil to health care. Even as subsidies for business interconnection to networks would remain implicit in public policy, the subsidies for the poor would be laid out as a line item in budgets ripe for political attack. As one example, when the CPUC proposed listing universal service funds as a line item on customer bills, reporters produced headlines about a "new" customer tax on toll service, creating a political focus on those subsidies even as those for business or upper-income users are ignored.[118]

What is yet more politically duplicitous is that even as the basis of local regional economies are torn asunder by the marketization of traditional regional economic anchors, these same market advocates promote the "decentralization" of government spending to local governments. This is doubly destructive to economic equity, since, beyond even the loss of political support for equity at the local level, networking technology is helping to undermine the fiscal survival of local government as its tax base is undermined by global commerce. In the next chapter I will outline the Internet's effect on regional government's tax revenue as a backdrop to the fiscal vice limiting the public information systems that governments are able to create for their citizens.

Local Government Up for Bid: Internet Taxes, Economic Development, and Public Information

Despite the financial crises of local government (exacerbated by the Internet) and the loss of economic institutions such as utilities and banks tied to regions, the quotation on the right reveals the model of local governance and activism promoted by New Democrats such as Will Marshall at the Democratic Leadership Council. It is a vision that sees cyberspace "enabling" local government power. And from Newt's Gingrich's mid-decade Contract with America to George W. Bush's "compassionate conservatism," Republicans have continued to frame their attacks on federal programs as an attempt to decentralize government in an era when the Internet is supposedly making federal bureaucracies inappropriate: "The question is no longer whether second wave governments are in remission, but rather what will the governments of cyberspace look like?" argued George A. Keyworth, chairman of the Gingrich-allied Progress and Freedom Foundation. "Massive, widespread decentralization is already, irreversibly underway."[2] The conservative and liberal

For today's progressives, there is no challenge more compelling than the need to replace a governance model developed for the Industrial Age—an era characterized by large, centralized institutions, public and private, that once guaranteed citizens a decent standard of living and personal security in exchange for their allegiance. . . . By dramatically lowering the costs of information and communication, microchip technologies break bureaucratic monopolies on expertise and diffuse knowledge—and therefore power—more broadly. The new model decentralizes decisions, expands individual choice, and injects competition into the delivery of public goods and services.
—Will Marshall, Policy Directory, Democratic Leadership Council[1]

sides of this consensus might disagree ferociously on many of the details of how and what was being decentralized (and share the hypocrisy in rhetorically ignoring what national regulatory apparatuses are being strengthened), but the consensus on the virtue of community economics and politics driving government is remarkably shared.

Yet the reality is that corporate businesses are already happily promoting the use of the technology as a way to disempower local government and make it serve the bottom line for business interests. If one wants to contemplate the rawest view of the tissue paper dispensability of local government for business in the new economy, there is no better example than in a 1996 *Forbes* article called, appropriately, "Buy the Government of Your Choice." Don't like local regulation? Looking for a lower-cost workforce with weaker labor protections? "Vote with your modem," wrote author Peter Huber, and go "shopping for a better government." Huber observed that most people have contemplated the instantaneous transfer of money across the globe as people and companies search for lower tax rates. But he sees the new opportunities for forcing governments to reshape government policy to serve global corporations as an even greater gain for companies in the new age of "cyber power." With the power of the new networked technology, Huber (approvingly) noted: "[A] manufacturing company can move jobs and capital around like pieces on a chessboard, shopping continually for the best-priced labor—and the best labor laws. Their managers will be able to move jobs almost as fast as governments can rewrite employment laws. At the margin, the managers of these transnational companies will adjust their portfolios of labor in much the same way as the manager of the Templeton Growth Fund trades stocks." Unless governments keep their taxes and "services" in-line with the global competition, Huber argues they will lose "market share to the competition."[3]

Networked technology is accelerating the competition between and within regions for tax dollars. While many local governments gained some temporary stability in the boom of the mid-1990s, after years in which cities such as New York and Los Angeles teetered on the edge of bankruptcy, the new technology of the Internet and the global economic changes accompanying it have threatened to deal a final body blow over time to the financial security of local governments. Local governments could once count on local economic development to produce local jobs whereby local employees could spend money in local stores, thereby generating local tax revenue for further development. This virtuous cycle has been fatally undermined by the new technology of cyberspace. Even as many states and local areas hope for increased revenue resulting from high-technology-based growth, it becomes harder and harder for local government to capture much of that growth in local tax revenue. There is an irony (or more specifically a strategy) to be found in the fact that conservative leaders promote the new technology as allowing greater decentralization, even as they hand responsibilities to local government increasingly unable to fund them.

This fiscal squeeze follows the pattern set by Proposition 13 and other property-tax-limitation measures that themselves responded to the earlier wave of increasing global speculation in local housing markets. The pressures of responding to the global economy has fractured the ability of local government to effectively push forward long-term economic development, and as rich communities have increasingly abandoned participation in and financial contributions to shared regional economic development, this has widened economic inequality between communities within regions. This "opt-out" by rich communities from shared investment through local government parallels the opt-out by the wealthy from the local banking, power, and phone systems that had once promoted some degree of equity within regions.

And if Huber is correct, the devastating effects of competition for tax dollars may end up being a sideshow to the much higher stakes of technological competition between local government bodies. Economic competition between regions is driving the most fundamental transformation of urban infrastructure since the rise of the interstate highway system and is driving the greatest internal restructuring of the functioning of local government since the emergence of integrated water, sewer and utility infrastructure at the turn of the century.

An invisible physical restructuring is occurring as networks of tax and government authorities interact with the new configuration of fiber-optic cables, satellites, microwave stations, and other means of data connection that reshape the virtual geography. These information crossroads of the new age are playing as key a role in economic booms and busts as railroads did in the days of the stockyards of Chicago. In some ways, the effects are even more fundamental since, unlike transit, telecommunications do not shrink space in some measured way, but instead make space irrelevant for certain economic functions of those connected directly to the high-speed telecommunications circulatory system. Through telecommunications networks, the financial markets of New York, London, and Tokyo are virtually adjacent, giving rise to a global financial market that places each of them more tightly at the center of its workings and employment. At the same time, new highly networked regions such as the San Francisco Bay Area find themselves able to "move next door" through networking investments.

However, such connections invisibly and selectively penetrate traditional physical geographies. In more traditional models of urban planning, cities were usually treated as unitary entities that could be enhanced through invest-ment by growth coalitions in the urban infrastructure, whereas the new model of telecommunications can tightly integrate subsections, even small city blocks, into virtual adjacency to global production, while other areas within a region can be left far off in the boondocks even while appearing to be in the traditional heart of urbanization.[4] This uneven investment in telecommunica-tions in different areas, both by private industry and by local governments, is helping to reinforce the polarization between regions and within regions. Cities react to the competition to serve the telecommunication needs of global corporations not by disappearing but by creating a grab bag of new technolo-gies for the elite while leaving those without with less, exacerbating inequality. The ideology of privatization has meant in practice that government in elite areas assists in the planning for upgrading the technology of economic centers for business, while leaving other areas with little support.

Even the hope that technology-assisted education, in the form of distance learning and other networked education technology, would help to diminish the gap between educational opportunities is giving way to the reality that technology investments for privileged students is just further reinforcing edu-cational stratification. Private business investments and support for schools in their suburban enclaves just further the economic disparity.

But the effects of networked technology have not been evidenced merely in the disparity of fiber laid or computers purchased but in the wholesale transformation of local government to serve the needs of business. The Pro-gressive Era, a period of government reform—the creation of a civil service and the integration of broad areas into integrated regional planning bodies—was itself a product of planning tied to new regional water systems and the drive for coordination of utility functions. The shape of local government is again being remade by the demands of the new telecommunications infra-structure.

The new technology seemed to promise local government a new opportu-nity to deliver services and expand economic development for its citizens. With local government as a vital source of information serving the whole community and with governments increasingly integrating their databases, in-formation would become a new fuel for the empowerment and integration of regional economies across the nation.

Unfortunately, these new information networks are often being used to di-vide communities against one another and empower global corporations with little permanent relationship to those communities. The selective market-oriented shape of current telecommunications, driven by both technology and federal government policy, is encouraging each jurisdiction to view its telecommunications policies, from hardware to Web sites, as one more element in the desperate attempt to attract and hold multinational corporations within city limits. This is occurring even as those same policies further undermine traditional methods of local economic planning. As in private industry, technology has emerged as one more tool to wield in downsizing local government employment.

As more and more of the operations of government are integrated into the Internet, cities are experiencing erosion of the barriers of information and of the need for day-to-day local relationships that had once supported a regional civic and economic culture oriented to local growth. Any attempt by local government to build up local businesses may be undercut by global businesses scooping up government bids off the Internet and underbidding local contractors based on volume sales. The Internet may be the tool that those global companies need to reach those governments and replace traditional firms that once had a "local advantage" in getting information at the town hall.

The information at town hall itself is becoming more a part of international commerce than a local tool for economic development as the lines between public information and private information brokers blur. It is an irony of the "information age" that most information-gathering divisions of government, both national and local, are suffering budget cuts—a factor that is pushing many local governments into privatizing information-gathering functions and into a drive toward profit-making enterprises to capitalize on that information. A further irony is that even as the integrated economic geography of regions fades under the onslaught of global commerce, the selling of information about the physical geography of regions, contained in so-called geographic information systems (GIS), is becoming one of the most profitable areas of government information. As selling government information for profit becomes a model for more local government, local government workers are losing out to information brokers who are often located far from their home cities. In all these trends, local governments increasingly find their fates more and more controlled by companies located far from their region. With local services provided by far-off companies and government information increasingly controlled by private companies, the public finds itself losing political control of many economic development decisions it had previously taken for granted.

The fact is that for government as for business, the local region is increas-

ingly not an economic unit unto itself but becomes merely a realm of negotiation between global actors. However, in the case of local government where
local officials negotiate with global business, the power of business to "shop"
for alternatives leaves local governments in desperate competition with one
another in a race to lower both taxes and standards—effectively leaving most
local community actors out of the power equation. Those who have more
power globally have a stronger hand for negotiating locally within a region,
while those unable to muster such power are out of luck. The end result is
more inequality within regions.

Prop 13 Meets the Internet: How State and Local Government Finances Are Becoming Road Kill on the Information Superhighway

The key to the fiscal crisis facing local governments is the Internet-driven
expansion of interstate retail sales of goods, ranging from computers to Christ-
mas sweaters, sales that remain untaxed because of a Supreme Court ruling
barring such state taxes on most interstate sales. That fact is good news for the
individual consumer (and often a critical sales advantage of mail-order and
Internet sales outfits) but is a growing catastrophe for the state and local governments that are dependent on sales-tax revenue.

Even before e-commerce took off, states by 1994 were already losing at least
$3.3 billion in revenue each year because of retail sales that had migrated to
mail-order businesses, as estimated by the U.S. Advisory Commission on Intergovernmental Relations (an agency that brought together representatives of
state governments to improve the effect of federal policy on the states).[5] With
the growth of the Internet and online sales, consumer access to a nationwide
and worldwide marketplace expanded exponentially. At the push of a button,
consumers increasingly have access to the lowest-priced goods nationwide,
and with the bonus of avoiding sales taxes, interstate sales have exploded over
the Internet, promising to leave state and local government in tatters.

California, at the heart of the new Internet technology, is likely to feel some
of the most severe effects of this change. Because of Proposition 13's limits on
property tax revenue, state and local governments in California are extremely
dependent on sales taxes to fund their budgets, so any increase in untaxed
interstate sales at the expense of local retail will be magnified there. In 1995
Wally Dean, mayor of Cupertino (the birthplace of Apple Computer),
summed up the shock his government colleagues would soon be feeling as
Internet sales took off in the next few years, undermining their traditional tax

and economic development goals: "The thing that scares us is that cities are run on local sales tax; if stuff is sold on the Internet, there's no sales tax. It's a house of cards for government finances. This could be the Achilles heel for state and local government. . . . We once built city government on local manufacturers and sales—you didn't think globally. This will mess with a lot of people's heads."[6]

How Real Is the Danger of the Internet to Local Taxes?

Retail sales online to consumers grew from an estimated $200 million in direct Internet sales in 1994 to $2.6 billion in 1997 to $45 billion in 2000. Business-to-business sales on the Net was estimated to be four times as much (although a much larger percentage of that commerce gets taxed by local authorities).[7] By 1996, 80 percent of businesses had a Web site. Pioneers ranging from Dell to Amazon.com highlighted the potential for industries in moving sales online. Amazon.com's ability to list 2.5 million in-print books (which it in turn orders from book publishers and warehouses on demand from retail customers) highlighted the advantages of online stores that can virtually bring together all the products a consumer might desire. Offering search engines, online reviews, and discounts, Amazon.com became a symbol of the promise of online commerce and the threat to local retail merchants.[8]

Even in cases in which the final sale may ultimately happen over the telephone rather than directly over the Net, an expanded online presence made it easier for many companies to expand and build trust in traditional mail-order operations. Mark Masotto of CommerceNet observed, "Clearly, you'll see more and more stories emerging of how putting information on the Internet is reducing the number of phone calls and number of brochures distributed. . . . The medium provides much more possibility to do interactive support—you can read and search information, immediately pull up the information you are interested in rather than looking through a whole catalog of information. It makes the person reading the information more effective in finding information."[9] Driven by an earlier generation of telecommunications and computer technology advances, the mail-order industry had been growing phenomenally for decades. Mail-order sales grew from only $2.4 billion in 1967 to more than $237 billion by 1993, extraordinary growth even when accounting for inflation.[10] With the Internet added, interstate commerce was guaranteed to boom.

At the same time, sales taxes have emerged as a increasingly important revenue source for state governments and often an even larger source of revenue for local governments. Beginning in the early 1980s, the federal government began to cut funding to the states, forcing state and local governments to pay for more and more services out of their own budgets. Sales taxes often became the revenue of choice. De facto, state governments substituted local sales taxes for federal income taxes, which had been cut in the early Reagan years. Fully forty-four states (and the District of Columbia) now impose taxes on retail sales, revenue that accounts for 25 percent of states' annual income. With a growing resistance to increases in income taxes and with tax limitation laws such as Proposition 13 making it harder to raise property taxes, sales taxes have become the most attractive way to raise local revenues.[11] By 1999, state and local governments were raising $203 billion from sales taxes.[12] For states, one-third of their total tax revenue was coming from sales taxes, which had been just 20 percent of revenue back in 1950.[13]

These two trends—more out-of-state sales and a greater dependence by local governments on sales taxes—are now on a collision course. By 2000, the U.S. General Accounting Office (GAO) estimated that states were losing as much as $9.1 billion from untaxed interstate sales, $3.8 billion from Internet sales alone. And by 2003, the GAO was estimating that they would be losing $20.4 billion, with $12.4 billion lost to e-commerce. This will amount to an estimated 8 percent loss in sales tax revenue.[14] A report released by the National Governors Association in 1997 cited the growing loss of sales tax revenue resulting from the new technology—as well as federal cuts in Medicaid—as one of the top budgetary threats to state government finances.[15]

For many local governments that suffered budget cutbacks throughout the 1980s and early 1990s, the effects could be even more devastating. While many cities in Silicon Valley became more flush with funds from the Internet-driven economic boom, this new stability hardly made up for the cuts suffered during the bad times. After slashing budgets by $293 million a year in the early 1990s, Silicon Valley's Santa Clara County finally balanced its budget in 1997 with an $8-million surplus[16]—hardly making a dint in restoring funds previously cut despite the economic boom. And the irony is that Santa Clara County is one of the California counties most vulnerable to lost sales tax revenue.

At the county level, Santa Clara County actually outpaces the larger Los Angeles, San Diego, and Orange Counties not only in the percentage of tax revenue coming from sales taxes but also in the total revenue from sales taxes, despite the larger population sizes of those other counties. Cities are even more vulnerable than counties, with many smaller cities receiving almost all tax revenue from the sales tax. It is not surprising that the mayor of Cupertino was ahead of the curve in worrying about this threat to his city's finances:

Cupertino depends on sales taxes for 81 percent of all taxes collected in the city, making it one of the most sales-tax-dependent cities in California. Even including nontax revenue sources such as state aid, fines, and service charges for utilities, Cupertino still depends on the sales tax for 45 percent of city revenues. However, in absolute terms it is clear that the large urban cities, among them Los Angeles, San Francisco, San Diego, and San Jose have billions in revenue at threat from the new technology.[17]

Why States Can't Collect Mail-Order Taxes: The Quill Decision

The obvious response to the loss of mail-order and Internet-based sales taxes would be for states to directly tax such sales. However, the Supreme Court in its 1967 *National Bellas Hess Inc. v. Department of Revenue* decision prohibited states from taxing out-of-state companies selling to state residents, basing its decision on the Commerce Clause of the U.S. Constitution. At the heart of that clause is to take regulation of commerce, including taxation, away from the control of local government in cases where the scale of that commerce has grown beyond the confines of one state. In the case of mail order, the view of the Court was that businesses operating in one state could not be taxed by another state merely because residents of that other state were buying the company's products through the federal mail system. With the explosion of mail order commerce and the ubiquity of catalogs, direct marketing and other changes in technology to reach customers, there had been some hope in states that the Supreme Court might alter what was considered "in-state" commerce, but in its May 1992 *Quill Corp. V. North Dakota* decision, the Supreme Court reaffirmed that mail-order firms were exempt from state sales taxes. By creating an extremely tough standard in defining "in-state" sales, technically called "nexus" in the law, the Supreme Court made it clear that Internet-based sales would be treated as out-of-state, tax-free transactions.

When the state of North Dakota attempted to impose state sales taxes on Quill, the state argued to the courts that the nature of direct marketing had created a "ubiquitous presence" of the company in the state through mail, telephone, and electronic solicitations, far beyond what the Supreme Court had envisioned when it banned interstate sales taxes in its 1967 decision. If states were to survive as fiscal units, the state basically argued, the courts had to recognize that the new technology made companies a part of the local economy just as much as if they had salespeople in the downtown mall. But in the *Quill* decision, the Supreme Court held to its "bright-line" rule that physical presence by company personnel in a state was required to trigger sales taxes.

The logic was that without such personnel present, the company was receiving no benefits from state services, so it need not pay taxes. Thus Quill would pay taxes in Delaware, Illinois, California, and Georgia, where it had employees, but in no other states.[18]

So, rather than the new technologies of direct marketing making companies more subject to sales taxes as they collapse geographic distances for their customers, the use of toll-free numbers, computer databases, and the Internet itself would allow direct-marketing companies to further dispense with the need for placing sales personnel, inventory, or showrooms within most states. Such technologies would actually help these companies avoid creating the physical presence that would trigger the nexus that would force them to collect sales taxes. As Internet Web pages located on servers in far distant states increasingly replace catalogs mailed to people's homes, it is clear that the physical connection between retailers and states trying to tax them will increasingly recede even farther.

Why Saving the Sales Tax Requires More Intrusive Government Regulation

Paradoxically, in the movement toward "local control" and "government decentralization" the increased dependence on local taxes and revenue in an increasingly global retail market is pushing governments toward policies of more burdensome regulation of business and more intrusive government for the individual in order to collect those out-of-state sales taxes. As the boundaries of local regions becoming increasingly artificial boundaries for government jurisdiction, even more jerry-rigged regulations are attempted as a way to salvage regional financial health.

In the *Quill* decision, the Supreme Court did leave open the option that, while the states could not unilaterally impose sales taxes on interstate commerce, Congress itself could establish such a tax and remit the proceeds to the respective states. Before Internet commerce took off, then-senator Dale Bumpers, Democrat from Arkansas, had authored the Tax Fairness for Main Street Business Act of 1994 to established that authority for states, but the bill failed in the face of opposition from the Direct Marketing Association and allied business and consumer groups, including the American Council of the Blind, Disabled American Veterans, and the National Alliance of Senior Citizens.[19] As late as 1997, local governments and representatives of the Direct Marketing Association were close to an agreement for firms to voluntarily collect sales taxes for states in exchange for limits on state audits and the right of those businesses to expand their presence within target states without invoking nexus for tax purposes. However, when the imminent deal was reported about in the *New York Times,* the affluent customers of direct retailers such as L. L. Bean generated such a volume of complaining phone calls to the retailers that they backed out of the deal.[20]

With the increasing hype over the Internet, movement in Congress began veering in the opposite direction, toward even further restrictions on state power over taxing such commerce. With the priority of encouraging growth of the Internet industry seen as more important than the fiscal needs of regions, Internet companies began promoting the Internet Tax Freedom Bill, sponsored by Congressman Chris Cox, Republican, of California, and Senator Ron Wyden, Democrat, of Oregon, to exempt online transactions from local taxes. As Cox aide Peter Uhlmann argued, the priority of Congress is to "help ensure that state, local and foreign tax policies don't interfere with the potential for economic growth over the Internet."[21] As with bank, utility, and telecommunications "deregulation," local power over economic development had to be reduced to serve the ambitions of industries turning an eye to global markets.

In the fight over the Internet Tax Freedom Bill, local governments represented by the National League of Cities and National Governors' Association fiercely criticized the federal government for preempting their tax powers and undermining local businesses that would see only an acceleration of their tax disadvantages, compared with mail-order businesses. Brian O'Neill, head of the National League of Cities, condemned Congress harshly: "This is unfair to Main Street business people. This is as un-American as it gets."[22]

While local governments worried that the bill would institutionalize tax losses from Internet sales, they were outraged that the ambiguity of the language banning taxes on Internet transactions would likely lead to the repeal of existing taxes on a range of local telecommunications. Most versions of the bill threatened to repeal, in twelve states, taxes collected on local Internet service providers. But the real worry was that the bill, by banning "indirect" taxes on the Internet, might be used by courts to strike down local taxes on telephone service, especially as more and more phone calls were projected to use Internet protocols in coming years. This would cost local governments billions of dollars and give further advantage to Internet-based telecommunications at the expense of local phone companies serving non-Internet users.[23] In the end, a more limited three-year moratorium on new Net taxes was finally passed in 1988 that just postponed the longer-term legislative debate on the future of local government fiscal survival.

Still, aside from the "Astroturf" pleas of shut-ins and the disabled, there is a moderately compelling argument against forcing out-of-state direct market-ers to submit to the administrative burden of being subject to the ever chang-ing tax laws of thousands of separate government jurisdictions. With forty-six states, the District of Columbia, and more than six thousand counties, cities, and school districts each collecting its own sales taxes, the complexity of track-ing tax rates in each area and dealing with local government authorities could easily overwhelm many businesses.[24] Some argue that the computers allowing the direct-mail boom could be used to ease the burden of calculating the tax costs, but the difficulties of dealing with so many separate government author-ities could not be completely eliminated.

Arnold Miller, treasurer of Quill, argued in the company's legal brief against the "untold hardship" of paying deposits, filing quarterly returns, and dealing with audits in so many jurisdictions. Miller had once worked at Sears Roebuck and Company, which through its stores had nexus in all states, and noted that Sears underwent at least five audits at any one time. Sears could endure the burden because it paid twenty-five professionals to deal solely with tax issues, a luxury that smaller direct marketing firms could not afford.[25] And while local tax issues were probably not the only reason, Sears discontinued its mail-order-catalog business in 1993 in favor of licensing its database of customers and addresses to specialty catalogs such as Hanover Direct of Weehawken, New Jersey, which escapes nexus in other states and thereby avoids sales tax burdens.[26] When Spiegel, the largest catalog direct marketer in the United States, acquired retailers Honeybee and Eddie Bauer, it suddenly was hit with nexus in thirty-four states. "You really do need a lot of computer power," noted Spiegel investor relations officer Debby Koopman. "For example, some states like Massachusetts and Connecticut exclude clothing mail-order sales up to a certain amount, say $75. Other states have one rate for shoes that are classed as clothing and another for shoes that are classed as athletic equip-ment."[27] The exact costs of forcing companies to collect sales taxes in all juris-dictions is unclear, but estimates place the amount as high as a 10 to 20 percent increase in operating costs to comply.[28] All this is aside from any extra costs of filing statements with all the different government jurisdictions.[29] States are involved in a process of "streamlining" systems to create a more uniform set of sales tax rules, but the process is difficult given the very different political priorities on which products should even be taxed.

The other alternative to having retail companies collect taxes is to have states directly tax consumers with a "use tax" in place of a sales tax. States can already legally do this, and they can step up their efforts to collect use taxes directly from end-consumers. Companies with resale permits in any state are already required in their routine sales tax audits to prove that they pay tax on everything purchased for their own use. And for individuals, some states are already using computerized records from U.S. Customs to bill residents for purchases made abroad that are subject to use taxes. The Software Industry Coalition, one of the main Silicon Valley voices in the sales tax debate, advocated that all states add a line to their state income tax forms specifically for sales tax on goods purchased out of state, thereby transferring the burden of sales tax collection (and possible audits by the government) from business to the consumer.[30]

Some have suggested that to collect such taxes from individuals, states could begin gathering information on sales directly from private sources such as individual credit card and checking account records. No state has dared to do this, but legislators may move in that direction if their sales tax revenues continue to fall. Some states, including California, prohibit such actions through strong privacy guarantees in their state constitutions. But in other places, we have the specter of a consumption-based equivalent of the Internal Revenue Service appearing to audit individual purchases.[31] This intrusion of government into peoples' private lives would be an ironic result of the decentralization promoted by conservatives in the name of "getting government off peoples' backs."

Sales Taxes and the Effects on the Poor

The other major problem with the increasing use of the sales tax as a revenue source is its disproportionate burden on the poor and working families. Beyond lobbying on behalf of their own economic self-interest, direct marketers trumpet the burdens on the elderly, the disabled, and poorer rural residents of taxing mail-order sales. While there is a certain cynicism in this "concern" by the Direct Marketing Association as it has trotted out allies from the disabled and elderly communities before the U.S. Congress, there is also a strong truth to their argument that taxing consumer sales affects the poor more than anyone else.

Study after study has shown the regressive nature of sales taxes as a revenue source. The most comprehensive study was by Citizens for Tax Justice, published in their report *A Far Cry from Fair*. In that survey of all taxes collected by local governments, the authors argued that "excessive reliance on sales and excise taxes is certainly the hallmark of regressive taxation." The report found that across the country, in 1991, the poorest 20 percent of families were paying 5.7 percent of their income in state sales taxes, while the richest 1 percent paid only 1.2 percent of their income in sales taxes—in other words, the poor paid *nearly five times the tax rate paid by the rich*. This contrasts sharply with the much more progressive state income tax. Nationwide, the average state personal income tax for a family of four is only 0.7 percent for the poorest 20 percent of residents and 4.6 percent of the income of the richest 1 percent.[32]

Because of the regressive nature of sales taxes, in those states that depend on them, such as Washington and Texas, the poor have the highest tax rates in the country. Adding in property taxes—which burden the poor as part of their rent—total state and local taxes in Washington State were 17.4 percent of the income of the poorest 20 percent. Contrast that with neighboring Oregon, which has a state income tax, where the poorest 20 percent paid only 9.8 percent of their income in state and local taxes. The results are clear: dependence on sales taxes leads to the heaviest taxation burdens falling on those least able to pay.

Many analysts worry that Internet sales are making this tax inequality worse, since upper-income taxpayers with computers have increasing access to a world of tax-free goods ordered over the Internet, while those with fewest resources are stuck with buying locally and paying sales taxes on their purchases. The Center on Budget and Policy Priorities has argued that untaxed Internet sales "create a vicious cycle leading to an ever more regressive sales tax. The erosion of the sales tax base resulting from on-line purchasing by businesses and affluent consumers would force states and localities to raise sales tax rates, encouraging more on-line buying, forcing further rounds of rate increases, until the lowest-income population groups unable to buy online would be left paying an ever-greater share of sales taxes."[33]

Technology, Suburbanization, and Prop 13

Although the economic losses and regressive tax burden resulting from dependence on sales taxes is a prime concern for regional economic planners, the deeper problem is the fracturing of the tax base as cities find themselves having to more desperately compete for retail outlet revenue rather than cooperating in expanding general growth. As cities become polarized because of this competition, it further increases inequality within regions, and given the regressive nature of sales taxes, increases inequality overall.

Before we turn to how this regional competition for sales tax is undermining economic development, it is important to understand the context of this problem in a longer history of regional polarization around tax policy and development. As a result of California's growth over the past decades, the polarization there was most intense, culminating in the passage of Proposition 13 in 1978 following an earlier round of technologically induced economic changes that skewed, then exploded, regional fiscal stability and planning.

In the postwar period, property taxes, not sales taxes, had been the key tax source for local governments. The economic expansion of the 1950s and 1960s not only created economic growth by increasing the number of homeowners, but it created the funding base for continued local expansion of services through this new class of property taxpayers. Homeowners, construction workers, community builders, and regional development as a whole supported one another in a virtuous cycle of expansion.[34] However, the economic and technological changes of the late 1960s and 1970s would undermine that virtuous cycle and fracture the political unity that had supported broadly distributed growth.

The same computer and communication technology that was allowing the new middle classes to take their money out of local banks and invest in global markets was also creating the global investment markets that prowled the country for local property investments as a hedge against the inflation of the 1970s. Investors in the United States and around the world were playing increasingly speculative games in the housing market, especially in the booming growth cities of California. Housing prices began to escalate wildly, setting the stage for the coming tax revolt of Prop 13 and its counterparts across the country. Whereas housing inflation in the 1950s and 1960s had been two-thirds of general inflation, in the 1970s that relationship reversed. In some areas of California, housing prices which had been increasing 2–3 percent every year in the mid-1960s, were increasing 2–3 percent *every month* by 1976.[35]

It was not only financial speculation that drove these housing prices upward but also a new dynamic of slow-growth politics and "suburban separatism" that began to dry up available development areas, increasing the premium on housing prices in those areas that were developed. Especially in the upper-middle-class communities of the new high-tech millionaires, slow growth ordinances began springing up and were copied on down the economic scale. By 1975, most cities and counties in California had some form of growth control policies, thereby vastly increasingly the value of uncontrolled land. Speculation exploded and in extreme cases, such as that of Orange County, almost half of all single-family homes built were bought by speculators.[36]

Mike Davis, in his book *City of Quartz*, contrasted the "Keynesian suburbanization" of the 1950s and early 1960s, when local finance supported local growth, with the "new Octopus" of giant developers pulling in financial backing from global markets. With the price of land rising so dramatically, many

of the old railroad companies and industrial concerns found their landhold-
ings to be their most valuable resource. Land development often became their
new economic focus. These new developers came into increasing political confrontation with the new upper-income suburbanites who were developing
their own strategies to maintain their incomes while severing their ties to general growth politics of the region. The goal of these new suburbanites was to
slow development in their own communities in order to preserve their quality
of life and escape the economic burden of providing services to new residents,
particularly poor residents of the region. This clash between developer elites
and suburbanite elites would culminate in the battle over Prop 13; in its aftermath, both elites would sever almost all remaining alliances with the working-
class and urban forces that had once fueled general growth politics.[37]

Beginning in the 1950s, wealth and racial divisions had fueled the creation
of an escalating number of municipal incorporations divided from urban core
areas. Previously, homeowner covenants and organizations had enforced racial
segregation while keeping most citizens within the same fiscal and political
jurisdiction. When the Supreme Court declared such covenants illegal in the
1940s, the old homeowner associations began to mobilize to find new strategies
for racial separation, strategies that would soon be joined with the goal of
fiscal separation from the poor as well. In the past, separate incorporation of
a municipality had been a possibility only for the wealthiest enclaves such as
Beverly Hills, but the passage in California of the 1956 Bradley-Burns Act radically changed the fiscal calculus of incorporation. Bradley-Burns allowed any
local government to collect a 1 percent sales tax exclusively for its own use, a
critical tool for suburban separatism whereby fringe areas with a shopping
center could now finance city government without needing much of a property tax. This was combined with new arrangements by local governments to
have counties contract (usually at cut rates) to perform basic services, leaving
the new towns with control of zoning without the fiscal hassles of managing
most services. As Davis argues, "Sacramento [the capital of California] licensed suburban governments to pay for their contracted county services with
regressive sales revenues rather than progressive property taxes—a direct subsidy to suburban separatism at the expense of the weakened tax bases of pri-
mate cities."[38] The first step toward local dependence on sales taxes had been
taken.

Upper-income homeowners began exiting cities in order to avoid paying
the standard taxes to support urban infrastructure. Davis notes the distinct
"gradient" of home values between each incorporation with lower-middle-
class, middle-class, upper-middle-class, and wealthy communities neatly di-

vided by the new jurisdictional lines of incorporation and zoning. With poor people and their need for services zoned out of these new towns, this fiscal zoning would help suck jobs out of the inner city to these minimal-service, low-tax areas. Racial and income divides would expand between these jurisdictions. In addition, federal and state spending on highways and other traditional urban spending would facilitate this fiscal succession by providing the critical infrastructure that once required regional growth alliances and planning. And by creating divisions between municipalities, capital investors interested in development could now more easily demand economic concessions from weaker fiscal units desperate for new revenue sources following the departure of the upper-income municipal residents.[39]

What is striking is that just as massive regulation was necessary for that same upper-income elite to secede from common banking and utilities systems in regions, it took strong government regulation to assist them in preserving their segregated residential enclaves. From providing them their own sales taxes apart from shared revenue streams to aiding them in delivery of basic services separate from regional systems, the state and federal governments nurtured these enclaves. And these upper-income homeowners, normally advocates of free markets in other aspects of the economy, would promote what conservative commentator George Will labeled "Sunbelt Bolshevism"[40] in their extensive system of land regulation, growth controls, and other zoning to undermine housing markets that might otherwise have brought "undesirables" into their municipal districts.

As the same time, a combination of government action (and refusal to take action) in support of the developer interests put the suburban separatists on a collision course with growth economics, leading to the further splintering of economic development resulting from Proposition 13. Even as money market funds and other new financial tools were leveraging personal savings out of local finance markets into speculative global markets, thereby naturally fueling intensified investments in real estate, the government actually began expanding subsidies for real estate, adding fuel to an already growing speculative fire. Through an alphabet soup of institutions—FNMA, GNMA, FHLMC, REITs— in combination with a range of tax advantages, the government was encouraging new flows of capital to bid up the price of housing throughout the 1970s.[41] The Federal Home Loan Bank Board was well aware that housing was being increasingly priced out of the reach of average homeowners, but it refused to do anything other than issue toothless warnings to the savings and loan institutions it governed not to lend to speculators who did not plan to reside in property they were buying. Such tighter regulation or a windfall profits tax on speculation could have gone a long way toward cooling the speculation that was turning housing from a prop of regional growth economics into a plaything for global investing.[42]

The result of the clash between speculation and suburban slow-growth controls was that in the four years before passage of Prop 13 in 1978, property taxes on California homeowners doubled. By 1978, the typical homeowner was paying four times as much for property taxes as for mortgage payments.[43] Compounding the indignity for property tax payers, the state government was running a budget surplus of $3 billion, which Governor Jerry Brown was neither spending on social programs nor returning as a tax cut, but instead sitting on as proof of his fiscal responsibility.

The Prop 13 results were not foreordained; the earliest roots of the tax revolt were among lower middle-class property owners feeling the economic squeeze; they were open to alliances that could have been more economically populist. But in both California and Massachusetts (where a similar measure, Prop 2½, was passed soon after Prop 13), initial attempts by progressive tax-reform activists to ally with those squeezed by these new global forces of speculation were abandoned in favor of alliances with developers and big business in what became the last hurrah in California of the old broad-based growth coalition. In the fight against Proposition 13, social-spending liberals and unions were joined by the broad economic elite of the state, from Bank of America to the California Taxpayers Association, the latter the main lobby for large corporations. The California Republican Party even refused to endorse Proposition 13. The alliance by progressives with the increasingly globally minded banks and developers, however, meant that no alternative solution to the tax pressures on lower-income homeowners was forcefully pushed.

This in turn left the way open for upper-middle-class homeowners in rich communities such as Sherman Oaks to give a more conservative bent to the tax revolt. Clarence Lo, in his classic study of the class dynamics of the Proposition 13 battle, describes how "upper-middle class homeowners drove down from the scenic hills of the Palos Verdes peninsula . . . back to the unwashed Toyota Tercels gridlocking Ventura Boulevard [where] they mingled with the K-Mart shoppers of Van Nuys. . . . Joining the less affluent in mass meetings, the homeowners of Rolling Hills Estates and Sherman Oaks eventually took the lead in organizing and shaping the entire tax limitation movement."[44] The new tax revolt was linked to anti-school-busing movements and other political campaigns that race-baited welfare programs. With the help of right-wing politicians such as Howard Jarvis, this alliance of suburban separatists would surge to an overwhelming margin of victory, 65 percent to 35 percent. Despite the racial overtones of the tax revolt movements, the reality of a broad-based problem with property taxes was shown by a victory in which even 42 percent of African Americans voted for the measure.

However, Proposition 13 would have devastating effects on local governments' financial stability, especially those in poorer inner-city areas, and would lead to the final fracture of any growth alliances between city and suburb and, as important, between the global economic investors and their traditional urban-union partners in regional growth alliances. Large corporations had opposed Prop 13, partly fearing that it would be followed by a round of increases in corporate and bank taxes to make up for the shortfall. When that populist reaction failed to appear, they began to enjoy their economic windfall from the tax revolt, and much of the corporate elite shifted political allegiances to the emerging Reaganite tax revolt nationally. Of the $5.5 billion in taxes cut by Proposition 13, $3.5 billion went to businesses and landlords, a model of corporate enrichment that would be replicated nationally. While particular battles over development would be fought between the suburban separatists and the corporate developers, these groups soon made peace over a shared enthusiasm about the mutual benefits they received from the tax revolt. (At the same time, the results of cumulative tax changes, including increases in Social Security taxes, meant that between 1977 and 1990, the poorest 90 percent of taxpayers ended up paying more, not less, in taxes than they had before the "tax revolt.")[45]

Sales Taxes and the Distortions of Economic Development

Lenny Goldberg, the head of the progressive California Tax Reform Association in the 1990s, has written: "The most noted irony of Proposition 13 is the extent to which it decimated the fiscal powers of local government and transferred power decisively to Sacramento—an irony because the major source of tax problem in 1978 was Sacramento, not local government."[46] After worries about local control, the post–Prop 13 result was, to take one example, a change from state government supplying less than 25 percent of school funding before the tax initiative to the state supplying more than two-thirds of school funding by the 1990s. Local government lost almost all fiscal power to leverage new growth, and the divisions between the fractured municipal jurisdictions made regional economic planning a near impossibility.

The shift from property taxes to sales taxes as the main source of local tax revenue created further distortions and perversities in how economic development affected communities. The inflexibility of Proposition 13 funding formulas—all property assessed at its 1975 value or whenever it last changed hands, adjusted for inflation by no more than 2 percent each year—meant that governments could not capture most of the results of growth as reflected in increasing property values. Since new housing developments often would not pay for themselves, especially over the long term as the inflation-adjusted value of taxes paid would fall, local governments began increasing up-front fees on construction—$3 billion a year in California in fees with an average of ten thousand dollars per unit. Essentially, while the old homeowners who pushed for Prop 13 would reap a massive capital-gains windfall, new home buyers (including any inner-city residents seeking to move to the suburbs) would have to "prepay" a large share of development costs. Since growth could not generate the tax revenue needed to sustain many of the social services and other amenities that once accompanied such growth, from schools to parks to museums, Proposition 13 further justified slow-growth policies. And since commercial property is covered by Prop 13, the measure breeds inefficient uses of property by businesses that survive only because they are paying so much less in property taxes than new businesses that have to pay dramatically higher taxes.[47]

With Californians paying less in property taxes (in real dollars) than they did back in 1977, and 75 percent less in property taxes than if Prop 13 had never been passed, local governments have had to increasingly depend on sales taxes to pay for social services of all kinds. This has led to a desperate competition between cities for the location of retail outlets, a competition that itself not only prevents strong regional cooperation but itself undermines revenue as cities financially subsidize such outlets. Even as Silicon Valley boomed, cities such as San Jose still found themselves subsidizing retail expansion as the simplest way to capture the fruits of that growth. The *San Jose Mercury News* highlighted the example of San Jose's offer to the electronics superstore Fry's Electronics of a no-interest loan amounting to a $1-million subsidy. The paper bemoaned the fact that "reliance on sales tax leads some cities to favor building superstores over industries that offer good-paying jobs. It discourages cities from adding housing, since more residents mean more city costs but not necessarily more revenue."[48]

Union activist and writer Greg LeRoy described in his book *No More Candy Store* how local and state government subsidies create a wild competition for the location of retail establishments with little evidence that such subsidies create any new jobs overall; they merely move them from one location to another. The tax revolt that started in 1978 only accelerated that trend of subsidies. LeRoy notes that in 1977, only nine states gave tax credits for R&D; by

1993, thirty-four states did. In 1977, only eight states allowed cities and counties to lend for construction, and now forty-five do; only twenty states gave low-interest, tax-exempt revenue-bond loans, now 44 do; only twenty-one states gave corporate income tax exemptions, now thirty-six do.[49] In the Federal Reserve of Minneapolis economic newsletter, the *Region*, in March 1995, Melvin Burstein and Arthur Rolnick (general counsel and director of research respectively for the bank) argued that in regard to the competition between states over economic subsidies,

> though it is rational for individual states to compete for specific businesses, the overall economy is worse off for their efforts. Economists have found that if states are prohibited from competing for specific businesses there will be more public and private goods for all citizens to consume. . . . In general, it can be shown that the optimal tax (the tax that distorts the least) is one that is uniformly applied to all businesses. Allowing states to have a discriminatory tax policy, one that is based on location preferences or degree of mobility, therefore, will result in the overall economy yielding fewer private and public goods.[50]

While six states have begun prohibiting cities from using tax subsidies purely to lure retail across municipal borders and some try to block subsidies to "footloose" companies, only one city, Gary, Indiana, has an ordinance that specifically denies tax abatements to projects that will relocate jobs from other cities. Unfortunately, the federal government has contributed to such wasteful relocation subsidies, since its biggest job programs, such as Industrial Revenue Bonds, the Department of Housing and Urban Development's Community Development Block Grants, and most Commerce Department programs, have no rules against using such funds to encourage relocation. Only two small-job subsidy programs have such rules, but states and cities can elude the rules by shuffling money from other federal sources to fund questionable projects.[51]

The competition for retail has created a ludicrous distortion of economic development patterns, as cities have had to desperately bid for successive waves of retail evolution. First, shopping in urban centers gave way to downtown retail in the suburbs. Then, downtowns began to weaken in the face of movement to concentrated suburban malls. Now, general-purpose department stores in malls are giving way to discount "big box" retailers such as Home Depot and Toys'R'Us. There was once some expectation that a residential population would generate proportionate retail revenues. Now, competing cities work to attract discount giants that suck in business from a whole region, oftentimes devastating the more dispersed retail that local governments depend on for financing their budgets. An extreme example is the small city of Emeryville which has attracted a large number of discount retailers. Emeryville now has more than five times the retail sales per resident as surrounding cities such as Oakland whose own retail has suffered from the competition.

Notably, the rules that leave out-of-state sales untaxed creates exactly the distorting economic policy that the Federal Reserve researchers denounced. Direct marketing through phone, cable, or the Internet takes this economic cannibalism to a new level. Cities and states are fighting to attract "call centers" to service direct-marketing companies, since such jobs are seen as nontoxic and "high tech." To cite one example, Oklahoma has done well in replacing lost oil patch jobs with telecom-based jobs, but the price has been massive subsidies to encourage companies to locate in the state. Oklahoma offers tax incentives, including through a law exempting business from sales tax on 800-numbers and WATS and private-line systems. There is one-stop environmental permitting, tax exemptions on distribution facilities, and major support for training and retraining workers. Data-processing firms get a five-year property tax exemption.[52] In pursuit of jobs, other states and local areas have created similar subsidies. In the end, they merely subsidize the flight of local retail to tax-exempt mail order.

Even though all local governments as a whole lose out in this competition, the hope for the individual areas is that jobs from call centers will be long lasting, and the gain in long-term jobs will offset the cost in local subsidies. But even that hope may wither in the face of new technologies. Bruce Lowenthal, Tandem's program manager for electronic commerce over the Internet, predicted that the Internet would eliminate the need for much of the work done by such call centers. The Internet would be an "interface" for finding out what customers need and letting them directly indicate what they want. Presently, "[s]uch 'interfaces' are done by data entry clerks," argues Lowenthal. "So many call centers may be replaced. You'll still need some people to deal with hysterical customers, but that's about it."[53] A large part of the industry of entry-level data clerks at call centers may melt away, leaving only a much smaller set of more specialized troubleshooters. And much of that may even move overseas, as companies increasingly use the new telecommunication technology to route calls to English-speaking workers in developed nations.[54]

Wiring Government and Communities

Still, the rush to develop the telecommunications infrastructure for call centers is merely one expression of government hopes worldwide to leverage technology for competitive advantage, both between regions and within them. In Asia, for example, Japan has had one of the most aggressive plans for expanding its technology infrastructure.[55] Korea, with an even tighter government planning system, coordinated by the Ministry of Information and Communication, has plans to directly spend $58 billion by 2015 to link all eighty major Korean cities and incorporate thirty thousand public organizations through fiber-optic cable. Envisioned in the plan is a high-speed broadband information highway across the country, integrated into the global information network.[56] At various levels of telecommunications privatization, European countries are more slowly moving forward to achieve the full digital upgrade of their infrastructures.

When Bill Clinton ran for president in 1992, he promoted federal investments in a national Information Superhighway as a crucial element of both his technology and his economic plan. However, with the defeat of Clinton's proposed infrastructure-jobs bill in 1993 and the chilling of attempts at broader initiatives after the Republican takeover of Congress in 1994, the federal government rapidly began backpedaling from most commitments to support local government efforts to upgrade technology infrastructure.[57]

Into this breach in the mid-1990s, very tentatively, stepped state and local governments and a mixed stew of business initiatives to collaborate (and in some case interfere) with such local government efforts. While the initial focus was on networking government offices, libraries, and schools, the efforts have also raised the issue of how to extend access to the general population for the "last mile" of high-speed connections to the government information being created in the public sphere.

One irony of the breakup of AT&T was that it actually increased state governments' involvement in running their own telecommunications systems, with many states like California creating quite elaborate phone systems connecting their state institutions. By 1996, California was spending nearly $2 billion annually on information technology and telecommunication systems, including seventeen separate telecommunications networks that grew in the wake of the AT&T divestiture in different departments and statewide agencies. Unfortunately, with different agencies working with different contractors, the state ended up with seventeen separate, often redundant telecommunication networks. These were pulled together somewhat through the creation of an integrated CALNET service, which by the mid-1990s was providing voice phone services to 85 percent of state agencies, as well as a number of cities and counties in the state.[58] With a new Department of Information Technology and an agency head dedicated to privatization, bids to privatize the system were being made by 1997 as telecommunications companies circled the statehouse looking to land the estimated $500-million five-year contract for connecting agencies.[59] Other privatization proposals were in the works as well by 1997, including a (failed) attempt to transfer the California State University technology infrastructure to a consortium run by Microsoft, GTE, Hughes, and Fujitsu.[60]

At the local government level, efforts to connect government offices to the Internet and enhance computerization were much more scattered. Much of the business interest in strengthening the public Information Superhighway has been in putting it more at the service of business in the economy, from speeding up permits for land development to making government data more available to the private sector. Through a wide range of institutions swirling around the business-led Smart Valley consortium and Pacific Bell's own involvement, businesses in Northern California played a leading role in setting the agenda for the public infrastructure and shape of the public Internet. In 1993 Pacific Bell launched its $35-million California Research and Education Network (CalREN) to promote innovative networking projects in the public sector.[61] The business goals of the project were bluntly noted by Syd Leung, CalREN's project manager: "The altruistic side is to help out California and fund these projects. On the business side, is to leverage as much public relations as possible. If we fund these other institutions, they can create new applications and we can leverage these into new business products."[62]

The strength of technology businesses in California intersected with the weakness of regional government, in what one regional official called the "Yugoslavia" of conflicting jurisdictions in the region. Without a regional pooling of resources, the squeezed local government budgets made cities open to any financial support and direction offered by local technology firms. This meant that most of the earliest cities to struggle onto the Internet, such as Palo Alto, did so with help from local technology firms, while other, poorer, low-tech towns were left further behind. In this scattershot approach to government access, high-tech companies were most interested in supporting the upper-income communities of their engineers and plant sites.

The experience in Northern California parallels the building of state information systems from North Carolina to Georgia to Texas, with each having its own mix of public and private involvement in putting their system together. North Carolina teamed up directly with three major phone companies (Bell-South, GTE, and Sprint Carolina) to use major new government markets to spur the creation of an overall high-speed fiber-optic system throughout the state, with a clear eye on spurring economic development.[63] Texas was an early pioneer in getting state agencies on the World Wide Web and quickly saw the advantages of moving state agencies such as the Department of Human Services from traditional networks to an intranet based on the IP protocol.[64]

Ironically, the most highly developed state information-networking system was to be found not in one of the traditional high-technology states but in one of the least expected places, Iowa. In 1989, Iowa had begun planning an improved technology network but had trouble finding businesses interested in its dispersed rural areas, so legislators approved the creation by the state itself of the Iowa Communications Network to link government agencies, schools, and libraries throughout the state by fiber-optic lines. By the mid-1990s, the state had spent roughly $200 million in bonded debt, with additional funds coming from the federal government in the form of grants, to wire veteran hospitals and other military-related facilities.[65]

The three factors facilitating the Iowa network seems to have been the size of the state (2.8 million people), state officials who were willing to take the leap forward, and the initial failure of the private sector to make serious bids for the project. There is an interesting contrast between the system that was produced through regionally coordinated response of Iowa (without much help from local technology companies) and the balkanized, haphazard system that emerged in the Bay Area under the influence of hundreds of technology companies trying to cut separate deals with different governmental authorities.

With broader regional and statewide high-speed access deployment stalled in many areas, individual cities (usually those serving high-income residents) began building their own high-speed networks. This often further undermined universal access to telecommunications in favor of regional polarization and inequality. In 1996, the city of Palo Alto, one of the first with a Web site, began investing $2 million in a twenty-six-mile superfast fiber-optic ring for Internet access to both connect its own government buildings and lease access to businesses, Internet providers, and telecommunication carriers. By the end of 1997, thirty-five Internet service providers, including WebTV, UUNET, Deutsche Telekom North America, and Hongkong Telecom, had set up shop in the central Internet Exchange for the fiber-optic loop. All this is aimed at making Palo Alto even more central to the economic and technological geography of the region and the nation.[66] Similarly, wealthy Anaheim, in southern California, is creating a public-private venture through the city-owned water and electric utility to invest $275 million in ninety-six fiber-optic loops with the goal of linking its three hundred thousand citizens and two thousand government offices to one another and the Internet at blindingly fast speeds. And unlike cities without the resources to make such investments and that expect to be spending large amounts on telecommunication services, Anaheim looks forward to earning $10 million for city coffers by the year 2010.[67]

Paradoxically, even as high-tech firms are promoting deregulation and the retreat of government, local publicly owned utilities are emerging as the most dynamic delivery systems for wide high-speed access—albeit only for upper-income communities. The results of market competition for telecommunications is more broadly following the model set in Scottsdale, Arizona, where universal cable and telephone service is giving way to piecemeal competition serving select communities.[68]

For most California cities, market competition in telecommunications is threatening to further shred city budgets. Because of a nineteenth-century law passed originally to encourage telegraph system expansion, telephone companies in California are allowed free use of public land in cities for the rights-of-way needed to lay wires and build facilities. This loophole has been expanded to include telecommunication services such as data networks and wireless services by more than sixty-five companies given free use of public land with no compensation to local governments. Worse, new state and federal legislation has extended restrictions on local government to cable companies, threatening to cost cities additional hundreds of millions of dollars in cable franchise and right-of-way fees.[69] Such legislation not only undermines cities' tax base but removes the leverage local governments have in extracting promises of universal service from many cable franchises.

Schools and Economic Equity

Even if local government networking and high-speed access to the home has been a slower, more haphazard process than originally envisioned, many pro-market advocates trumpet the "American Success Story," in the words of the Progress and Freedom Foundation, of computers and networking in the public schools. The Progress and Freedom Foundation itself cites studies showing that 98.6 percent of schools already have computers, with 64 percent of all schools already connected to the Internet. Like President Clinton and most of Silicon Valley, they also cite efforts such as "Net Days" in which as many as one hundred thousand volunteers across the country have spent a day installing equipment and wiring schools. By the end of 1996, more than 250,000 volunteers in forty states had installed wire in more than fifty thousand classrooms.[70] Under state telecom regulations, schools have been given discounts of between 20 and 90 percent on telephone and Internet access.[71] FCC access fees, amounting to $1.7 billion annually, were directed at public schools and libraries starting in 1998 (cut back from an initially proposed $2.25 billion per year).

While all this is accurate, the numbers are deceptive in indicating how technologically sophisticated most schools really have become, and the numbers hide the fact that technology spending is largely reinforcing inequality in education. The respected *Education Week* newspaper surveyed the broad problems in implementing technology in schools nationwide and, in particular, rated California in 1997 as one of the most technologically backward school systems in the country. California ranked forty-third in the nation in providing up-to-date computers, with only one computer available for every fifty-two students. This is added to the fact that California already ranked dead last in the ratio of books and librarians to students.[72]

Beyond the overall inadequacy of the spending is the fact that dollars are going overwhelmingly into equipment, with little funding for training teachers and other school personnel to make that spending effective. A 1997 report by the Benton Foundation, one of the most engaged funders of nonprofits in the technology field, has highlighted the fact that whereas federal guidelines advocate that 30 percent of technology budgets for schools go to staff development, only 5 percent of school technology spending went to training.[73] The corporate donations of equipment and quick-shot volunteerism of Net Days only add to the mismatch between equipment spending and lack of training for teachers.[74] Another four-year study of the impact of the Internet in schools noted that in poorer schools, high-achieving students were usually the only ones getting access to the technology, which exacerbated inequality in education between high-achieving and low-achieving students.[75]

These recent studies highlight the way in which the new round of Internet-related spending is repeating the mistakes of the 1980s, when computers were first introduced in mass numbers and technology acted to heighten differences in classroom experience between rich and poor schools. Back in 1992, *Macworld* magazine famously devoted a issue to the rising inequality in the classroom, directly contradicting the then glowing statistics: "What we found is a false dependence on statistical analysis and a reality so discouraging that it made us question how this situation has remained unremarked on for so long."[76] An article in that issue emphasized the role of lack of training in increasing inequality between schools using the new technology effectively and those left behind:

> Inner-city and rural school districts rarely have the skills or funds to maintain their machines. These districts lack the training and social support to use computers effectively. In most cases, computers simply perpetuate a two-tier system of education for rich and poor. Nearly every school in America owns personal computers. But without expertise to use and maintain them, thousands of machines lie fallow. When

computers are grafted onto dismal, underfunded schools that lack appropriate staff support, students and teachers rarely use them effectively.[77]

The volunteerism favored by many corporations and strongly promoted by Silicon Valley organizations such as Smart Valley merely adds to this inequality. For example, Smart Valley and Joint Venture have focused corporate contributions overwhelmingly on high-tech suburban schools, even as the poor city of Oakland, across the Bay, has been largely ignored. Volunteers with technological expertise are most available in upper-income communities; thus, events such as "Net Day" reinforce the advantages of schools serving middle- and upper-middle-class kids. After the 1996 Net Day, the *San Jose Mercury News* noted that its computer analysis of the effort showed that "by significant margins, schools in the most affluent neighborhoods have attracted more volunteers than the less affluent." The paper noted that schools in richer communities nationwide already had access to the Internet at twice the rate of schools in poorer communities, and Net Day bolstered that trend.[78] Many poorer schools did not even try to participate, knowing that they lacked the staff support to follow up. "When we've got leaky roofs, filling a room with technology seems stupid," argued Kevin Gordy, lobbyist for the California School Boards Association, at the time. "It's like serving filet mignon to someone who's starving."[79]

Unfortunately, even public spending on technology in schools has been modeling itself on this "volunteer" approach. California's governor Pete Wilson strongly promoted state government funding of a foundation run by John Detwiler, an investment banker and advisor to Wilson on education policy. The foundation refurbished computers and donated them to schools. Critics, including state legislative analyst Elizabeth Hill, noted that the program locked out many schools, since the Detwiler foundation required each school to find a locally donated computer in order to receive one from the foundation. The result was that in the first six months of the program, more than two-thirds of the computers funded by this state program went to schools in Silicon Valley.[80]

The use of prison labor by the program further widened to the gap in training in local schools. Along with providing public funds, the state also provided prison labor to assemble and refurbish computers to be donated to schools. Many schools offered to refurbish the computers themselves in lieu of having to find a locally donated computer, but the Detwiler program denied this option. This moved needed training funds and potential personnel out of the school districts. The foundation's dismissive attitude toward school personnel was accented when the program's executive director, Diana Detwiler, emphasized the productivity of the prison workers, adding contemptuously, "Maybe it's because they don't get any summer vacation." The program received $10 million from the state in fiscal 1996–97, in a process that Wilson sought to expand as a model for the state.[81] New funding for the program was blocked in the summer of 1997 when former employees publicly reported that the foundation had been misrepresenting the success of the program to legislators.

Unfortunately, many more states are following the California model, and volunteers, corporate donations, and lack of spending on teacher training will only continue to reinforce inequality between richer and poorer school districts.

Technology and the Transformation of Civic Government

The technological polarization between cities within regions is only one piece of the broader transformation of civic government occurring under the impact of the new technology. Moving information online and the computerization of public information are part of wide-ranging changes in the nature of local government. The balkanization of regional telecommunications infrastructure is reflected in the move to privatization and "entrepreneurial government" lionized by New Democrat thinkers such as David Osborne and Ted Gaebler. In their book *Reinventing Government,* Osborne and Gaebler see this new entrepreneurial government as a successor of Progressive Era civil service bureaucracies. While hailing the new commitment to "empowering citizens," they also admit that much of the momentum for government transformation is derived more from budget constraints like Proposition 13 and the need to forge public-private collaborations out of budgetary desperation.[82]

That the fracturing of telecommunications infrastructure is coinciding with the fracturing of traditional city and regional government is appropriate given the origin of regional government in the early twentieth century due largely to the need to manage emerging energy, water, and sewage networks.[83] One of the impulses for universal infrastructure back then were new understandings of the bacterial spread of disease and the commitment to extend water and sewage systems to all residents as a way of assuring general public health.[84] Since elite computer users do not see the spread of computer viruses as caused by unequal technology access, we have lost that factor in pushing forward universal networked infrastructure.

A stronger parallel to the varied trajectories of telecommunications infrastructure development is the varied historical municipal approaches to transportation systems. Analyst Anthony Sutcliffe has argued that both the degree of infrastructure development and its universality have been closely tied to how and who was able to harness the social product of infrastructure networks. Transit by itself never generated the profits necessary for its expansion, so the key to its development lay with the capture of the external social product generated by transit. In Europe, comprehensive transit systems emerged as a result of municipal ownership of peripheral land whose increase in value funded further transit expansion. In the United States, government ended up ceding land speculation and the profits generated by urban sprawl to private interests, thereby undermining integrated transit networks in favor of suburbanization.[85]

Much as the dismantling of urban mass transit systems in the United States helped facilitate suburban land speculation and subsequent profits, the dismantling of telecommunications utilities has blocked the ability of government to direct the social gains of telecommunications to universal expansion of the network, thereby facilitating the process of private gain in telecommunications "sprawl." If anything, as noted in Chapter 6, the effect of the growth of Internet commerce will be a significant loss in revenues for local government and the further erosion of programs aimed at economic inclusion and regional integration between rich and poor communities.

In the postwar period of growing suburban sprawl, the main countervailing force working against municipal ruptures between suburb, city, and later areas of exurban growth was the direction of federal funds into regional planning bodies. If specifically local growth was not to be harnessed for regional integration, the overall national boom was redirected through block grants and other programs into political forms that restrained municipal centrifugal forces. The 1980s and 1990s, however, would see both the dismantling of integrated regional telecommunications systems and the political liquidation of most of the federal programs that encouraged regional integration. The result is what urban planner Allan Wallis has called the era of the "networked" region, in which different municipalities are held together through a complex web of relationships brought together by the active involvement of public, nonprofit, and increasingly business sectors.[86]

Into this regional vacuum has emerged the "gated community" business politics for regions, detailed in Chapter 4, as global companies located in localities shape regions into useful research and production nodes in the global production system. The reality for local governments has been the movement from growth coalition politics to a politics of selling cities and regions in competition with both neighbors and regions across the world. Telecommunications and the Internet have emerged as a critical element in transforming civic culture in service to business needs.

Budgetary Desperation and the Virtual Downsizing of Government

When the Association of Bay Area Governments (ABAG) began its abagOnline Internet system as an off-budget "skunkworks" project in 1994, it saw its role as encouraging local governments to play catch-up and bring government processes up to the needs of local businesses and the public. With a five-hundred-thousand-dollar grant from the National Telecommunications and Information Administration, it hoped to help promote and integrate emerging online government sites in the region. Terry Bryzinsky, head of information services for ABAG, explained: "ABAG perceived that local government has not fundamentally changed the way it's done business for many years, while business for its own survival has embraced change. So local government has fallen behind. Now, we see great financial pressure on survivability for local government, and pressures on local government in delivering services. So ABAG saw a need to help local government adopt technologies that promise greater efficiency."[87]

While uneven, corporate financial support, from sources including Pacific Bell and Smart Valley, the latter's funds aimed at hooking up schools to the Internet, played a role in pushing governments online in the region. As important, the same pool of expertise and technology that Saxenian highlights in the private sector was "in the air" in the region and also played a strong role in early public movement onto the Internet. Warren Slocum, county assessor of San Mateo County, noted, "One day you meet the Pac Bell guy who invented the pager system, so the next day you are talking about paying your assessment taxes by touch-tone phone. It's the informal contacts that make a difference."[88] However, many governments might as well be "in Tanzania," in the words of ABAG's Bryzinsky, as far as technological expertise was concerned.

In the end though, local governments' fundamental reluctance has been driven more by budgets than by cultural resistance. With few resources available for any untraditional projects, it was often nearly impossible to launch new initiatives stemming from the new technology—one reason the federal grant for ABAG made its role more prominent. The State Employment Development Department's (EDD's) Kurt Handelman expressed a common irritation at complaints of government slowness in adopting new technology:

"Contrary to public perception, we don't have extra money. Every dime from the federal government goes to specific places. There are no slush funds. So we have to know how to make it profitable before we do it."[89]

Even in the heart of Silicon Valley, Mayor Wally Dean of Cupertino noted that "[g]overnments in the state of California are so severely cash-strapped that they can't take the first step. To make the commitment to get a city on-line, you need elected officials to have the political guts to do this and the budget to do it." Dean explained that he was able to sell his initial version of a quasi-Internet system called CityNet to his fellow city councillors only based on its fiscal advantages. With the CityNet system, Dean argued that the city had been able to layoff 11 percent of city staff without affecting city services. "We're starting to learn how to move information around directly. . . . Our city hall is paper generated—when something is generated, you have a staff person file it, pull it out for users, etc. If you can eliminate the labor charge and do it electronically, it becomes a very cost-effective way of running a city."[90]

Other governments around the country have sought similar cuts in government costs as a prime reason for increasing networking throughout the state. North Carolina's corrections department has harnessed its high-speed networks for videoconferencing for arraignments and hearings, saving an estimated seven hundred dollars for each hearing using this method, in mostly rural areas.[91] In Texas, the state government pioneered one of the most developed public online systems in the early 1990s. As one example, the Texas Department of Human Services launched a three-year project to relocate six hundred statewide offices from a host-controlled network that supported state welfare benefit services to an Internet protocol–based system. The entire project was expected to generate $9 million in savings over the 1993–98 period.[92]

Increasingly, budget savings are being made through the outsourcing of technology and privatization. The federal Office of Management and Budget (OMB) has assumed overall oversight of information technology spending in recent years and sees lean budgets and the need to replace downsized federal employees as a key reason for technology investments. "With the number of federal employees declining and resources more scarce, there is certainly a significant incentive for the government to become a world-class user of technology," said John Koskinen, deputy director for management at the Clinton OMB. At the federal level, privatization of technology spending has been rapidly increasing to the point where two-thirds of the estimated $26 billion of yearly information technology spending is being done through private contractors.[93]

The reshaping of government information systems is seen by many as a deliberate reorganization of government with the goal undermining the power of unions and traditional government-led growth advocates. San Mateo County assessor Warren Slocum was a strong advocate for moving Internet operations out of the county and onto ABAG computers, less as an attempt at regional integration than as a way to undermine the county's unions in its technology division. "Unions and Democrats would argue that we should be employing all these people. Should it be done by the county itself or through shared resources like ABAG? This government has a foot in many different worlds and it's a transitory world, and he who controls the public policy debate on this issue controls the future. The debate is much more than the Internet, related to whether the county should be building hospitals when the private hospitals have vacancies."[94]

Marketing Government to Global Corporations

The most immediate applications of the new technology for local government just seem to heighten, or at least accentuate, the vulnerability of local towns and cities to the forces of the global economy. In *The Condition Of Postmodernity*, geographer David Harvey describes a vision of local regions, under the pressures of new technology, as unable to establish long-term differentiation of space between areas. Instead, we see a desperate search for short-term exploitation of lower wages or tax benefits for business—a race that the loss of tax revenues to online commerce only escalates. Marketing regions to attract the relocation of business becomes a high premium for local government.[95] The way in which all information becomes an almost undifferentiated mass on the Internet may add to this problem, but the World Wide Web accentuates the desperation of cities and towns in promoting their "unique" attraction to business development by global companies.

If downsizing government employment is the strongest initial selling point for information technology, local governments see the competition for economic development as a key reason for getting on the Internet. If regions are losing their purpose as planners and growth promoters, they are increasing their role in marketing to global corporations. From the beginning, ABAG's Bryzinsky viewed making the Bay Area competitive as a central purpose behind encouraging local governments to put information online:

> The economic health of the region is definitely tied to government actions and policies. For example, if we have different ports in the

region competing for business, Long Beach wipes us out because they have a uniform marketing operation. . . . So regional government is definitely involved in international competition. We saw the Internet as a key part of the battlefield . . . with cities on-line gaining economic competitiveness for attracting business, for encouraging international opportunities in world trade. If Nissan announced they were looking for a new site, you can bet there would be a lot of on-line lobbying.[96]

He expects that cities without online access will be left out in the international competition for jobs, noting, "Companies would rather do business with modern governments rather than governments from the stone age."

Smart Permits: Blurring the Line Between Government and Business

One crucial aspect of using the Internet to make cities more attractive to global corporations is allowing them to process land use permits electronically. Smart Valley initiated an online project called "Smart Permitting," since a stock worry for governments has been fear of losing jobs to other areas because of companies seeking faster permit approval.

Much of the literature about new flexibility in corporate alliances has highlighted the way technology has blurred the boundaries between the internal structures of different firms. In the same way, online permitting is beginning to blur the boundaries between the internal structures of private business and the government. The Smart Permitting project coordinated efforts across twenty cities around Silicon Valley with private industry to set the software standards for how permits would work online. The process has largely followed a design completed by Santa Clara County's Manufacturing Group, a consortium of high-tech manufacturers who pushed for agreement between towns on uniform regulations.

Electronic permitting is expanding across the country, including the kinds of public-private collaborations developed in Silicon Valley. In 1995, Minnesota's Pollution Control Agency began a software-based system of permit applications in collaboration with American Management Systems, Incorporated, a systems integration firm based in Fairfax, Virginia. As the company sells the software to other states, the state of Minnesota will receive royalties from the software sold.[97] In addition, other states and cities are moving from software to Internet-based permits and application review.

The darker side of this process is that it puts pressure on other local governments to adopt the new standards and application processes built into software systems or else face much higher comparative costs in processing permits if they stick to their own standards, chosen by local voters. And the emphasis on the speed of processing permits threatens the time for review—which is the major chance for democratic input in permit decisions. Even as it increases efficiency, online permitting threatens to move standards for permits from local voters over to quasi-private consortia designing standards not so much in accordance with local democratic decisions but in the name of software efficiency and standardization, which may ignore the long-term health of local citizens.

For Silicon Valley, this danger is not abstract but a matter of history. The semiconductor boom of the 1960s and 1970s has left a legacy of more Super-fund contamination sites (twenty-nine) there than in any other area in the country, with more than 150 groundwater contamination sites. The small city of Sunnyvale alone has six Superfund sites. It was only when citizens and grassroots groups formed organizations such as the Silicon Valley Toxics Co-alition and Communities for a Better Environment in the early 1980s that the industry, formerly thought of as "clean," began to be investigated and pollu-tion restrained.[98]

Activists see those gains threatened by the push for "streamlining" permit processes and moving permit information online. Greg Karras, a senior scien-tist at Communities for a Better Environment, worried that "not everyone is online and many of the people most affected by this pollution are not online: people of color, low-income folks." He also scoffs at the idea that industry's goals in modifying permit processes is merely to make them more consistent and efficient, since lawsuits by companies blocking regulations has been the major cause of "the permit process being lengthened, delayed, and weakened." High-tech cities such as Sunnyvale run subsidized waste-treatment plants for the electronics industry and have sided with polluters in lawsuits, adding to the inconsistent and weakened regulations for other cities around the region. "This is in exact opposition to the streamlining bullshit that the industry mouths and it is because of lawsuits by the polluters. This all makes it harder for government to do its job."

Karras cites the danger to democracy of electronic permitting in the exam-ple of regulatory changes in 1997 that proposed weakening water pollution protections in the region, changes that Karras labels "the worst thing for the SF Bay in the history of the Clean Water Act." Where in past years, the propos-als would have been mass mailed to community organizations and interested citizens for comment, in 1997 the Environmental Protection Agency posted the proposals on its Web site as all the public notice needed. Karras himself did not track down the proposed changes until fifty-five days into the legal sixty-day comment period for challenging the regulations while poor and low-tech community organizations in affected towns only heard about them when Karras himself called them. "If it can happen like this for something this big, imagine how it will work for individual permits?"[99]

How Online Purchasing Is Undermining Local Development

Along with the effects of downsizing employment and speeding permit approval, one promise of the electronic networking for local government is cutting the costs of its own purchases, especially important as more goods and services are outsourced. The problem for local government is that virtual purchasing of goods will likely have the same impact on local economic employment as online consumer purchases have had on local tax revenues, namely, the undermining of integrated local economies.

Government contracts are the classic example where government corruption and government activism for local economic development have gone hand in hand. By steering contracts toward local firms, government officials nailed down political support but could also help out local businesses that needed a leg up. Governments might be targeting up-and-coming firms that are crucial to a neighborhood's development or the officials might just be lazy and call the same two or three firms and remind them when a contract was about to be made available by the city.

The Internet and other online channels promise to change that, opening the process to smaller bidders who might never have heard of the bids before but also opening up bids to multinational corporations that would never have troubled themselves before to track information on the myriad of local governments. ABAG and other government groups around the country have followed the model of the Los Angeles Metropolitan Transit Authority (MTA), the largest local government purchasing agency in the country and one of the first to establish an electronic bulletin board posting procurement needs.

After the electronic bulletin board was established in 1993, with its contracts posted electronically, the agency saw a 7 percent overall reduction in costs on $170 million of business. The MTA had traditionally mailed out procurement contracts to interested businesses once a month, but that was useful only for long-term construction contracts. As outsourcing of personnel services had increased, monthly contracts were not processed fast enough to keep up with agency needs. With contracts online, potential private contractors could keep up-to-date instantaneously; on hard copy, 30 percent of bids had been worthless because dates were outdated by the time people received the bids.

At the same time, online bidding made more complicated contracts easier to administer, since updates and new specifications can be posted online as well. "It used to be that you had to rewrite and mail out upgrades and worry about whether everybody gets the bids at the same time," noted the creator of the MTA system, Cary Paul Peck, who headed vendor relations for the MTA. The new system allowed bidders to enter a code and keep track of changing specifications over time.[100]

So far this seems all to the good: lower prices for government, more access for a wider range of businesses. The rub is that a wider range of businesses also can include businesses with no local connection to the local economy. On the positive side, a small earth-moving company in one economically depressed part of a state may get to bid in another part to drum up business. However, it also makes it easier for the giant multinational construction firm Bechtel to scout out the largest and the smallest contracts in the state, tracking down contracts and displacing smaller local businesses, whereas it would have previously been too expensive to spend the time keeping track of all the contract possibilities.

While the purpose of the system was to expand universal access, Peck acknowledges that it is really the sophisticated users who most heavily download bids: lobbyists, engineering firms, larger companies. While small companies with "Tandy [computers] and a schlocky modem crawl on-line," most of the traffic is from large corporations such as Bechtel.[101] ABAG's Bryzinsky hopes the contract exchange gives small companies a better chance to weather local economic storms: "If I was a small company in Redding, and we were an earth-mover company and in a depressed area—the Internet makes it more feasible for that small company to find out about Cal Trans jobs all over the state." But he also admits, "It also makes it easier for Bechtel . . . and Bechtel has the most resources to divert to finding work."[102]

Local government trying to support local economic development may increasingly see global businesses underbidding local contractors. Where the upside of Internet access had been more businesses trying to bid, the downside may end up being the narrowing of actual successful bids to a few global companies. A visit to any local government convention right now will reveal that it is inevitably dominated by tables set up by new global companies seeking to supply everything from trash collection to food service for local jails. The Internet may be just the tool that those global companies need to reach those governments and replace traditional firms that once had a "local advantage" in getting information at the town hall.

The other potential issue is the downsizing of government in favor of private global information "brokers." Whereas the Los Angeles MTA has moved forward on full public access to procurement information, California state government was slow to move from an all-paper state-bidding system, which costs $42 million for its printing bill alone each year. Instead of moving directly to online access, the state was looking in the 1990s to sell its information to a private company that would repackage the information electronically and sell it back to anyone interested in getting information on available government contracts.[103] In effect, this would transfer jobs out of state government procurement offices and into the hands of private companies that would then profit from reselling public information back to the public. (We will return to the issue of government information being privatized later in the chapter.) As selling government information for profit becomes a model for more local governments, not only will local businesses lose employment opportunities but in addition local government workers may lose out to information brokers located far from their cities, further cutting the link between local government spending and local job creation.

Many local agencies involved in job placement for the unemployed, especially for the long-term unemployed, complain that business refuses to cooperate in efforts to use new technology to list jobs and help the unemployed find work. Most businesses prefer to find all but elite employees through secondary channels to avoid being inundated with new applicants. This without question is linked to the rise of contingent labor—temps, self-employed contractors and participants in other employment schemes that take businesses out of much of the direct-employment market.[104]

Kurt Handelman, of the state employment department, from a statewide policy vantage point, noted the worry of policy makers that the hiring process would become too visible. Policy makers might see the possibility of destabilization of job markets that have thrived on limited information. "You don't want to advertise that there are two hundred lumberjack jobs in Humbolt County to a jobs-starved state, and have caravans going up the state. There could be a bad influence on migration and stability."[105] This illustrates the point that although the technology may facilitate openness, there are significant elite interests in maintaining stability in job markets and there are business interests in maintaining limited information for low-wage job seekers.[106]

The reality is that state and federal employment agencies have bowed to the will of employers in not consistently collecting and publicizing information

on job openings for low-wage workers. So instead of the dream of an electronic "hiring hall" for regional labor markets, we see a core of elite engineers
networking for jobs electronically while the vast majority of low-skill workers
exist in an almost invisible peripheral job market. Rather than the new technologies making the job market more visible, they have assisted in the growth
of temporary agencies that manage that peripheral market more effectively for
the corporations, which can thereby avoid dealing directly with most of their
peripheral employees.

Databases and the Privatization of Public Information

The choice of government not to collect and distribute information on job
openings goes to the core politics of networked technology and government:
what information will be collected, who will have access, and at what price
will it be available? How those questions are answered is increasingly shaping
how government itself is organized internally and in relation to other political
bodies and the private sector. Government is one of the central sources of
information in society, and one would expect that that role would be enhanced
in an age deemed the "information economy," yet local governments are
finding the role as much a burden as an opportunity.

Even for information that government has traditionally collected, basic bureaucratic resistance had restricted access to data long before the Internet even
was imagined. Control of information is power within government; most
agencies fight tooth and nail against losing that control and many see the
Internet as a decisive threat to their control. Despite laws mandating access to
government information, studies in 1997 by the nonprofit Center for Investigative Reporting and subsequently by the Associated Press found stiff resistance
from California's state government to releasing a range of information they
held in electronic form.[107]

In political terms, this kind of resistance is understandable, given the nature
of bureaucratic infighting. Michael Duffner was the first Webmaster for ABAG
and subsequently was hired to help run the city of San Francisco's Web site;
Duffner argues, "There are many competent people in government who
haven't lost sight of what government should do. But they also have to protect
their own interests or they won't survive."[108] Many officials worry that if information is shared, the public will use it in the short-term without recognizing
the ultimate source of that information. The result will be (and has been)

budget cuts that ultimately undermine the ability of those officials to keep providing the original information.

Within Northern California, how to effectively share information has intersected with the "Yugoslavia" of divided city, county, and agency bailiwicks. ABAG is one agency that has been slowly, too slowly in some critics' view, trying to get local governments to integrate their information through its Web site and information exchange systems. Argues ABAG's Bryzinsky, "It is a challenge to ABAG to get governments together to avoid a Tower of Babel. We may not have picked the best standard, but it's hard. But there's nothing that we can't rearrange and move; we aren't pouring concrete here."[109]

One of the more successful approaches to government cooperation regarding data exchange has been in the city and county of San Diego, which as far back as 1984 formed a nonprofit, public-benefit organization, the San Diego Data Processing Corporation, to handle government data in a partnership called the San Diego Regional Urban Information System (RUIS). RUIS now serves seventeen departments in the city and county covering 2.6 million residents and forty-two hundred square miles. The partnership has been reluctant to move forward on selling any data until public access has been assured but budgetary pressures are mounting.[110]

The stakes are high across the country in the battle for control of government information and the software to create integrated databases. Even before the advent of the Internet, spending on versions of the so-called geographic information systems (GIS) totaled more than $3.5 billion by 1991. Roughly a quarter of the market was selling information systems to the Pentagon and another half to mostly federal government offices, but geographic information software makers see their real growth in local government and leveraging government information to private industry markets. By the end of the decade, analysts were projecting up to $100 billion in software and system sales.[111]

But with everyone from utility companies to fast-food franchisers to rental companies interested in using government data in locating and managing their branches, the industry has been exploding in recent years. Insurance companies are looking to use GIS data to calculate how far a customer's home is from a fire hydrant or police station and to deduce the real risks in the daily commute. Information from a range of government sources can be combined with private sources of information to create detailed customer profiles on each individual, an important tool for banks and other companies engaged in "data mining" to manage customer marketing. Significantly, all three major credit bureaus and others in the information field have acquired or merged with GIS mapping companies to forge new partnerships in producing data-mining software to track individuals across the country.[112]

The Invasion of Privacy and the Threat of Economic Discrimination

Although many people worry about bureaucratic resistance to releasing information, others worry that too much is being released. Surveying plans in California to connect justice system information with social services with transit and Department of Motor Vehicles information, Victor Pottoroff, deputy director of the California State Association of Counties, comments that soon people "probably will be able to punch up too goddamn much information about me."[113] San Mateo assessor Warren Slocum relates the story of police calling him up and complaining that his office was selling property information that some company was using to publish the home addresses of judges. Slocum worries about the implications: "If anyone wants to get this information, someone can come into this office on the computer and find it. Women living alone—we don't want people knowing they have three bedrooms and living alone."[114]

As what was previously government-only information is made public and integrated with private information sources, many analysts are worried about the systematic invasion of privacy as previously disparate pieces of information about individuals are combined to provide marketers with a comprehensive profile of each person. When P-Trak, a database service offered by the information company Lexis-Nexis, provided Social Security numbers on individuals nationwide, a frenzy of public horror swept the country, but many analysts noted how limited the P-Trak service was in light of the comprehensive information available to companies in other, less public database vendors. For set fees, it is now relatively simple to find out whether any specific individual owns a plane or has ever been a felon, when he or she last moved, and what his or her driving record looks like, as well as to obtain an employment background check—all gleaned from public records.[115]

Access to database information that the Federal Bureau of Investigation was once denied by law has now been made available to anyone willing to pay for it as governments have marketed information to private data vendors. Most legislation protecting privacy is so weak that little has slowed this ongoing encroachment into the private lives of people. While marketers trumpet the advantages in delivering tailored offers to customers, legal scholars highlight the growing imbalances of power in an information economy in which individual preferences are so perfectly surveyed by global corporations. "Through the use of data banks, the state and private organizations can transform themselves into omnipotent parents and the rest of society into helpless children," wrote Paul Schwartz and Joel Reidenberg, two American lawyers, in a book commissioned by the European Union to study American data privacy laws.[116]

The danger, however, is not only at the individual level. As states and local government increasingly market themselves to global capital, the aggregate data released on individuals within a region has the potential to radically expand geographic discrimination and economic redlining of all forms. One of the most advanced data integration projects has been in Washington State, where in the name of expanding its "competitive environment," the state has implemented information projects to integrate information databases within a web of public-private partnerships. Charles Cook is a former director of a U.S. Congress Office of Technology Assessment study of the original conception of a National Research and Education Network and now privately studies emerging data networks in the economy. He has highlighted the dangers of programs like Washington State's in which databases of community information about "at-risk" four-year-olds and juvenile offenders are being linked to public information systems:

> Someone has obviously decided there is a public purpose to be served in creating a database of at-risk four-year-olds and another of violent offenders. Missing from the Washington State scene is any widespread public understanding either that these and other data bases exist. Also absent is any reasonable means of challenge by the public to state agency use of them. For example, what if [the state] were to decide that when a business had narrowed its choice to four potential locations, the final step towards maximum competitiveness for that business would be to show it the population density of at-risk four-year-olds and violent offenders in each of the sites? The business would surely choose to locate in an area with as few undesirables as possible. The potential of information technology to be used in building economic ghettos in Washington State is not generally known let alone an item on the public policy agenda.[117]

The Lure and Losses for Governments of Selling Information

In an ideal world, the choices between expanding access to public information and protecting the privacy of individuals in the electronic age would be balanced on the basis of a careful scrutiny of gains and losses for democracy in our society. Unfortunately, the more relevant context has been the budgetary desperation of local and state governments that view the selling of public information as a revenue alternative in a time when other tax sources are being undermined by global pressures.

Federal projects promoting "smart highways"—containing toll booths that collect tolls by means of electronic sensors—are also being used to raise revenues for local government by selling "information of commercial value" to entities such as auto insurance companies willing to purchase the data.[118] Even as ABAG in Northern California has opposed some public sale of information, it has tried to raise revenue (somewhat unsuccessfully) from its public contracts exchange and other ventures. San Mateo's assessor Warren Slocum saw the detailed information on property sales collected by his office as a potential revenue bonanza if sold to real estate companies.[119] Across the country, local governments have turned to selling public information as a new revenue source.

Free-speech advocates have responded by demanding that such public information be released at no charge rather than be limited to those with the funds to pay for such services. Some have pursued lawsuits and others have turned to legislation to block the sale of information to private resellers. In California, one of the first victories of free-speech advocates was a legislative mandate to put all legislative proceedings on the Internet, thereby ending a longtime monopoly on such information in electronic form held by a private company.

Local governments in California have fought furiously against similar mandates for local government to release all information onto the Internet, arguing that they could not afford to format their information in electronic form without private sale of the information, so the public would not have access in any case. This is the bind local governments find themselves in: if they have to provide the information in electronic form for free, not only do local governments forego the added revenue of commercial sales, they also incur new costs in preparing the information in a usable form.

The danger of this economic bind is not only to local government budgets but also to the existence of information being available in any form. Many local governments faced with increasing economic pressures have discontinued various kinds of public information collection altogether. Don Wimberly, as head of a public-private Bay Area consortium for selling government information, sees laws banning governments from selling information as a mistake that will only speed the loss of public information. He cites the example that in the early 1980s, Santa Clara County made the financial decision to stop collecting some land information history, including the tract maps created by land surveyors:

> There are lots of examples of this; they don't create the information, don't collect it, or they throw it out. We need to create an incentive for local governments keeping and sharing their information. You're a city manager; it's going to cost you thirty thousand dollars to generate this information, but you can only charge you for what it costs to copy it. It gets you nothing if you don't capitalize on this. So if it's not your number one priority, you'll drop that data collection, which is what Santa Clara did.[120]

Across the country, governments have been discontinuing data collections, and Alaska and other states have virtually abolished their data centers. This trend has reached the federal government; the Census Bureau has cut its operations and pared back what information it gathered in the year 2000 census. Officials in the bureau have already cut back on the publication of and reports on data from previous surveys.[121] Other federal government information centers are following suit to such an extent that many analysts who depend on government information are seeing an emerging crisis. In his 1996 address, the president of the National Association of Business Economists, Maurine Haver, argued, "Resources for research on how to measure the increasingly complex service sector and high tech industries have simply not been available." He noted that statistical agencies have maintained present programs by reallocating research dollars to information collection and production, but termination of information collection programs had already begun. Surveys including the U.S. Industrial Outlook, the Quarterly Plant and Equipment Survey, and others had been terminated, with more to follow.[122]

Michael Duffner, from his experience in running Web sites for both ABAG and the city of San Francisco, is scornful of those advocating the resale of public information as the solution. His view is that if one vendor is reselling the information, the city probably failed to get its full value in any case, while citizens end up paying a premium for access to privatized information that should have been available for free, since they paid for it as taxpayers. The relationship corrupts information-collection agencies and invariably brings in less revenue than the costs of collecting it in the first place. The entire approach ignores the long-term gains from adequate freely available information while concentrating on its short-term value, a recipe for the public to undervalue information: "It's like government making timber sales at a massive loss, selling off assets below cost and selling the value of future tourism and quality of life. Information is a similar kind of thing. When you do sell information, you are never going to get what it cost to acquire it, so if you sell it you will decide it's not worth it and get out of the business. You should just see how information like census data drives the economy and give it away."[123] The danger inherent in the information age is that as detailed, invasive information on individuals is increasingly in the hands of the global corporations, the public is increasingly denied information on the broad social and economic trends in the economy that might give them the chance to collectively act to reassert authority over those global forces.

Public-Private Partnerships and the Privatization of Democracy

As governments privatize information sales and even its collection, the end product becomes the loss of democratic control as information critical to democratic decision making becomes the property only of those with the wealth to pay for it. Increasingly, vital government information is being distributed and sold exclusively by private companies. The Reporters Committee for Freedom of the Press is one organization that has been raising alarms bells about the impact of this trend on democracy. They point to Mississippi, where a single publisher has been given exclusive rights to distribute and sell the electronic version of the state's laws; or to the federal government's sale to private companies of once public data from the National Institutes of Health, the Social Security Administration, and the FCC.[124] Such practices force both the public and the government itself in many cases to pay for access to information critical to democratic discussion and that was created and collected at taxpayers' expense.

The most notorious example of the costs of privatizing control of information is in the area of federal legal decisions. The federal government in the early 1970s created an electronic database, called JURIS, of all federal court decisions and much of its administrative law. In 1983, it contracted out maintenance of JURIS to longtime legal publisher West Publishing. When its contract was not renewed, West demanded that all records in the database created during its contract be expunged to prevent public advocates from using the Freedom of Information Act to create a publicly available legal database to compete with West's own private electronic databases. By dismantling JURIS, West forced the federal government to pay it millions of dollars each year to access the publisher's database of the government's own case law. At the same time, this monopoly control of legal citation by West burdens those in the legal system with few economic resources to pay the high costs of West's electronic services.[125]

The Bay Area Shared Information Consortium (BASIC) information system being developed in Silicon Valley by a private-public consortium is taking this privatization trend to a new and, from a privacy viewpoint, much scarier point. At one level, the goal is simply to take the scattered information-privatization deals and integrate them into a one-stop information spot selling information from both private and public sources, from governments to utility companies and credit bureaus. By integrating these information sources, any business that is planning to locate in a city will have a comprehensive view of everything from zoning regulations to how many customers likely to be interested in their particular product are located close to their proposed location.

There is the obvious danger for democracy in a situation in which the information infrastructure of the Bay Area would be controlled by a business-led consortium whose information, by definition, would not be subject to Freedom of Information review. However, BASIC has taken the trend toward information privatization one step further by combining all this electronic geographic data with satellite imagery of the physical geography of the region, in a project called the Bay Area Digital GeoResource (BADGER), made possible by the participation of NASA's Ames Center and Lockheed in the consortium. Largely as an offshoot of missile defense Star Wars funding, Lockheed and other companies pushed through declassification of satellite imagery in 1994 in order to resell those images in commercial ventures.

With satellite imagery available that can show photographs with a scale of one meter per pixel, an observer can clearly see buildings, trees, and even cars. By combining this data with BASIC, the idea is to create a digital atlas distributed over the Internet by means of which utility companies can visually survey vegetation encroachment on their wires or fast-food outlets can locate stores based on market data and observation of the physical flow of cars at any one point. There will no doubt be advantages for government (at a price) in surveying the effects of suburban sprawl, but the real gains will be for commercial enterprises intending to market their goods or to locate within any region.

"This is the Home Shopping Network for geographic data," in the view of David Milgram, Lockheed's principal investigative researcher on the project. "If I have a property map and a recent satellite image, I could identify where the swimming pools are and which are dirty. I could then identify the homes that have swimming pools and finding out who's property it's on, and sending a mailing to people about my swimming pool cleaning service. Or see whose roofs needs reshingling and tell people I have a deal for them."[126] While not legally licensed yet, there is an obvious next step: satellite imagery that can see down to the scale of individual people, whose movements could be tracked throughout the day by satellite, thereby opening up a whole new kind of surveillance-based marketing within regions.

Prostrate Governments in the Information Age

That governments are assisting in the marketing of its citizens is the clear end product of the desperate economic marketing of their regions to global corporations on the prowl for low costs and compliant governments. With local government budgets teetering on the brink of insolvency, information technology has emerged more as a tool for marketing and downsizing public employment than as the promised vehicle for open government and citizen empowerment.

The loss of local control over sales and telecommunications taxes adds to the general fracturing of local economic development caused by the interaction of technology and the increasingly global economy. All this should coerce a reevaluation of the push to "decentralization" of government responsibilities to local government. Such responsibility makes little sense in a world in which multinational corporations often outpower whole states in total assets and can pit local governments against one another in the competition for jobs and local revenue. Although much information-age rhetoric harkens back to images of small firms and decentralization, the reality is of soon-to-be-trillion-dollar corporations straddling the globe. Even modest-sized enterprises operate more and more on a global basis. Faced with such a disparity in power and the fracturing of such governments' ability to cooperate easily, local governments can hardly be expected to devise fair and efficient systems of taxation that can deliver the social goods and economic development needed. Under such decentralized revenue approaches, the poor and working class face increased tax burdens. The rise of national and global commerce calls for national and even global revenue approaches. While the microchip may be getting smaller, the plane of economic activity encouraged by this technology is national and global.

Modern progressive government was born at the turn of the century as the need to manage industrial growth intersected with the need to manage the growing networks of power, telephone, and water infrastructure. As those integrated utility networks have been torn apart by "deregulation," new telecommunications infrastructure is increasingly polarized, with highly developed high-tech suburbs and downtown urban enclaves at one end and poorer towns left behind with inadequate access and unequal schools at the other.

In order to meet the demands of global corporations, the very structure of local government is being transformed through outsourcing, the privatization of information, and the general blurring of the lines between government and business functions in areas ranging from local contracting to development permits to employment policy. Where regional governments once saw their role as promoting general growth in a region, that function is increasingly lost in favor of a desperate marketing to global corporations as each city or region bids with a combination of tax breaks and compliant government for corporate location choices. The exposing of the privacy of citizens has become an integral part of that bid for corporate location as once-private government data on individuals is increasingly sold to offset budgetary crises (themselves caused at least partly by the globalizing effects of networked technology.)

Despite the rhetoric of New Democrats or Gingrichite "Third Wave" advocates, the local region has become not the venue for empowerment of individuals but the space where the power of global capital so outmatches the power of any other actors as to largely eclipse democracy itself. Whether with the threat of disinvestment or full-scale flight, global corporations are increasingly able to design government regulations and information infrastructure to suit their needs at the expense of local empowerment and equality within the region.

The reality is that if the new technology holds any promise of empowerment for the individual and less well off communities, it is at the national and global level. It is only at such a level that the combined power of working families, working through both government avenues and through labor unions, has any chance to match the power of global corporations.

Already, officials in local governments are using the Internet to strengthen their communication with one another and for lobbying at higher levels of government. "We can also use information technology to track changes in federal and state legislation and regulation," notes Victor Pottoroff of the California State Association of Counties, "since those things change on almost a daily basis." Groups such as the National League of Cities have become increasingly savvy in fighting against such bills as the Cox-Wyden Internet Tax Freedom Bill, which would further erode local government revenue-raising ability. Because of their mobilization, city and other local government authorities were able to prevent the bill's passage in 1997 and significantly water it down in 1998.

But the real impact of the Internet is among community organizations and unions, which increasingly find their power at the local level stymied by global corporate power and have turned to allied organizations nationally and globally to restore some balance to the power equation. If anything blocks a revitalization of grassroots power through use of networking technology, it is the ideology of decentralization promoted by many conservative apostles of the new technology. In the concluding chapter in this book, we will look at that decentralization ideology in the context of the arguments made so far and explore the real potential for community mobilization through use of the new technology, to create a truly national and even global response to centralized corporate power in this new era.

Conclusion: The Death of Community Economics, or Think Locally, Act Globally

When in the 1990s, Newt Gingrich and other conservatives promoted political decentralization and New Democrats framed much the same politics in terms of "reinventing government," it was hardly surprising that this bipartisan consensus helped drive market competition in telecommunications and the transfer of government responsibilities to local government in a host of other programs. All this added to the crisis increasingly facing most local governments. What is remarkable, though, is how much the drive and desire for decentralized economics and politics was often shared even by the strongest grassroots progressive and left-wing opponents of the governing establishment. Although critical of corporate power and skeptical of achieving the form of decentralization they desire under present economic power relations, many progressive activist organizations, especially those tied to an environmental vision, often promote the goal of decentralization as ardently as those advancing the rhetoric of the establishment.

One typical manifesto from mid-decade came from a coalition called the General Agreement on a New Economy (an explicit response to the General Agreement on Tariffs and Trade [GATT]), which was led by a range of scholars associated with left institutions, among them the Institute for Policy Studies, the Worldwatch Institute, the National Jobs for All coalition, and other key research organizations. In their manifesto, they argued for the goal of "[a] new form of federalism rooted in communities where local goals and programs for a sustainable future can be chosen through a fully participatory process, with regional and federal functions that can help integrate local efforts into a cohesive whole capable of ensuring overall sustainability, equity and full employment" and the "use of private and public initiatives and resources to achieve these local goals, with supplemental federal funding provided to communities and regions in a manner that accounts for differences in local resources and severity of problems."[1] They added the conditions of restraint of corporate power and, in practice, most progressive activists would defend the federally based programs that they ideally would see decentralized. But the convergence with the stated goals of the establishment is remarkable and shows the reserves of vision in the United States that decentralizers draw on in promoting the ideal of community-based decision making in the economy.

Even some left activists saw their similarity with conservatives and even explicitly aligned themselves as the "left wing" of a Third Wave politics. Carl Davidson, a Chicago activist and leader of various socialist organizations over the years, is managing editor of Chicago-based *cy.Rev: A Journal of Cybernetic Revolution, Sustainable Socialism, and Radical Democracy,* which published an article he co-wrote titled "The Third Wave and the Republicans: Is Newt Gingrich a Closet New Leftist?" Davidson traced the emergence of Gingrich's politics and rhetoric back to not only the corporate Right but also the 1960s Left: "Is Gingrich a hidden 1960s new leftist in 1990s conservative clothing? Not only is he using some of our old slogans, he also appears to be invading our political space. After all, it was only after the antiwar and civil rights movements of 30 years ago that it became possible to badmouth the White House and the federal government the way it's being done today." As a national officer of Students for a Democratic Society in the late 1960s, Davidson spoke with more authority than most when he, with some irony, hailed the fact that "Newt Gingrich is leading the most successful attack on the capitalist state since the 1960s. Tearing apart bureaucracies, desanctifying authority, delegitimizing the corporate liberal political system, decentralizing power closer down to the grass roots." Even as Davidson attacked the hypocrisy of Republican defense spending and pro-corporate power in Gingrich's agenda, he argued that the core of the rhetoric of decentralization was a natural result of the telecommunications Third Wave: "There is nothing wrong with returning many government programs to the state or municipal level. Local politics is more accessible to grass roots' movements. A good amount of decentralization, moreover, is an inevitable consequence of telecommunications and its impact on the economic base of society. Gridlock in Washington is partly a result of federal bureaucracies being too distant and too clumsy to handle regional and urban realities."[2]

Given the descriptions given in this book of the disempowerment of local government and the splintering of regional economies as useful units for economic planning, what are we to make of this decentralization consensus spanning the ideological spectrum from the far Right to the far Left poles?

At one level, we can see it less as a response to the new technology than, as Davidson notes, a continuation of New Left and other traditional American politics that has always sought utopia in local and small-town values. The technology becomes an excuse to dress the politics of nostalgia in new cybernetic clothes. Any political position can label any disliked spending and social structures as "too centralized" and as obsolete in the "new" economy. The Right can target the social welfare state while the Left can target corporate subsidies and rail against corporate power. Conservatives get the best of this cynical politics, since corporate power, as we have detailed, remains centralized; but all political actors find sustenance in this cyberpolitics of nostalgia. Progressive activists end up hyping a program of "community economics" that resembles nothing so much as a longing for the old "growth coalition" regional politics, a politics that is increasingly defunct as cybertechnology allows the selective interpenetration of regions, which erodes the growth linkages of regions.

But beyond nostalgia, there are real contradictions in the new technology and economics of the networked world, which make the promise of decentralization alluring, even if dead wrong politically (especially for those promoting egalitarian values). We have federal government technology spending driving economic revolutions, even as local private entrepreneurs appear to commercialize the new technology, seemingly out of their garages. We have a globalization of production at the same time as singular regions such as Silicon Valley loom ever larger on the economic landscape. We have the dramatic globalization of business alliances even as regional business consortia become more critical in ensuring production standards in our information-rich economy. We have the market-driven dissolution of regional economic linkages tied to banking, power, and telecommunications infrastructure, while simultaneously the appearance of new technologies and companies expand local choices for participating in the global economy. Across the spectrum, we have the rise of market competition and "deregulation" of industries at the same time as expanded regulations permeate both the domestic and international economy.

Out of this confusion of where the dynamic engine of growth originates in the economy, it is hardly surprising that people find resonance in directing political focus on the aspects of the economy that are situated within the comfortable scale of the local. There was a compelling logic to the period between the Depression and the Cold War, when national-scale industries dominated the land and were counterbalanced by national-scale legislation that shaped and restrained their actions within regions of the country. As networking technology has allowed business to globalize, the federal government has seemed incapable of fully coping with the changes, while global institutions seemed too murky a focus for democratic management, so the local region has emerged as the only handhold, however slippery, for activist energy in using the government to steer citizens' economic fate.

In turn, we have had the peculiar regionalization of global business organization, its breaking up of concentrated national factories into carefully constructed regional niches that expertly harvests what author Saxenian has labeled "regional advantage" in the service of the global corporation. All this has contributed to a certain myopia of decentralization and the idea that the community is now the natural scale for government action. Progressives may not want regions to compete against other regions and would prefer cooperation, but they are still lulled by the illusion that integrated regions exist and that they have common fates. The ready participation of business in ever increasing "public-private" partnerships and consortia at the regional level further validates those who reify regions as economic units.

Pivotal to the relationships explored in this book is the core role of federal action in its support of economic growth in regions such as Silicon Valley. The Internet and networked technology offer a prime example of this, even as federal policies were combined with the dissolution of the economic ties between richer and poorer members of such geographic regions. As described, it is a relatively bleak story as inequality grows and local political institutions find themselves overburdened by responsibilities they cannot cope with. But there is also a positive side to the story, for even as local egalitarian economic development becomes a nearly impossible concept in the new economy, new opportunities for managing the national and global economy are opening up. And even as the grassroots rhetoric of many community organizations emphasizes local economics, increasingly the new technology is being used in practical ways to strengthen global organizing for economic justice. While the latter was often sporadic and mostly unsystematic for most of the 1990s, the use of information technology has been expanding within the corporate world for well more than thirty years, while most unions and community organizations have begun to turn to electronic networking only in the past half decade. To assume ever increasing economic inequality is to ignore the fact that one side of the economic class divide is only now beginning to practice use of these tools and mastery is years down the road. With the explosion of activism that led up to and followed the WTO confrontation in Seattle in 1999, we may be seeing the beginning of a new alternative politics that can be achieved as the community and labor side of the economic conflict is as pervasively networked globally as is business. Only as this becomes more and more of a reality will we be able to measure the true political and economic balance of power.

Community Power in the Age of the Internet

Described in this book is the rather grim situation of power and inequality as the age of the Internet begins. Within local economic politics, it is the global business class that exercises most power, since its easy threats to move away production and other jobs benefiting non-elite workers chills political debate on economic development. At the national level, both the ideology of decentralization and the raw economic power accumulated as a result of these trends have generated for political policies that have accelerated the trends of economic polarization.

What is important to recognize, as is emphasized in this book, is that there is nothing in the technology that makes activist government irrelevant. If anything, the opposite has been the case, as "market deregulation" has led to escalating caseloads for national regulatory commissions and legal dockets. Intellectual-property laws and trade deals have put government action at the center of determining the dispensation of economic fortunes between different interest groups. New international regimes of trade agreements, international tribunals, and economic coordinating bodies have appeared. There is no technological reason why the underlying laws and regulations governing use of the new technology could not favor economic equity in a way that supports all communities rather than leaving them at the mercy of global gamesmanship by multinationals.

What has changed with the new technology is therefore not the degree of government intervention but of who it benefits and how power operates. And where the technology has played its most critical role is in strengthening the political hand of business to use rapid communication to coordinate far-flung geographic empires and increase its bargaining leverage in local and even national political fights.

But while the strengthening of the power of multinational companies has been the first result of the new technological developments, to merely extrapolate an ever increasing gap in power and expanding inequality would be to ignore the long-term dynamics of technological change and social struggle over geographic space. As the economic historian Immanuel Wallerstein has documented, an early response to industrial technology was seen in the flight of capitalists from the old cities of the Middle Ages, areas that were controlled by craft and guild rules. They built factories in traditional rural areas where a new geography of company-driven urbanization would define the economic landscape.[3] Although that change would empower a new economic class and introduce new power inequalities, it also would inspire in the following centuries new organizational responses. In particular, national trade unions would emerge and push for the centralization of government policies over working and living conditions, which in turn would create the modern social democratic conditions in which "growth coalitions" had originally thrived.

A crucial element in the rise of such national associations as a check on the excesses of industrial economic power was the new communication technology of that day, namely the printing press and the newspaper. Alexis de Tocqueville, as noted in his *Democracy in America,* saw the newspaper as more than merely an avenue for speech; for him it was also a conduit for civil associations formed by average citizens. As he wrote in 1839:

> In a democracy an association cannot be powerful unless it is numerous. Those composing it must therefore be spread over a wide area, and each of them is anchored to the place in which he lives by the modesty of his fortune and a crowd of small necessary cares. They need some means of talking every day without seeing one another and of acting together without meeting. So hardly any democratic association can carry on without a newspaper. . . . Newspapers make associations and associations make newspapers.[4]

In a time when the economics of newsprint and the rise of a mass, nonassociational press has undermined the traditional political newspaper, the rise of Internet communication as a tool of association may very well promise a counterbalance to the rising power of the economic elite. Although recent in its application and still reaching only a fraction of those rooted in community because of the "modesty of their fortune," political and economic organizing over the Internet is slowly emerging as a critical tool for amalgamating the power of community members and workers across the country and the globe. It is in many ways a dramatically more powerful means of "talking every day" and "acting together" than the newspaper every was.

The effects of the new technology on community organizations and unions is only tentative so far, so the stories concluding this study can only be anecdotal beginnings, but they do show the emerging possibilities that counterbalance the grim economic and political trends that existed as long as networking technology was the exclusive preserve of business. What is remarkable is how quickly the Internet has been moving from a high-tech toy to a day-to-day tool for organizing—often at a pace that leaves many activists bewildered. Its use not only offers new ways to exercise power within regions and on a global basis but also promises to reshape what democracy itself will mean for mass organizations. Like the economic spin-offs that are centered in the technology regions of Boston and Silicon Valley, many of these initial political uses of the Net have their roots in those regions.

Public consciousness of the Internet as a political tool for progressive political change first came about in 1995. Not surprisingly, the earliest stories emerged off of college campuses, where the density of Internet connections was already relatively high. The event that caught the national media's attention was a March 29 national rally on one hundred college campuses across the country protesting against the Republican Contract on America. Remarkably, the Boston-based Center for Campus Organizing—the lead organizer of the national rallies—had only a dozen or so campus contacts when it announced the day of action, only a month and a half before it was to take place. But using the Internet, it quickly spread the word, gained agreement on a set of principles endorsed by the network, and created a national event that was covered by media ranging from *USA Today* to the *Nation*. The March 29 action followed closely on the heels of protests against California's anti-immigrant Proposition 187, protests that had spread nationwide by means of electronic mail lists dedicated to the issue and sponsored by the Center for Community Economic Research at UC-Berkeley.[5]

Rich Cowan, a former MIT student who headed the Center for Campus Organizing in the mid-1990s (and still works with community organizations through the Organizers Collaborative),[6] emphasizes that the Internet is no complete substitute for face-to-face meetings, but he does believe that the Internet is a "vehicle for groups to exchange strategies and introduce new people to successful organizing strategies and the lessons they've learned."[7] This is the power of the Internet: to allow activists to almost instantly share what is happening in their region, redefine national goals proposed by others, and come to a consensus on dates and forms of joint action. From the fight against Proposition 187 to the March 29 rallies to national marches that would follow on May 6 against the Contract on America, furious on-line organizing became a fixture on the Internet in the spring of 1995 in a way that had never been seen before.

Getting Online: Nonprofits and the Net

Off campus, national nonprofits began slowly connecting online in the mid-1990s, and many local community organizations began to see the Net as a key to connecting communities divided by distance and media disinterest. Many began getting online for prosaic economic reasons. James Johnson, an organizer in Sacramento for the Oakland-based Center for Third World Organizing (CTWO), got online in 1995 to share his work on police brutality and accountability with other organizations from Portland to Denver to Rhode Island to South Carolina. The move to electronic mail was a deliberate one by CTWO, which supplied computers to its affiliates and offered basic computer training to organizers. Johnson noted that before e-mail, "CTWO organizations used to spend fifteen thousand dollars per year just on UPS and postage and those costs dropped significantly. Where before you were looking at ten dollars per pop, now we just e-mail it."[8] With the rapid-fire exchange of information, community groups can more rapidly make the political case that local issues, whether lead poisoning in paint or police brutality, are actually national trends.

Empty the Shelters (ETS), another national organization advocating for low-income communities, made an even deeper commitment early on to conducting internal communications on the Internet. With ETS offices in Chicago, Philadelphia, San Francisco, Oakland, and Atlanta, staff and activists decided that using the Internet would be the most practical way to coordinate their work on homelessness and poverty issues, given their limited national budget. ETS bought each activist center a modem and an online account and set up an online discussion conference for communication. The organization went a step further and began holding meetings online and posting all communications on their online conference. It even stopped faxing out information in order to force everyone to use the equipment that had been purchased.

The decision to move much of the organizational communication online was based on raw economics: "We had seven thousand dollars in the budget," noted ETS coordinator Emilee Whitehurst, who runs the Oakland office, "and we made the decision to invest in accounts and modems instead of putting the money into a national meeting. We had a debate and the choice was to bypass a meeting last December (in 1994)." Whitehurst admitted that her own technophobia inhibited her and she did not adopt the technology as quickly as other offices, but she points out that lack of training was a problem as well. Without money in the budget for training, Whitehurst emphasized, merely investing in technology will not work. And she noted that face-to-face meetings will still remain vital: "There are certain things you can do online and some things you should reserve until you meet in person. We waited too long to get on-line but we have to be wise in how we use it and for what decisions."[9]

Pat Bourne, who as its director helped run an early computer training and networking program at San Francisco State University called SFUNET, emphasized that despite the benefits of networking, the transition to the new technology will always be a battle for many groups in low-income communities. SFUNET concentrated on working with groups focused on the problems of homelessness, and Bourne worries about those being left behind: "People in the computer area have no idea the budget constraints these non-profits are under," complains Bourne. "Just getting $60 for a modem is hard. The lack of resources is unbelievable—the lousy office equipment, the pressure they are under. This is why I wince when people talk about Windows or the Web and talk about, hey it's only $800 for a [PC]." Each advance in technology raises the bar a bit higher and threatens to leave behind those struggling online: "There was a point where any piece of junk could use the computing power of the host computer. But with graphics and the Web, it demands more power from the person's computer."[10] Beyond the costs of equipment are the costs of keeping staff trained in the technology. For nonprofits dependent on the sweat and expertise of an important volunteer or staff person, turnover can also be a devastating blow.

Still, nonprofits across the spectrum, from environmental to peace organizations, increasingly began to use electronic communication as the decade progressed. Helping that effort was the emergence of a whole new class of activists devoted specifically to assisting other organizations, especially those serving low-income communities, in getting on the Internet. Dubbed "circuit riders," these new activists act as contingent consultants, traveling from organization office to organization office, installing needed software, training staff, and helping in crises. Groups such as TechRocks, a project supported originally by the Rockefeller Family Fund foundation, have not only developed cadres of circuit riders to help low-income organizations get over the "digital divide" but have even moved into the arena of software programming to develop tailored programs such as databases to specifically serve the needs of nonprofits.[11] Given the dramatic gains in lowered communication costs if that initial technical assistance is given, TechRocks and other groups feel that they have helped turn organizations' small initial investments in the technology into large gains in organizing success.

Unions and the Electronic Targeting of Business

Many organizations, particularly unions, have begun using the Internet not only to communicate with allies but also as an integral part of their organizing weaponry against opponents.

A central feature of the new economy, especially in places such as Silicon Valley, has been the outsourcing and subcontracting of many jobs, particularly those held by less-skilled, and often minority, employees. This has allowed many high-tech firms to create the illusion of being "good employers" even as large chunks of their effective workforce, whether in subcontracted assembly plants or in support services, work for substandard wages in awful conditions. One use of the Internet has been to cross the social distance between the segmented workplaces of elite core workers and the sites where peripheral workers are being organized.

Justice for Janitors, one of the most militant union campaigns waged in the country, led by the Service Employees International Union, took on and won a broad effort to organize the janitorial workforce of Silicon Valley—making it one of the only segments of the industry that has been unionized in recent years. Besides carrying out militant street action, the union used the Internet to publicize its campaigns against Apple Computer, Oracle, and Hewlett-Packard on a global basis to tarnish the companies' images as "model" employers. Additionally, they used electronic bulletin boards to directly inform engineers and programmers of the working conditions of those workers who, unseen by the elite employees, cleaned the latter's offices every night, putting pressure on those companies to recognize the union.

Oakland-based Local 2850 of the Hotel Employees and Restaurant Employees (HERE) International union, which covers much of the emerging Highway 580 extension of Silicon Valley, began using the Internet even more directly in their organizing campaigns. In its initial foray onto the Internet, the union publicized retaliatory firings at the Lafayette Park Hotel and in the process generated hundreds of letters, phone calls and e-mailed support letters from activists and organizations around the country. Its aim was to target a small chain of luxury hotels known as the Western Lodging Group, of which the Layfayette Park was a part. As the campaign evolved, a key tactic of the union became targeting corporate customers of the hotel who regularly used it to house visiting clients.

One of the hotel's largest corporate customers was PeopleSoft, a computer software company that paid the hotel to house employees and corporate partners who came into town. After PeopleSoft refused to relocate to another hotel, Local 2850 took the simple step of highlighting negative economic facts taken from the company's own legal filings, then posted the fact sheet they compiled to computer-oriented newsgroups on the Internet. The result was almost instantaneous; the company received a barrage of letters from worried customers and investors, while PeopleSoft claimed that its stock declined by more than $63 million in value following reactions to 2850's electronic postings. Soon after, PeopleSoft announced that it was moving its customers and other visitors to another hotel.

Other companies have been similarly targeted. Through their actions, Local 2850 managed to use the Internet to link the struggle of some of the most peripheral workers in the Silicon Valley production system—those who clean the rooms of clients—to the global capital investors in Bay Area firms. The next step was for 2850, along with a broader network of affiliated hotel union locals, to build online connections to target the global hotel-reservation system of travel agents and event planners to discourage them from booking customers in targeted hotels. By means of a well-publicized Web page and electronic messages sent to key people, images of picket signs and chanting protesters have been used to steer potential customers thousands of miles away from booking rooms in hotels being targeted for organizing.[12]

One of the most ambitious attempts to employ the new information technology in taking on the global firms of the Bay Area was launched by the South Bay Central Labor Council, encompassing 110 affiliated unions representing one hundred thousand workers in much of Silicon Valley. Amy Dean, head of the Labor Council based in San Jose, sees the future of the labor movement in efforts to organize the vast array of contingent workers created in the region. Having strongly backed the Justice for Janitors campaign, Dean sees it only as the beginning: "The janitors were just the first among the contingent workforce. This is going to involve everybody from janitors to technical writers to software gypsies and testers to quality assurance engineers. When we talk about doing windows in the valley, we're not just talking about the janitors who clean them, but the software engineers who write them."[13] Along with a community-based policy center called Working Partnerships that the Labor Council launched as a direct counterpoint to the business-led Joint Venture, the Labor Council's Web site is seen as a tool to be employed over time to create a community for contingent workers who rarely meet and can then be organized into economic power.

One aspect of the Internet that is strengthening unions but is less welcomed by many national union and organizational leaders is its use in increasing democracy within national organizations. The Internet promises a major upheaval in democratic debate as local areas gain greater voice and connections laterally outside the hierarchical structures of organizations. One case in point is the American Federation of State, County, and Municipal Employees (AFS-CME) where an electronic mail list called PUBLABOR in the mid-1990s became an early forum for debates on democracy within the organization. Local union leaders and activists who had never had a chance to talk face-to-face were able to share sharp criticisms of the national union leadership for alleged failure to support locals and for instances of undemocratic leadership. Complaints over use of union dues in one local would lead to other participants chiming in to describe similar or different experiences in their region of the country. Others in AFSCME would staunchly defend the national leadership amid the raucous debate and in-depth analysis of the organization's successes and failures.

AFSCME member Katie Buller created and was the first moderator of the PUBLABOR list. She received some pressure from her local AFSCME leadership to use her position to tone down criticism of the top leadership, but noted, "What is AFSCME gonna do? Throw me out for creating a forum for free speech?" She adds that International union staff were generally supportive of her work on the list, arguing that "AFSCME is often a target because it is the biggest of the public employee unions." She praises the organization for working to become "Net literate."[14]

What AFSCME is facing is becoming increasingly chronic across larger organizations, ranging from unions to the NAACP to environmental organizations: chunks of the "mass" membership and local leaders get online and gain the ability to initiate dialogue on the direction of their organizations outside the often tightly controlled annual or biannual conventions. But most of the union leaders saw assisting the expansion of Internet use as only strengthening their organizations over the long run as local concerns could more easily become national issues. National unions were soon issuing regular email updates to members, helping locals establish Web sites, and most dramatically, negotiating union contracts to provide subsidized computers and Internet access to members. In 1999, the United Auto Workers negotiated a union contract with Ford Motor Company to provide a personal computer, printer, and Internet service to each of its 350,000 workers for as little as five dollars a month. In 2000, Local 32B-32J of the Service Employees International Union, representing largely immigrant janitors, would negotiate a similar provision for twenty-six thousand workers to receive subsidized computers and training, a model for other unions around the country serving low-income workers that they could follow in addressing the digital divide and strengthening internal communication.

The Internet and the Globalization of Economic-Justice Organizing

Beyond facilitating national organizing, the Internet is hastening the process by which global alliances challenge privatization and economic trade deals that ignore the concerns of environmentalists and labor. On News Year's Day, 1994, the day that NAFTA was implemented, the Zapatista guerrilla army announced its existence to the world and launched a series of raids that focused attention on their demands for land reform and the rollback of neoliberal economic policies in Mexico. From the beginning, the Internet was used to convey to the world the ideas and messages of the Zapatistas and their mysterious leader, Subcommandante Marcos. The Mexican government clashed with the rebels, then spent the year in uneasy negotiations seeking a peaceful settlement of the conflict.

As the peso's value plunged in 1994, many nervous investors blamed the Zapatistas for undermining the country's economy. Although serious analysts noted that the operations of a small army in a marginal state like Chiapas was not itself significant, perception could create reality in the volatile financial markets. Finally, an analyst for Chase Bank of New York on January 13, 1995, published a memorandum declaring, "While Chiapas, in our opinion, does not pose a fundamental threat to Mexican political stability, it is perceived to be so by many in the investment community. The government will need to eliminate the Zapatistas to demonstrate effective control of the national territory and security policy."[15] Within a month, under pressure from both Chase and other international financial forces, Mexican president Carlos Salinas launched a military assault on the Zapatistas.

Initially, the Chase memo received no coverage in the U.S. media except in a small newsletter put out by *Nation* columnist Alexander Cockburn and his then collaborator Ken Silverstein. But then the political networks that had been distributing information on the Internet for the Zapatistas picked up on the newsletter and redistributed it with the comments from the Chase analyst. A firestorm of angry e-mail and phone calls threatening boycotts and other sanctions against the bank swept over Chase headquarters. In the first mention of the story in the major media, Chase publicly disassociated itself from the analyst and his recommendations. As Mexican papers expressed outrage at Salinas's capitulation to U.S. financial interests, Salinas was once again forced to back away from the war against the Zapatistas. A few months later, in May, the Mexican foreign minister, José Angel Gurria Trevino, paid a backhanded tribute to the technological sophistication of the Zapatistas and their allies when he tried to dismiss the rebellion as just a "war of inks, of writings, and a war on the Internet."[16]

The conflict in Chiapas became for many progressive activists a paradigm of a new global politics—a conflict rooted in a specific region, yet appealing for global economic changes that would make reform possible. It was a war waged by some of the poorest and least technologically sophisticated people on the planet who were using cutting-edge technology to bypass the biases of the global mass media. Even as the electronic flows of money crashed through regional economies and increased economic inequality, progressives took heart that the new electronic flows of information were giving the world community of activists a new tool for coordinated response.

If the electronic financial tools had come out of those square miles of Silicon Valley, many of the electronic tools carrying the Zapatistas' messages were also flowing from that same acreage, largely in the form of the Association for Progressive Communications (APC), a global consortium of dozens of national electronic networks, with its headquarters at its U.S.-affiliate, the Institute for Global Communications (IGC). IGC was the center for initial progressive networking in the United States with more than seven thousand progressive environmental, peace, and labor organizations using IGC for their Internet access by 1997; through the APC, IGC members could connect directly with more than fifty thousand nonprofits in 133 countries.[17]

Originating as electronic bulletin boards connected by Apple II's in the early 1980s, before the Web even existed, two electronic networks, EcoNet and PeaceNet, arose to link environmental organizations and peace groups, respectively. These networks, based in Menlo Park, grew out of the rich left-wing organizing tradition of the 1970s that had briefly fused with the emerging personal computer revolution in venues such as the Homebrew Computer Club. Lee Feldenstein, the red-diaper baby who had helped launch Homebrew, used profits from his shares of Osborne Computer to build an early networking project called Community Memory, until Osborne itself collapsed. Other projects, among them the Peoples Computer Company (a magazine, not a manufacturer) and for-profits such as DYMAX, all located in Menlo Park, would spread the gospel of technology to politicos in the area.[18] Out of this milieu arose EcoNet and PeaceNet, which in 1986 merged to form IGC (later adding components involving labor, conflict resolution, and women's organizations). In 1988, IGC embarked on a major effort to spread electronic networking internationally, especially in third-world countries and among indigenous peoples. At the same time, IGC had donated its software to Green-Net, a progressive electronic network in the United Kingdom, and IGC began assisting in the creation of electronic network affiliates around the world. In 1990, seven founding national networks created the Association of Progressive Communications, which had fifty affiliates by 1997. APC would be chosen by United Nations agencies and conferences to host discussions and resources, a recognition of their critical international role.[19]

The rise of the Internet did lessen IGC's role as often the exclusive electronic connection that activists might have with other activists around the world. Unable to compete as an ISP, at the end of the decade IGC relinquished its direct-service role to private companies, but it maintained a progressive "portal site" linking a wide-ranging set of organizations for activists seeking to obtain and exchange information.[20] Through this role, it continued building networks connecting areas that the Internet still has yet to deeply penetrate and, more important, helped construct the political and social infrastructure that made links possible. In many ways, IGC and APC thrived early on as centers for international linking of community organizations that had never achieved regular contact in the past. One group cited by IGC staff is the Environmental Law Community, which used electronic networking to strengthen an existing organization and extend its reach internationally. IGC program manager Michael Stein painted this picture of the system's efficacy in conferencing and e-mail:

> Their staff [was] running around the world with Powerbooks and they're driving the technology. E-law's John Bonine flew to Mongolia at the request of a small law cooperative and won the struggle against a mining concern about three years ago. Now, there's a computer network in Mongolia called MAGIC that joined APC and it was through his efforts. There's someone who used the electronic medium to drive community building. They're sold on IGC and APC, so wherever John goes Powerbook in hand, they promote it.[21]

Stein also cited the Pesticide Action Network, based in San Francisco, which has helped build a stronger worldwide movement to promote pesticide reform. It worked hard initially to get all thirty-five of their steering committee members online around the world, creating affiliates in Africa, Asia, and Central America. Going to endless trouble to get modems and computers to their affiliates around the globe, "they really made use of the computer to enhance capacity building in their movement."[22]

As the Internet and the World Wide Web has grown, other individual groups have pioneered its use to spread information and aim at global corporate targets. The Rainforest Action Network (RAN) approaches environmental issues from a perspective broadly connecting issues of human rights, environmentalism, and international economics—an approach that made extensive networking across different constituencies a requirement. Early on, RAN's Web site achieved a reputation for containing dynamic content and information resources, attracting more than thirty thousand visitors a week by 1997. One of its most dramatic efforts on the Web was its campaign to curtail logging by the Mitsubishi Corporation. Direct action by activists was combined with an Internet campaign in which supporters could inundate Mitsubishi headquarters with faxes automatically sent from the RAN site upon request. As noted by one researcher who did a broad profile of RAN's Internet tactics, "It is fair to say that Rainforest Action Network's dynamism, inclusiveness and cognizance hinge in large part on its effective use of the Internet."[23]

However, if any organization early on showed the political promise of the Internet as a counterbalance to corporate power, it was the Minneapolis-based Institute for Agriculture and Trade Policy (IATP). As NAFTA, the World Trade Organization, and other international trade agreements and institutions assumed ever larger roles not only in governing trade but also in limiting environmental and labor rights worldwide, the IATP pioneered the use of the Internet to keep activists around the globe updated with the information needed to track and fight for a more just trade system in the world. The IATP was established in the 1980s by organizations throughout the Americas to create a multi-national progressive counterbalance to those advocating neoliberal rules on trade. "In 1988 and 1989," explained IATP executive director Mark Ritchie, "we saw how to use the computer information technology to summarize the information we were collecting and distribute it cheaply to establish a common information base in a wide area of organizations. We spawned a whole new genre of publications and genre sharing."[24] A co-founder of the Fair Trade Campaign, working against NAFTA and GATT, IATP would play a critical role in educating environmentalists, unionists, and community organizations about the need for not only an alternative trade policy but also international organizing across borders to achieve those goals over the long-term.

IATP updates would become a weekly and sometimes daily information source with global news summarized and then redistributed to activists around the world. Bulletins would be sent over the Internet to cities around the world, then faxed by organizations there to allies without Internet connections—drastically cutting the communication costs and expanding the information distributed to hundreds of organizations whose members had never heard of IATP or the Fair Trade Campaign. These information bulletins were supplemented by online archives and strategy conferences located at IGC in which trade policy and organizing strategies could be broadly debated. While neither NAFTA or GATT were defeated in the U.S. Congress, the organizing campaign against them forged new links domestically between environmentalists and labor unions, while a new vista of international organizing appeared, linked to common struggles involving trade issues.

Beyond its organizational savvy with electronic communication, IATP was pushing itself (within its low-budget means) to the technological edge. "We've experimented with having people electronically scan articles in other countries," noted Ritchie, "then send them up to us on the Internet. Then we run them through language translation software. It doesn't give perfect translations, but it allows us to monitor information that would be otherwise inaccessible to us."[25] Not only does IATP use the Internet for electronic discussions, it also began experimenting with conference calls directly over the Net—bypassing the phone companies—as early as 1995. IATP produced a weekly television and radio show for a number of years and began using satellite time to beam it around the world. What is remarkable is that IATP achieved its quite penetrating level of worldwide communication with a relatively small staff—a feat made possible only through the leveraging of the Internet to expand both its sources of community information, which it then publishes, and its reach to those who need its day-to-day information.

The Internet may not have single-handedly defeated "fast track" authority—the ability of the president to negotiate treaties immune to congressional amendments—in the fall of 1997 but it played a key role in bypassing a mainstream media devoted almost unanimously to free trade in traditional corporate terms. No less a supporter of free trade than the *New Republic* has noted the gulf in viewpoint between "an insular and patronizing pro-globalization establishment" and what the magazine labeled "resentful and suspicious anti-globalization populists." The magazine noted that the next battle on globalization, the Multilateral Agreement on Investments (MAI), was already probably lost without a single major article about the agreement in major newspapers due to "MAI paranoia [which] has ricocheted through the Internet." The magazine worried that the elite was ignoring this growing grassroots organization on the Internet at its peril.[26] By March 1998, the *Economist* ratified this loss in an article titled "The sinking of the MAI," which stated, "Labour and environmental groups want high standards written in for how foreign investors should treat workers and protect the environment. Their fervent attacks, spread via a network of Internet websites, have left negotiators unsure how to proceed."[27]

Activists committed to their local community have begun to flip on its head the old 1960s mantra Think globally, act locally. With power at the local level at the mercy of global corporate competition, activists on the Internet increasingly see global alliances as, ironically, the only way to preserve local sovereignty in any meaningful form. In that paradox of the electronic age, it is the technological and global elite that increasingly pays rhetorical homage to "community" (gated, in most cases) and local government, while it is local community and union activists who increasingly see the global village as their political home.

Seattle and Alternative Globalizations

The events surrounding the 1999 World Trade Organization (WTO) meeting in Seattle were the culmination of the emerging political use of the new technology and of the new politics being formed in the wake of new global corporate political structures flowing themselves in many ways from the technology. The central importance of Seattle was that what had been localized struggles and single issue politics were reconceived as a common struggle against the current shape of the emerging global structuring of politics and the economy. Broadly labeled an "antiglobalization" movement by many supporters and critics, it might better be understood as a challenge to the frustrations of democracy that the particular form that globalization promoted by the WTO, the International Monetary Fund (IMF), and the World Bank has taken in recent decades. In this, the movement was an amplification of the more specific frustrations expressed about the limitations of technological possibilities and local democracy that this book has documented.

The most surprising thing about Seattle was that the mainstream media and elite politicians were so taken aback by the scale of the protests. From the Zapatistas to the defeat of MAI, the signs of a new politics centered on trade and what was broadly labeled "globalization" were increasingly evident. And on the electronic networks of unions, environmentalists, and community groups, the upcoming Seattle confrontation was the topic of a constant stream of discussion. More than fifteen hundred organizations globally signed onto an anti-WTO petition circulated on the Internet by the Nader-aligned Public Citizen, a major organizer of the People for Fair Trade Network, which helped sponsor the major legal rallies and demonstrations in alliance with the major environmental groups and labor unions. A RAND study had just recently denounced the threat of such "NGO swarms," organizing by e-mail, as "multi-headed, impossible to decapitate" political opponents to governments.[28] The Direct Action Network, the organization that would direct most of the nonviolent street blockades, was sending out e-mail calls for a "festival of resistance" in the streets of Seattle.[29] Possessing the flavor of calls to come to Chicago in 1968 and to Woodstock in 1969, the question of whether people "were going" was in the electronic air.

Seattle in many ways also represented the closing of the historic divide separating those two earlier events, whereby the political Left and countercultural activists had moved into increasingly separate streams of activity in the 1970s. With Seattle, Naderite political organizers would mingle freely with ex–Grateful Dead followers who had gone from swapping tapes in concert parking lots to swapping songs on Napster and were now swapping news of debt crises in Bolivia over the Internet. In the streets of Seattle, the city of grunge music would meet the city of software programmers and the city of Boeing union workers.

Not trusting the mainstream media to cover an event it had barely anticipated, Web designers, film innovators, software programmers, and activist writers combined to produce a new institution, the Independent Media Center (IMC), or Indymedia. IMC Web sites would act as a central exchange for news and articles about demonstrations and events, along with allowing activists to comment on them and strategize the next action. During the Seattle protests, handheld cameras, combined with the new technology of Webcasting, would allow activists to bypass traditional media as they broadcast film of the street protests directly onto the Internet. Built as a volunteer collective, the model of the Seattle-IMC offered a model that would spread rapidly from city to city into a network from Chiapas to Russia to Australia to Brazil.[30] In its annual listing of underreported stories, Project Censored, which salvages stories underreported in the mainstream press, proclaimed the formation of the IMC network as a sign of one of the few trends challenging the increasing concentration of media ownership in a few corporate hands.[31]

Bay Area Activism Meets Global Politics

As the technology of Northern California was used to recruit demonstrators and to project the story to the world, another element of the region, its vibrant, raucous progressive politics, was projected onto the world stage as well. The decade of the 1990s had seen an escalating series of local political battles— over issues ranging from education funding to immigrant rights to affirmative action to local labor struggles—that had inevitably faced the limits of the capacity of local institutions and government to solve the underlying issues at stake. Despite being the heart in many ways of "act locally" political inclinations, the "think globally" tendency began pushing for a broader stage for action, especially as trade and development policies left domestic and global concerns on intersecting trajectories.

While the Direct Action Network had leaders in its loosely knit structure from around the country, the heart of early leadership came from the Bay Area, exporting tactics and training from direct action in that region to Seattle and to protests that took place the following year in Washington, D.C.; Philadelphia; and Los Angeles. More formally, the nonprofit Berkeley-based Ruckus Society had been sponsoring direct-action trainings for three years before Seattle, producing what the media dubbed a "graduate school for protesters." With a $370,000 budget, its director, John Sellers, who started his career with Greenpeace, had produced two thousand graduates of its weeklong camp trainings and 120 trainers teaching direct-action tactics across the country. "We don't sign our actions," said Sellers, "but it's likely that our alums are in there" when mass action occurs.[32] Before Seattle, the weeklong camp added, along with its normal repertoire of climbing stunts and coordinated movement, an expanded training in "digital activism" from online security and encryption to designing effective anti-WTO Web sites.[33]

Many of the most prominent spokespeople for the Seattle movement came from the Bay Area, connecting international justice issues to grassroots organizing at home. Annuradha Mittal, executive director of the Institute for Food and Development Policy (better known as Food First!) noted the explicit connection between work that the organization did in fighting welfare reform in the United States and the networks it was building against WTO and IMF policies, which she saw as impoverishing developing nations. A native of India who saw fighting poverty in the United States as a key battleground, Mittal envisioned the goal of the Seattle protests as expressing "moral outrage against a system that is working against the working poor—the working poor in America and the working poor overseas. . . . Until we fight this battle right here in this country . . . nobody is going to give a damn about people's right to food in Mexico or in the Philippines."[34]

Global Exchange, a San Francisco–based organization that had played a large role in organizing student antisweatshop networks across the country through speaking tours and e-mail lists, was a strong backer of the Direct Action Network. One of its co-directors, Medea Benjamin, became a major spokesperson for the movement at Seattle and even managed, before being arrested, to waltz into a delegates-only WTO summit meeting, shake hands with WTO director general Michael Moore, and walked to the lectern to begin politely upbraiding the assembled delegates for not factoring human rights and environmental and labor issues into their trade considerations.[35] Benjamin, a former economist for the World Health Organization and other United Nations agencies, passionately argued for the connection between corporate abuses abroad and in the United States. She became the 2000 Green Party candidate for U.S. Senate in California and played a lead role in grassroots organizing during the electricity crisis, arguing months before state politicians took any action that in an "era of nearly unchecked corporate greed, 'deregulation' will inevitably lead to outrageous prices."[36] This emerging consciousness of the intimate connection between local and global issues is the hallmark of new community and NGO organizing.

Workers of the World Unite?

That the community groups and young street activists were joined by twenty thousand union members arriving in buses from across the country represented the bridging of a political divide of another sort. This new "Teamsters and Turtles" coalition opposing the WTO signaled a radical change in the approach to foreign policy on the part of national union leaders. While most of the unionists who made the bus ride there had probably heard about Seattle in old-fashioned ways, it was quite new-fashioned changes in global corporate structures that drove this new alliance by labor against the WTO and United States international policies.

If midcentury had seen local growth coalitions between community business boosters and union leaders, the same period of the Cold War had seen an even more intimate alliance of the national AFL-CIO labor federation with the foreign policy of the United States. For decades, the foreign policy centers of the national union federation were funded heavily by the Central Intelligence Agency and later in more regular fashion by the National Endowment for Democracy. Many progressive activists had been alienated by the labor federa-

tion's support for the Vietnam War and its support for often right-wing regimes in Latin America and around the globe.

But as trade issues began to predominate over the older Cold War ideology,
AFL-CIO leaders began to question whether their foreign policy strategies,
which often undermined the more militant trade unions of the developing
nations, was also undermining labor power at home. When John Sweeney
became AFL-CIO president in 1995, he dissolved the existing union international institutions and substituted a new center with new leadership that was
focused on linking union struggles abroad with organizing at home. "If there's
an organizing drive [in the United States] with a multinational with operations
in Brazil," noted Ron Blackwell, the labor federation's new director of corporate affairs, "we need to know what the operation there is like. It helps us act
as if we were global." The labor federation Solidarity Center now does union
training in countries from Cambodia to Bangladesh to Central America and
along the way has worked closely with the campus antisweatshop movement,
in day-to-day organizing parallel to the efforts of the more dramatic Seattle
demonstration alliance.[37]

In concert with the end of automatic AFL-CIO support for U.S. foreign
policy, a new, much closer relationship developed with the International Confederation of Free Trade Unions (ICFTU), essentially the global equivalent of
the AFL-CIO. The ICFTU had itself become far more inclusive in the wake of
the end of the Cold War, by the end of the 1990s representing more than 200
million union workers globally, as national union federations from developing
nations joined, federations that were previously excluded or had refused to
participate over decades because of Cold War divides.

Against the media image of first-world unions facing off against developing
nations over inclusion of labor clauses in international agreements, the ICFTU
endorsed such clauses, arguing in its official statement, "A workers' right provision would strengthen the political authority of the WTO and break, rather
than build, barriers to world trade."[38] Although there were some dissenting
unions that preferred to strengthen the enforcement powers of the International Labor Organization, the vast majority of national trade union federations, including those in South Africa, Turkey, and Malaysia supported the
ICFTU position on incorporating labor standards into the WTO. In Seattle,
G. Rajasekaran, head of the Malaysian Trades Union Congress, attacked his
government's position, arguing, "The governments of developing countries
try to water down labor standards because they've been convinced by companies that they need to do so to attract investment. But that's wrong: Labor
standards and economic development are mutually supportive."[39] When labor

rallied in Seattle, speakers from Mexico, South Africa, the Caribbean, China, France, and elsewhere announced that the rejection of the current WTO regime was a global response by workers worldwide.[40]

Local Resistance Versus Global Transcendence

There were obvious tensions and contradictions among those mobilizing together in what has been dubbed the "antiglobalization" movement, including caused by the name to actually designate the movement. Given that the movement was explicitly based itself on global alliances and use of the technological possibilities of global society, hardly any of the groups involved supported autarchic national economies and the ending of trade, despite that caricature of the movement promoted by some elite observers. What the protesters opposed was the specific system of global governance backed by the multinational corporations and international institutions such as the WTO and IMF, a system that was less the naturalistic result implied by the term *globalization* and much more the planned political result implied by Bush Senior when he referred to a "New World Order," an ominous enough term that it was quickly retired.[41]

Yet there are open questions about the degree to which any alternative global system can preserve the diversity of cultures and local economic systems that are presently being undermined versus looking to new mechanisms that may sacrifice some of that diversity for universal standards that can at least deliver greater economic equity. This internal dissent is hardly surprising given the complexities of both substantive global policy choices and tough strategic options in a global system with few of the traditional democratic mechanisms available. Some activists see possibilities of reform of the system through labor and environmental standards, while others view dismantling the whole regime of unelected trade bodies as the first step needed to build an alternative democratically based global system. This strategic divide was captured in the oft-quoted slogan "Fix it or nix it" among unions and NGOs opposing new trade rounds.[42]

If anything focused the ire of the groups protesting globally, it was the lack of democratic accountability in this new global system, where decisions are made by unelected tribunals, through secretly negotiated treaties, and by economic institutions such as the World Bank, where voting is allocated on the basis of the wealth of the shareholding nations. *New York Times* columnist Thomas Friedman attacked the protests by contrasting the meeting of "elected officials" with "self-appointed . . . union activists";[43] however, the very idea that a single representative from a country could encompass the diversity of interests in each nation is exactly what was contested by the protesting organizations.

If anything is most pronounced in the world, it is the increasing polarization of interests *within* nations between the poor and the wealthy and within communities divided by race, caste, and language. The very expansion of global multinationals and the use of technology across far-flung regions has eroded the idea of a world shaped largely by nation-states with singular "national interests." Nation-states, like smaller regions themselves, can no longer be treated as unified political actors (if they ever could), so interests under the system that are in fact in conflict within particular state regimes have to be understood on a global basis. The new global regime has been designated simply "globalization" by some, but others have sought a more precise political term, whether "neoliberalism," "postimperialism,"[44] or in the evocative phrase of one set of authors, "Empire."[45]

This last term, the title of a book by Michael Hardt and Antonio Negri, was used by the authors to capture both the democratic deficit and the claims to universalism in the new global system. Embraced as a modern "Communist Manifesto" for the new movement, their book argued that the tensions between local and global were an inevitable by-product of a global system that produced local differences for its own economic and political ends. This new "Empire" operating beyond the nation-state itself is a "regime of the production of identity and difference" that "sets in play mobile and modulating circuits of differentiation and identification." Empire as a system is designed to trap resistance in local conflicts that cannot transcend the very system producing the conflicts, leaving local actors frustrated while handing power to the global actors operating at the global level.[46] This global analysis parallels the frustrations of regional economic planning documented in the present volume, while arguing that new organizing at the international level is pointing to the creation of an alternative transcendence of the present system of global empire.

In a sense the exploding density of new NGO on the global scene is itself a product of this new system, as the lack of traditional democratic venues forces the substitution of nonprofits operating across national borders to represent interests that cannot to be captured within nation-state politics.[47] New technology allows people to communicate and organize to not only lobby their own state or even other states, but also to mobilize across borders to unite global interests that are currently disenfranchised. And beyond such global lobbying, the technology allows local actors to marshal support through boycotts, sympathy strikes, or other forms of pressure against global corporate actors who had previously had nearly free rein to dictate terms within local politics. Beyond the distractions of the smoke and broken windows of particular protests, this is the most radical change embodied in the new post-Seattle politics.

Conclusion

What this all promises is a growing struggle in each region as the positive and negative trends of the information age clash. As global companies create new strategic alliances using the elite human resources of each region in turn, labor and other organizations are marshaling the tools of the information age to organize their communities and the contingent workers pushed to the fringes of economic and political power.

As he surveyed early nineteenth-century America, de Tocqueville cast a critical but pleased eye on the democratic possibilities of his age. But the rise of industrialization did trouble him in the effects it was likely to have on the quality of "association" that he saw as so critical to his view of democracy. He despaired that "[a]n industrial theory stronger than morality or law ties [the industrial worker] to a trade, and often to a place, which he cannot quit. He has been assigned a certain position in society which he cannot quit." He worried that as the lives of workers narrowed and their cultural vision narrowed, a new manufacturing "aristocracy" was arising that had abandoned all loyalty to territory or place and that "there is no true link between rich and poor." With no loyalty to the individuals they employed or to the territory in which they lived, claimed de Tocqueville, this new industrial class of employers represented the greatest threat to democracy in his survey. Although he hoped they would not flourish, De Tocqueville argued: "I think, that, generally speaking, the manufacturing aristocracy which we see rising before our eyes is one of the hardest that have appeared on earth. . . . For if ever again permanent inequality of conditions and aristocracy make their way into the world, it will have been by that door that they entered."[48]

While de Tocqueville's fears were realized in the later part of the nineteenth century, there had been hope after decades of struggle by unions and political organizations that a new form of liberal social democratic state had been established in the mid-twentieth century. A key aspect of that vision was the sense, as John Logan and Harvey Molotch described it, of a business class that had tied itself economically to the regions in which it did business through cross-class alliances and growth coalitions to build the economic infrastructure that benefited all in our economy. More recently, however, we have seen the new globalization of economic commerce and the emergence of regions not as shared democratic spaces but as merely useful venues for the economic management of global technologies. In the Internet we see contrasting visions: on the one hand, rootless corporations promote global commerce as local tax revenues disappear into cyberspace and, on the other, new grassroots organizing is practiced by unionists and environmentalists building a new paradigm of promoting global economic rules aimed at the needs of local communities.

Thus the growth of the Internet embodies a radical change in urban space whereby global links in communication and organization become the key to exercising any local power. Internal alliances within a region will always matter, but local power is inevitably flowing to those who can use the new information technology to deploy global power in support of those local needs. That is the challenge of communities in the new information age.

Notes

Chapter 1

1. Nicholas Negroponte, *Being Digital* (New York: Vintage Books, 1995), 165.
2. Alvin Toffler, *The Third Wave* (New York: William Morrow, 1980), 246.
3. David Gordon, *Fat and Mean: The Corporate Squeeze of Working Americans and the Myth of Managerial "Downsizing"* (New York: Free Press, 1996), 19–20.
4. *AFL-CIO Working America: The Current Economic Situation,* available [online], <http://www.aflcio.org/cse/mod5/situation1.htm>.
5. *AFL-CIO Working America,* <http://www.aflcio.org/cse/mod5/situation3.htm>.
6. *AFL-CIO Working America,* <http://www.aflcio.org/cse/mod5/situation4.htm>.
7. William Julius Wilson, *When Work Disappears: The World of the New Urban Poor* (New York: Alfred Knopf, 1996).
8. James Johnson, Sacramento-based organizer for CTWO, interview by author, March 29, 1995.
9. See Gordon, *Fat and Mean,* 61–94. Gordon describes the "stick strategy" used by corporate America to increase profits at the expense of worker wages and benefits.
10. See Thomas Ferguson and Joel Rogers, *Right Turn: The Decline of the Democrats and the Future Of American Politics* (New York: Hill and Wang, 1986); Sidney Blumenthal, *The Rise of the Counter-Establishment: From Conservative Ideology to Political Power* (New York: Times Books, 1986). Blumenthal, a close advisor to Hillary Clinton, would see his views caricatured in the phrase "the vast right-wing conspiracy" to denote this view of ideological mobilization by the Right.
11. See Robert Reich, *The Work of Nations: Preparing Ourselves for Twenty-First Century Capitalism* (New York: Alfred Knopf, 1991) and *The Future of Success* (New York: Alfred Knopf, 1991); Robert H. Frank and Philip J. Cook, *The Winner-Take-All Society: Why the Few at the Top Get So Much More Than the Rest of Us* (New York: Penguin Books, 1996).
12. Saskia Sassen, *Cities in a World Economy* (Thousand Oaks, Calif.: Pine Forge Press, 1994); Bennett Harrison, *Lean And Mean: The Changing Landscape of Corporate Power in the Age of Flexibility* (New York: Basic Books, 1994).

13. Antonio Gramsci, "Americanism and Fordism," in *Selections from the Prison Notebooks of Antonio Gramsci* (New York: International Publishers, 1992).

14. Alan Lipietz, *Mirages and Miracles: The Crises of Global Fordism,* trans. David Macey (London: Verso, 1987).

15. Jeff Gates, "Statistics on Poverty and Inequality," *Global Policy Forum* (1999), available [online], <http://www.igc.apc.org/globalpolicy/socecon/inequal/gates99.htm>. For an in-depth analysis of the recent research detailing expanding global inequality between and within nations, see also Robert Wade, "Global Inequality: Winners and Losers," *Economist,* April 28, 2001.

16. See Alvin Toffler, *The Third Wave* (New York: William Morrow, 1980); Jeremy Rifkin, *The End of Work: The Decline of the Global Labor Force and the Dawn of the Post-Market Era* (New York: G. P. Putnam's Sons, 1995); Reich, *The Work of Nations.*

17. Benjamin R. Barber, *Jihad vs. McWorld: How the Planet is Both Falling Apart and Coming Together—and What This Means for Democracy* (New York: Times Books, 1995); Thomas L. Friedman, *The Lexus and the Olive Tree* (New York: Farrar, Straus and Giroux, 1999).

18. Wilson, *When Work Disappears.*

19. Mike Davis, *City of Quartz: Excavating the Future in Los Angeles* (New York: Vintage Books, 1990), 153–219.

20. Reich, *The Work of Nations,* 282–300; Manuel Castells, *The Informational City: Information Technology, Economic Restructuring, and the Urban-Regional Process* (Cambridge, Mass.: Basil Blackwell, 1989).

21. Michael Piore and Charles Sabel, *The Second Industrial Divide: Possibilities for Prosperity* (New York: Basic Books, 1984).

22. Michael Best, *The New Competition: Institutions of Industrial Restructuring* (Cambridge: Harvard University Press, 1990), 204–5.

23. Best, *The New Competition,* 235.

24. Best, *The New Competition,* 235.

25. Annalee Saxenian, *Regional Advantage: Culture and Competition in Silicon Valley and Route 128* (Cambridge: Harvard University Press, 1994).

26. Richard Gordon, "Collaborative Linkages, Transnational Networks, and New Structures of Innovation in Silicon Valley's High-Technology Industry," report no. 4, "Industrial Suppliers/Services in Silicon Valley" Silicon Valley Research Group, University of California, Santa Cruz, January 1993.

27. Harrison, *Lean And Mean,* 8.

28. Harrison, *Lean And Mean,* 24.

29. See Stephen Hymer, "The Multinational Corporation and the Law of Uneven Development," in *Economics and the World Order,* ed. J. Bhagwati (New York: Free Press, 1972), 113–40.

30. Manuel Castells, *The Rise of the Network Society* (Cambridge, Mass.: Basil Blackwell, 1996), 171.

31. Reich, *The Work of Nations,* 89.

32. Peter Evans, *Embedded Autonomy: States and Industrial Transformation* (Princeton: Princeton University Press, 1995), 216–17.

33. David G. Becker and Richard L. Sklar, eds., *Postimperialism and World Politics* (London: Praeger, 1999); Scott R. Bowman, *The Modern Corporation and American Political Thought: Law, Power, and Ideology* (University Park: Pennsylvania State University Press, 1996).

34. Reich, *The Work of Nations*, 234.

35. Saxenian, *Regional Advantage*, 34–37.

36. David Harvey, *The Urban Experience* (Baltimore: Johns Hopkins University Press, 1989), 19.

37. Harrison, *Lean And Mean*, 26.

38. Piore and Sabel, *Second Industrial Divide*, 156.

39. Castells, *The Informational City*.

40. Harvey Molotch, "The City as Growth Machine: Toward a Political Economy of Place," *American Journal of Sociology* 82 (1976): 309–32; expanded into John R. Logan and Harvey Molotch, *Urban Fortunes: The Political Economy of Place* (Berkeley and Los Angeles: University of California Press, 1987).

41. Harvey, *The Urban Experience*, 206.

42. Harvey, *The Urban Experience*, 259.

43. David Harvey, *The Condition of Postmodernity* (Cambridge, Mass.: Blackwell, 1989), 293.

44. Reich, *The Work of Nations*, 252.

45. Reich, *The Work of Nations*, 254.

46. Evans, *Embedded Autonomy*.

47. Saxenian, *Regional Advantage*.

48. Kenneth Flamm, *Creating the Computer: Government, Industry, and High Technology* (Washington, D.C.: Brookings Institute, 1988), 252–53.

49. Antonio Gramsci outlined the concept of hegemony and consent in his more pluralistic view of Marxism in essays such as "The Modern Prince"; and its role in the economic sphere in his "Americanism and Fordism." See Gramsci, *Selections from the Prison Notebooks of Antonio Gramsci* (New York: International Publishers, 1992).

50. Manuel Castells argues that the Soviet system collapsed because of its incapacity "to assimilate and use the principles of informationalism" to mobilize its workers. See Castells, *The Rise of the Network Society*, 13.

51. See Jürgen Habermas, *Legitimation Crisis* (Boston: Beacon Press 1975).

52. Charles Sabel, "Learning by Monitoring: The Institutions of Economic Development," in *The Handbook of Economic Sociology* (Princeton: Princeton University Press, 1994), 137.

53. Peter Evans, "Introduction: Development Strategies Across the Public-Private Divide," *World Development* 24 (June 1966): 1033–37; and "Government Action, Social Capital, and Development: Reviewing the Evidence on Synergy," *World Development* 24 (June 1966): 1122.

54. Robert D. Putnam, with Robert Leonardi and Raffaella Y. Nanetti, *Making Democracy Work: Civic Traditions in Modern Italy* (Princeton: Princeton University Press, 1993), 156.

55. Best, *The New Competition*.

56. Evans, *Embedded Autonomy*.

57. Best, *The New Competition*, 44–45.

58. W. Brian Arthur, "Increasing Returns and the Two Worlds of Business," *Harvard Business Review*, July–August 1996.

59. Paul Krugman, "Complex Landscapes in Economic Geography," *American Economic Review* 84 (May 1994).

60. Ann Markusen, Peter Hall, and Amy Glasmeier, *High Tech America: The What, How, Where, and Why of the Sunrise Industries* (Boston: Allen and Unwin, 1986).

61. See Doreen Massey, Paul Quintas, and David Wield, *High-Tech Fantasies: Science Parks in Society, Science, and Space* (London: Routledge, 1992).

62. Ann Markusen, Peter Hall, Scott Campbell, and Sabina Deitrick, *The Rise of the Gunbelt: The Military Remapping of Industrial America* (New York: Oxford University Press, 1991).

63. Peter Hall, "The Geography of the Post-Industrial Economy," in *The Spatial Impact of Technological Change,* ed. John Brotchie, Peter Hall, and Peter Newton (London: Croom Helm, 1987); R. J. Buswell; R. P. Easterbrook, and C. S. Morphet, "Geography, Regions, and Research and Development Activity: The Case of the United Kingdom," in *The Regional Impact of Technological Change,* ed. A. T. Thwaites and R. P. Oakey (New York: St. Martin's Press, 1985); Castells, *The Rise of the Network Society,* 88–89. See also Evans, *Embedded Autonomy*; Alice Amsdem, *Asia's Next Giant: South Korea and Late Industrialization* (New York: Oxford Unviversity Press, 1989).

64. Charles Sabel, "Bootstrapping Reform: Rebuilding Firms, the Welfare State, and Unions," *Politics and Society* 23 (March 1995), 5–48.

65. Reich, *The Work of Nations,* 154.

66. Piore and Sabel, *Second Industrial Divide,* 287.

67. Castells, *The Informational City.*

68. Harrison, *Lean And Mean.*

69. Massey, Quintas, and Wield, *High-Tech Fantasies.*

70. Steven K. Vogel, *Freer Markets, More Rules: Regulatory Reform in Advanced Countries* (Ithaca: Cornell University Press, 1996), 3.

71. See Nathan Newman, "Net Loss: Government, Technology, and the Political Economy of Community in the Age of the Internet" (Ph.D. diss., University of California, Berkeley, 1998), 551.

72. Pam Woodall, "The Beginning of a Great Adventure: Globalisation and Information Technology Were Made for Each Other," *Economist,* September 23, 2000.

Chapter 2

1. "Survey of the Internet: The Accidental Superhighway," *Economist,* July 1, 1995, 4.

2. Clive Thompson, "The Al Gore Internet Is Not As Crazy As It May Seem," *Newsday,* October 15, 2000.

3. "What That TCI–Bell Atlantic Merger Means for You," *Fortune* 128 (November 15, 1993): 82–90.

4. Ellen Messmer, "Industry Heavyweights Team Up to Design Information Superhighway," *Network World* 10 (December 20, 1993): 20.

5. See Alfred Chandler Jr., *The Visible Hand: The Managerial Revolution in American Business* (Cambridge: Harvard University Press, Belknap Press, 1977), chaps. 3–5.

6. Steven Levy, "How the Propeller Heads Stole the Electronic Future," *New York Times Magazine,* September 24, 1995, 59.

7. Howard Rheingold, *Tools for Thought: The People and Ideas Behind the Next Computer Revolution* (New York: Simon and Schuster, 1985).

8. The history in this chapter is derived from a wide range of sources detailed in the Bibliography; invaluable sources are Katie Hafner and Matthew Lyon, *Where Wizards Stay Up Late: The Origins of the Internet* (New York: Simon and Schuster, 1996); along with Rheingold, *Tools for Thought.* Other sources used include Steven Levy, *Hackers: Heroes of*

the Computer Revolution (Garden City, N.Y.: Anchor Press, 1984); Vinton G. Cerf, "Computer Networking: Global Infrastructure for the Twenty-First Century" (1995), available [online]: <http://cra.org/research.impact>; Henry Edward Hard, "The History of the Net" (master's thesis, Grand Valley State University, 1993); Robert Hobbes, "Hobbes' Internet Timeline v2.5" (1993–96), available [online]: <http://info.isoc.org/guest/zakon/Internet/History/HIT.html>; Tim Berners-Lee, *Weaving the Web: The Original Design and Ultimate Destiny of the World Wide Web by Its Inventor* (San Francisco: HarperSanFrancisco, 1999).

Also see Bernard Aboba (with Vinton Cerf), "How the Internet Came to Be," in *The Online User's Encyclopedia* (Reading, Mass.: Addison-Wesley, 1993); Bruce Sterling, "Short History of the Internet," *Magazine of Fantasy and Science Fiction,* February 1993; Christos Moschovitis, Hilary Poole, Tami Schuyler, and Theresa Senft, *The History of the Internet: A Chronology, 1843 to the Present* (Santa Barbara, Calif.: ABC-CLIO, 1999); Simson Garfinkel, *Architects of the Information Society: Thirty-Five Years of the Laboratory for Computer Science at MIT* (Cambridge: MIT Press, 1999); Michael Hiltzik, *Dealers of Lightning: Xerox PARC and the Dawn of the Computer Age* (New York: HarperBusiness, 1999).

9. Evans, *Embedded Autonomy,* 49.

10. There is a more extensive history of this evolution of professional governance of the Internet in Robert E. Kahn, "The Role of Government in the Evolution of the Internet," *Communications of the ACM* 37 (August 1994): 15–19.

11. Hafner and Lyon, *Where Wizards Stay Up Late,* 248.

12. Hafner and Lyon, *Where Wizards Stay Up Late,* 248.

13. Larry Press, "Seeding Networks: The Federal Role," *Communications of the ACM* 39 (October 1996): 11–18, available [e-mail], <lpress@isi.edu>.

14. See Steven Baker, "The Evolving Internet Backbone" (history of the Internet computer network) *UNIX Review* 11 (September 1993): 15.

15. Joseph Gottlieb, "Education: The Right Formula," *Network World* 5 (May 30, 1988): 28–30.

16. Moschovitis, Poole, Schuyler, and Senft, *The History of the Internet.*

17. For other information on this later development of the Net, see Kahn, "The Role of Government," 15–19.

18. Gary H. Anthes, "Commercial Users Move onto Internet," *Computerworld* 25 (November 25, 1991): 50.

19. William Schrader and Mitchell Kapor, "The Significance and Impact of the Commercial Internet," *Telecommunications* 26 (February 1992): S17.

20. Press, "Seeding Networks: the Federal Role."

21. See National Academy of Sciences, *Evolving the High Performance Computing and Communications Initiative to Support the Nation's Information Infrastructure* (Washington, D.C.: National Academy Press, 1995), for a broad discussion of the economic benefits derived for private industry from the existence of government research.

22. Ellen Messmer, "America Online Buys Internet Service Provider," *Network World* 11 (December 5, 1994): 4.

23. Dan Shea, "MCI Focuses on the Internet," *Telephony* 227 (November 28, 1994): 6.

24. Paula Bernier, "The Science of High-Speed Computing," *Telephony* 228 (May 1, 1995): 7, 15.

25. R. Scott Raynovich. "*LAN Times* Talks to PSINet's William Schrader About Internet Connectivity and the Competition." *LAN Times* 12 (October 23, 1995): 8; and Kahn, "The Role of Government," 15–19.

26. Jim Balderston, "GTE, Uunet Merge Internet Services," *InfoWorld* 18 (July 15, 1996): 12.

27. "BBN to Acquire Barrnet from Stanford," *Cambridge Telecom Report,* June 27, 1994.

28. "BBN Planet Launches Internet Services Nationwide," *Cambridge Work-Group Computing Report,* March 20, 1995.

29. D. C. Denison. "Woburn, Mass.–Based Internet Services Firm Genuity to Cut Jobs," *Boston Globe,* May 3, 2001; Sean Buckley, "Back Talk," *Telecommunications,* April 1, 2001.

30. Michelle V. Rafter, "A Transcontinental Undertaking Company Follows Railroads in Building a National Network," *Chicago Tribune,* December 1, 1997; Toni Mack, "Empty Pipes—the Delicious Dilemma for Phil Anschutz and Joe Nacchio: Fill Qwest Communications' Capacious New Network or Sell It Off," *Forbes,* November 30, 1998; Tom Mc-Ghee, "Corporate Opposites Attract, *Albuquerque Journal,* August 30, 1999.

31. "Bulletin," *R & D,* February 1, 2000; Wirbel Loring, "Architects of the Internet Russ Hobby: Advancing to Internet2 over 'Bones,'" *Electronic Engineering Times,* September 27, 2000; Dawn Bushaus, "Project Provides a Breeding Ground for Developing Multicasting, Quality-of-Service Solutions," *InformationWeek,* March 20, 2000; Martin Kady," Energy Gives Qwest Breakthrough Deal," *Denver Business Journal,* January 14, 2000.

32. Karl Polanyi, *The Great Transformation: The Political and Economic Origins of Our Time* (Boston: Beacon Press, 1944), 141.

33. W. Brian Arthur, "Increasing Returns and the Two Worlds of Business," *Harvard Business Review,* July–August 1996.

34. "A World Gone Soft: A Survey of the Software Industry," *Economist,* May 25, 1996.

35. Tim Studt and Jules Duga, "Strong U.S. Economy Drives Continued R&D Growth," *R & D,* January 1, 1999; Scott Stern, "Impact of Basic Research on Technological Innovation," *Congressional Testimony by Federal Document Clearing House,* September 28, 1999.

36. Louis Uchitelle, "Companies Spending More on Research and Development," *New York Times,* November 7, 1997.

37. See Louis Uchitelle, "Corporate Outlays for Basic Research Cut Back Significantly," *New York Times,* October 8, 1996; and National Academy of Sciences, *Evolving the High Performance Computing.*

38. National Academy of Sciences, *Evolving the High Performance Computing.*

39. William Jackson, "Next-Generation vBNS Gets Three-Year Extension," *Government Computer News,* June 5, 2000; Wirbel, "Architects of the Internet."

40. Berners-Lee, *Weaving the Web,* 196–97.

41. See "DARPA Official Hits Policy," *Chilton's Electronic News* 37 (July 29, 1991): 4; Gary H. Anthes, "Feds Ax High-Tech Point Man: DARPA Move Seen as Slap at Industry Funding, *Computerworld* 24 (April 30, 1990): 1; Brian Robinson, "Bush, DOD Attacked for Firing Fields," *Electronic Engineering Times,* no. 588 (April 30, 1990): 1; Stan Baker, "Industry Wonders: What's Next?" *Electronic Engineering Times,* no. 588 (April 30, 1990): 84; Jack Robertson, "Transfer of DARPA Head Ignites Furor," *Electronic News* 36, no. 1807 (1990): 1, and Brian Robinson, "DARPA: We'll Risk It," *Electronic Engineering Times,* no. 534 (April 17, 1989): 19.

42. Martin Likicki, James Schneider, David Frelinger, and Anna Slomovic, *Scaffolding the Web: Standards and Standards Policy for the Digital Economy* (Science and Technology Policy Institute Rand, 2000), 34.

43. For the trials of the NTIA in the Clinton administration, see Judith Hellerstein, "The NTIA Needs to Rethink Its Role in the New Telecommunications Environment," *Telecommunications* 30 (August 1996): 22.

44. Lisa Armstrong, "A New Tune for the Internet Pipers?" *Communications International* 21 (December 1994): 8.

45. Jamie Murphy and Charlie Hofhacker, "Explosive Growth Clogs Internet Backbone," *New York Times*, July 29–30, 1996; and Jamie Murphy and Brian Massey, "No Shortage of Bottlenecks on Information Superhighway," *New York Times*, July 29–30, 1996.

46. William Jackson, "Next-Generation vBNS Gets Three-Year Extension," *Government Computer News*, June 5, 2000.

47. Denise Pappalardo, "WorldCom Adds to Its 'Net Riches," *Network World* 14 (September 15, 1997): 6.

48. Scott Kirsner, "WorldCom/MCI: Good Giant or Bad?" *Wired News*, October 2, 1997; Seth Schiesel, "MCI Accepts $36.5 Billion Worldcom Offer," *New York Times*, November 11, 1997.

49. "AFL-CIO, Church Groups, Urge FCC Reject Worldcom-MCI Merger," *Communications Daily*, August 13, 1998. See two research reports by the labor-backed Economic Policy Institute for the basic positions of the MCI opponents: Jeff Keefe, *Monopoly.com: Will the WorldCom-MCI Merger Tangle the Web?* (Economic Policy Institute, 1998); Dan Schiller, *Bad Deal of the Century: The Worrisome Implications of the WorldCom-MCI Merger* (EPI, 1998).

50. Bell Atlantic, Filing in response to WorldCom-MCI merger, Federal Communications Commission, Washington, D.C., January 5, 1998.

51. Andrew Leonard, "Gobbling Up the Net," *Salon*, June 12, 1997.

52. Scott Rosenberg, "WorldCom Buys Up the Internet; The Net Becomes WorldCom's Fiefdom," *Salon*, October 9, 1997.

53. "Telecoms: FCC Warns Future Mergers After Worldcom/MCI," *Network Briefing*, September 16, 1998; Jeannine Aversa, "Feds Clear Way for Takeover of MCI," Associated Press, September 15, 1998.

54. Neil Weinberg, "Backbone Bullies Beneath the Internet's Happy Communal Culture, a Cadre of Giant Carriers Is Mercilessly Squeezing Every Last Dime It Can out of Smaller Players; Users Are Picking Up the Tab," *Forbes*, June 12, 2000.

55. John Borland, "Net Blackout Marks Web's Achilles Heel," CNET News.com, June 6, 2001.

56. "In Search of the Perfect Market," *Economist*, May 10, 1997.

Chapter 3

1. Richard Barbrook and Andy Cameron, "The California Ideology" (1995); reprinted in *Crypto Anarchy, Cyberstates, and Pirate Utopias*, ed. Peter Ludlow (Cambridge: MIT Press, 2001).

2. Louis Rosetto, "Response to the Californian Ideology" (Hypermedia Research Center, 1996), available [online], <http://www.wmin.ac.uk/media/HRC/ci/calif2.html>.

3. Everett Rogers and Judith Larsen, *Silicon Valley Fever: Growth of High-Technology Culture* (New York: Basic Books, 1986). Along with Everett and Larsen's book, other works with good early histories of Silicon Valley include Thomas Mahon, *Charged Bodies: People,*

Power, and Paradox in Silicon Valley (New York: New American Library Books, 1985); Michael Malone, *The Big Score: The Billion-Dollar Story of Silicon Valley* (Garden City, N.Y.: Doubleday, 1985); and Dirk Hanson, *The New Alchemists: Silicon Valley and the Microelectronics Revolution* (Boston: Little, Brown, 1982).

4. Malone, *The Big Score,* 15.

5. Malone, *The Big Score,* chap. 12.

6. Hanson, *The New Alchemists* and Mahon, *Charged Bodies,* 160.

7. Mahon, *Charged Bodies,* 153.

8. Everett and Larsen, *Silicon Valley Fever,* 35–36.

9. Saxenian, *Regional Advantage.*

10. Malone, *The Big Score.*

11. Hanson, *The New Alchemists,* 79.

12. Mahon, *Charged Bodies,*

13. Anna Slomovic, *An Analysis of Military and Commercial Microelectronics: Had DoD's R&D Funding Had the Desired Effect,* A Rand Graduate School Dissertation (Santa Monica, Calif.: Rand, 1991).

14. Saxenian, *Regional Advantage,* 42.

15. Larry Press, "Before the Altair: The History of Personal Computing," *Communications of the ACM* 36 (September 1993): 27–33.

16. Howard Rheingold, *Tools for Thought: The People and Ideas Behind the Next Computer Revolution* (New York: Simon and Schuster, 1985), chap. 9.

17. Steven Levy, *Hackers: Heroes of the Computer Revolution* (Garden City, N.Y.: Anchor Press, 1984). Levy's book is one of the most in-depth stories of how the social ethic of free and open computing formed in the bowels of MIT's computer lab and the way it crossed the continent to California.

18. Paul Saffo, "Racing Change on a Merry-Go-Round: MIT Management in the Nineties Program Reports Industry Overall Is Not More Productive Because of Computing Technology," *Personal Computing* 14 (May 25, 1990): 67. Saffo details the revolutionary vision of Engelbart and how little modern business has engaged with the full thrust of Engelbart's vision.

19. Rheingold, *Tools for Thought,* 214.

20. Steven Baker, "The Evolving Internet Backbone: History of the Internet Computer Network," *UNIX Review* 11 (September 1993): 15.

21. Rheingold, *Tools for Thought,* 199.

22. Jim Warren, "We, the People, in the Information Age: Early Times in Silicon Valley," *Dr. Dobb's Journal* 16 (January 1991): 96D.

23. Michael Hiltzik, *Dealers of Lightning: Xerox PARC and the Dawn of the Computer Age* (New York: HarperBusiness, 1999), 18–20.

24. Rheingold, *Tools for Thought,* chap. 10.

25. Levy, *Hackers,* 142–43.

26. Hafner and Lyon, *Where Wizards Stay Up Late,* 151–52.

27. Press, "Before the Altair."

28. Rheingold, *Tools for Thought,* 203.

29. Levy, *Hackers,* chap. 10.

30. Stephen Manes and Paul Andrews, *Gates: How Microsoft's Mogul Reinvented an Industry—and Made Himself the Richest Man in America* (New York: Simon and Schuster, 1993), chap. 12.

31. See Ann Markusen, Peter Hall, and Amy Glasmeier, *High Tech America: The What,*

How, Where, and Why of the Sunrise Industries (Boston: Allen and Unwin, 1986); Doreen Massey, Paul Quintas, and David Wield, *High-Tech Fantasies: Science Parks in Society, Science, and Space* (London: Routledge, 1992). Richard Florida and Martin Kenney also emphasize the lost public funds by local authorities that have ended up merely subsidizing local business in university-business "partnerships" (*The Breakthrough Illusion: Corporate America's Failure to Move from Innovation to Mass Production* [Basic Books, 1990]).

32. E. J. Malecki, "Public Sector Research and Development and Regional Economic Performance in the United States," in *The Regional Impact of Technological Change*, ed. A. T. Thwaites and R. P. Oakey (New York: St. Martin's Press, 1985).

33. Florida and Kenney, *The Breakthrough Illusion;* see chap. 2 for the main part of this argument.

34. See Hanson, *The New Alchemists.*

35. From an interview in Mahon, *Charged Bodies,* 77.

36. Florida and Kenney, *The Breakthrough Illusion,* 18.

37. Malone, *The Big Score.*

38. M. J. Taylor, "Organizational Growth, Spatial Interaction, and Locational Decision-Making," *Regional Studies* 9 (1975): 313–23.

39. Manes and Andrews, *Gates.*

40. Everett and Larsen, *Silicon Valley Fever,* 63–65.

41. R. Gordon and L. Kimball, "The Impact of Industrial Structure on Global High Technology Location," in *The Spatial Impact of Technological Change*, ed. John Brotchie, Peter Hall, and Peter Newton (London: Croom Helm, 1987).

42. Florida and Kenney, *The Breakthrough Illusion,* 74.

43. Gordon and Kimball, "The Impact of Industrial Structure," 168.

44. Mark Hall and John Barry, *Sunburst: The Ascent of Sun Microsystems* (Chicago: Contemporary Books, 1990). Written by two former employees, this book is a broad look at the internal developments of Sun in its first eight years and is the source of much of the information about Sun in this section.

45. Brent Schlender, "Whose Internet Is It, Anyway?" *Fortune* 132 (December 11, 1995): 120–42.

46. Saxenian, *Regional Advantage,* 134–35.

47. See Hafner and Lyon, *Where Wizards Stay Up Late,* 250.

48. Darryl K. Taft, "Opportunity Lies in Government Downsizing," *Government Computer News* 11 (February 17, 1992): S14.

49. Thomas R. Temin and Shawn P. McCarthy, "Sun Chief Sells Unix and Minces No Words," *Government Computer News* 12 (December 6, 1993): 20.

50. Hall and Barry, *Sunburst,* 151.

51. Kate Button, "Hub of the Solar System: Sun Microsystems' Scott McNealy's Attitude Towards Cloning," *Computer Weekly,* December 3, 1992, 39.

52. Peter Clarke, "European Spacemen Launch Free Sparc-Like Core," *Electronic Engineering Times,* March 6, 2000.

53. Saxenian, *Regional Advantage,* 134.

54. "Bigger Faster, Better Ears," *Economist,* December 7, 1996, 59–60.

55. Lawrence M. Fisher, "Routing Makes Cisco Systems a Powerhouse of Computing," *New York Times,* November 11, 1996.

56. For good accounts of Cisco's rise, see Julie Pitta, "Long Distance Relationship," *Forbes* 149 (March 16, 1992): 136–37; and Joseph Nocera, "Cooking with Cisco," *Fortune* 132 (December 25, 1995): 114–22.

57. Steven Baker, "Fiber to the Desktop," *UNIX Review* 13 (1995): 19–25.

58. Susan Kerr, "3Com Corp. (The Datamation 100)" *Datamation* 38 (June 15, 1992): 141.

59. Tim Stevens, "Multiplication by Addition," *Industry Week* 245 (July 1, 1996): 20–24.

60. Fisher, "Routing Makes Cisco a Powerhouse."

61. "Bigger Faster, Better Ears," 59–60.

62. Jerry Michalski, "O Pioneers!" *RELease 1.0* 94 (January 31, 1994): 5.

63. Tim Stevens, "NCSA: National Center for Supercomputing Applications," *Industry Week* 243 (December 19, 1994): 56–58; and Kimberly Patch, "Spyglass Takes on Mosaic Licensing: Will Focus on Support and Security," *PC Week* 11 (August 29, 1994): 123.

64. Ellis Booker, "Spyglass to Commercialize Future Mosaic Versions," *Computerworld* 28 (August 29, 1994): 16.

65. Ellen Messmer, "Spyglass Captures Mosaic Licensing," *Network World* 11 (August 29, 1994): 4.

66. Accounts of Netscape's start-up are from Albert G. Holzinger, "Netscape Founder Points, and It Clicks," *Nation's Business* 84 (January 1996): 32; Eric Nee, "Jim Clark," *Upside* 7 (October 1995): 28–48.

67. Nee, "Jim Clark."

68. Tom Steinert-Threlkeld, "The Internet Shouldn't be a Breeding Ground for Monopolies—Mosaic Communications' NetScape Giveaway Could Be Prelude to Market Dominance," *InterActive Week* 1 (November 7, 1994): 44.

69. "University of Illinois and Netscape Communications Reach Agreement," *Information Today* 12 (March 1995): 39.

70. Messmer, "Spyglass Captures Mosaic Licensing."

71. Nee, "Jim Clark."

72. Steve Lohr, "Spyglass, a Pioneer, Learns Hard Lessons About Microsoft," *New York Times*, March 2, 1998.

73. Morris Edwards, "The Network Computer Comes of Age," *Communications News* 33 (July 1996): 44–45.

74. Warren, "We, the People."

75. Stewart Alsop, "Please Save Me from My Editors," *Fortune* 134 (October 28, 1996): 203–4.

76. Edward F. Moltzen, "IBM Puts 20% of R&D into networking," *Computer Reseller News,* no. 705 (October 14, 1996): 290.

77. Zina Moukheiber, "Windows NT—Never!" *Forbes* 158 (September 23, 1996): 42–43.

78. Brent Schlender, "Sun's Java: The Threat to Microsoft Is Real," *Fortune* 134 (November 11, 1996): 165–70.

79. Peter Lewis, "Alliance Formed Around Sun's Java Network," *New York Times,* December 11, 1996.

80. Alex Lash, "JavaOS Headed to Consumer Gear," CNET News.com, March 25, 1998.

81. "Sun Spots Its Chance," *Network World* 13 (August 12, 1996): I33; Deborah Gage, "Sun Launches New Servers, Support Boost," *Computer Reseller News,* no. 681 (April 29, 1996): 67, 70; and Moukheiber, "Windows NT—Never!"

82. John Markoff, "Several Big Deals Near for Sun's Java Language," *New York Times,* March 24, 1998.

83. "A New Player in Set-Top Boxes," *Business Week*, March 12, 1998.

84. Alan Deutschman, "The Next Big Info Tech Battle," *Fortune* 128 (November 29, 1993): 38–50. "The House That Larry Built," *Economist*, February 19, 1994, 73; and Kim S. Nash, "NCube Piggybacks on Oracle Plans," *Computerworld* 28 (January 31, 1994): 33.

85. Deborah DeVoe, "nCube to Unveil Multimedia Servers," *InfoWorld* 18 (February 19, 1996): 38.

86. Adam Rogers, "Getting Faster by the Second," *Newsweek*, December 9, 1996, 86–89; Mark Hall, "CIO Goes Beyond the Cutting Edge," *ComputerWorld Canada*, March 23, 2001.

87. Alexander Wolfe, "Cray Enters Race for Teraflops Computer," *Electronic Engineering Times*, no. 922 (October 7, 1996): 1, 16.

88. Richard Williamson, "After Ruling, nCube Must Make Changes," *Interactive Week*, October 3, 2000.

89. Deutschman, Next Big Info Tech Battle."

90. Petit \Charles, "Praise for California at Science Summit—but Leaders Worry About Low Test Scores," *San Francisco Chronicle*, May 29, 1996, A15.

91. Paul Saffo, "Racing Change on a Merry-Go-Round: MIT Management in the Nineties Program Reports Industry Overall Is Not More Productive Because of Computing Technology," *Personal Computing* 14 (May 25, 1990): 67.

92. Hanson, *The New Alchemists*, 37.

Chapter 4

1. Joint Venture: Silicon Valley, *An Economy at Risk: The Phase I Diagnostic Report*, report prepared by the Center for Economic Competitiveness (SRI International, 1992).

2. *The Joint Venture Way: Lessons for Regional Rejuvenation*, report produced by Joint Venture: Silicon Valley Network, San Jose, 1995.

3. See Julian Bright, "The Smart City: Communications Utopia or Future Reality?" *Telecommunications* (international edition), 29 (September 1995): 175–81.

4. Kathy Leong Chin, "Siliconnected Valley (Silicon Valley Companies Form Smart Valley Inc. Non-profit to Build Giant Information Network)," *InformationWeek*, no. 438 (August 16, 1993): 22.

5. Joint Venture: Silicon Valley, *An Economy at Risk: The Phase I Diagnostic Report*, report prepared by the Center for Economic Competitiveness (SRI International, 1992).

6. Patricia Schnaidt, "Cruising Along the Super I-Way: Cyber-Pioneers Look to the Information Superhighway to Carry Education, Health Care, and Commerce Applications," *LAN Magazine* 9 (May 1994): S8.

7. "CommerceNet Moves East," *Telecommunications* (Americas edition), 29 (February 1995): 12; and Jim Brown, "More Public/Private Sharing in Mass.," *Network World* 12 (May 15, 1995): 45.

8. Monica Greco, "AMTEX Bears First Fruit," *Apparel Industry Magazine* 56 (April 1995): 32–38.

9. Ellen Messmer, "Ford Testing a 'Private Internet' to Link Suppliers," *Network World* 12 (March 13, 1995): 14; and "Automakers Drive Private IP Network," *Network World* 13 (September 30, 1996): 33.

10. Michael McRay, interview by author at CommerceNet offices, June 5, 1995.

11. McRay, interview, June 5, 1995.

12. David Bank, "Smart Valley to Get $8 Million; Goal of U.S. Grant Is to Help Turn Internet into Electronic Marketplace," *San Jose Mercury News,* November 24, 1993: 8D.

13. Tom Masotto, Working Catalogs program manager, interview with author at CommerceNet offices, June 13, 1995.

14. McRay, interview, June 5, 1995.

15. Randy Whiting, interview by author at HP's Palo Alto office, July 21, 1995.

16. Karen Greenwood, Smart Valley project manager for BAMTA, interview by author at Smart Valley office, September 28, 1995.

17. Paul Kutler, interview, November 15, 1995.

18. Bill Davidow, interview by author at Davidow's Palo Alto office, November 11, 1995.

19. Mack Hicks, interview by author at Hicks's San Francisco home, June 22, 1995.

20. Whiting, interview, July 21, 1995.

21. Mason Myers, interview by author at BAMTA office, December 9, 1995.

22. Joint Venture, *Index of Silicon Valley 1997: Measuring Progress Towards a 21st Century Community,* prepared by Collaborative Economics for Joint Venture: Silicon Valley (1997).

23. Eric Benhamou, CEO of 3Com, interview by author at 3Com office, July 21, 1995.

24. Joint Venture, *Index of Silicon Valley 1995; 1996; 1997.*

25. Bruce Lowenthal, interview by author at Tandem office, July 21, 1995.

26. Peter Sinton, "Businesses Wired on Intranets," *San Francisco Chronicle,* September 26, 1996.

27. Francois Bar, "Configuring the Telecommunications Infrastructure for the Computer Age: The Economics of Network Control" (Ph.D. diss., University of California–Berkeley, 1990).

28. Jeffrey Garten, "Why the Global Economy Is Here to Stay," *Business Week,* March 12, 1998.

29. Whiting, interview, July 21, 1995.

30. Lowenthal, interview, July 21, 1995.

31. Mary Brandel, "On-line Catalogs Are Booting Up," *Computerworld* (*Electronic Commerce Journal* supplement), April 29, 1996, 5; and "AMP Unit Targets Online Commerce," *Electronic Engineering Times,* no. 919 (September 16, 1996): 24.

32. Bob Tedeschi, "The Net's Real Business Happens .Com to .Com," *New York Times,* April 19, 1999.

33. "Boring, Boring, Boring: Business-to-Business E-Commerce Is a Revolution in a Ball Valve," in the *Economist*'s "In Search of the Perfect Market: A Survey of Electronic Commerce," May 10, 1997.

34. Hicks, interview, June 22, 1995.

35. McRay, interview, June 5, 1995.

36. Dan Goodin, "The Nonprofit Millionaires," *Industry Standard,* February 26, 2001.

37. Greenwood, interview, September 28, 1995.

38. Bennett Harrison, *Lean And Mean: The Changing Landscape of Corporate Power in the Age of Flexibility* (New York: Basic Books, 1994), 23.

39. Lowenthal, interview, July 21, 1995.

40. Michael Krantz, "Doing Business on the Net is Hard Because the Underlying Software Is So Dumb. XML Will Fix That," *Time,* November 10, 1997; Constance Loizos, "Ecologies of Scale," *Red Herring,* February 1, 1998.

41. Goodin, "The Nonprofit Millionaires."

42. Christopher Marquis, "U.S. Approves Formation of Supply Web Site for Automakers," *New York Times,* September 12, 2000.

43. Goodin, "The Nonprofit Millionaires."

44. Tim Clark, "Pushing E-Commerce Standards," CNET News.com, March 19, 1998.

45. Much of the history of MCC in this chapter comes from David Gibson and Everett Rogers, *R&D Collaboration on Trial: The Microelectronics and Computer Technology Corporation* (Boston: Harvard Business School Press, 1994). They also give a good overview of the role of consortia in modern electronics. See also Peter Burrows, "Craig Fields's Not-So-Excellent Adventure," *Business Week,* no. 3362 (March 14, 1994): 32; Otis Port and Peter Burrows, "R&D, with a Reality Check," *Business Week,* no. 3355 (January 24, 1994): 62–65; "MCC Digs Out of the 'Celestial Sandbox of Computer Science,'" *Electronic Business* 18 (May 18, 1992): 67–68; John Carey and Jim Bartimo, "If You Control . . . Computers, You Control the World," *Business Week,* no. 3170 (July 23, 1990): 31; and Sam Verhovek Howe. "Austin Rides a Winner: Technology," *New York Times,* January 31, 1998.

46. Joint Venture: Silicon Valley, *An Economy at Risk,* 51–56.

47. See John Seely Brown and Paul Duguid, *The Social Life of Information* (Boston: Harvard Business School Press, 2000), 161.

48. *The Joint Venture Way,* 118.

49. Stephen Birmingham, *California Rich: The Lives, the Times, the Scandals, and the Fortunes of the Men and Women Who Made and Kept California's Wealth* (New York: Simon and Schuster, 1980).

50. Saxenian, *Regional Advantage.*

51. Anna Lee Saxenian, "Contrasting Patterns of Business Organization in Silicon Valley," Working Paper no. 535, Institute of Urban and Regional Development, University of California, Berkeley, April 1991.

52. "Chip Makers, U.S. Announce Venture," Associated Press, September 11, 1997.

53. Richard Florida and Martin Kenney, *The Breakthrough Illusion: Corporate America's Failure to Move from Innovation to Mass Production* (New York: Basic Books, 1990), 105.

54. David Gibson and Everett Rogers, *R&D Collaboration on Trial: The Microelectronics and Computer Technology Corporation* (Boston: Harvard Business School Press, 1994), 520.

55. Greenwood, interview, September 28, 1995.

56. Gibson and Rogers, *R&D Collaboration on Trial,* 215.

57. Gibson and Rogers, *R&D Collaboration on Trial.*

58. Davidow, interview, November 11, 1995.

59. McRay, interview, June 5, 1995; "The Age of the Cloud: A Survey of Software," *Economist,* April 14, 2001, 26.

60. Eran Gross, interview by author at BAMTA office, November 15, 1995.

61. Lowenthal, interview, July 21, 1995.

62. Tom Abate, "Internet Infighting," *Upside Today,* September 30, 1995.

63. Likicki, Schneider, Frelinger, and Slomovic, *Scaffolding the Web,* 30–32.

64. Brenda Sandburg, "The Standards Bearers: When It Comes to the Internet, Patents Are Made to Trade—Even for Mighty Microsoft," *Recorder,* March 10, 1999.

65. For a full view of how "transaction costs" of relationships create hierarchies in the economy, see Oliver Williamson, *The Economic Institutions of Capitalism: Firms, Markets, Relational Contracting* (New York: Free Press, 1986).

66. Whiting, interview, July 21, 1995.

67. "Microsoft Steps Up Calif. Presence," Associated Press, March 26, 2001.

68. Michael Noer, "Stillborn," *Forbes,* March 19, 1998.

69. Richard Shim, "Dell Throws Its Weight Behind DVD + RW," CNET News.com, June 28, 2001.

70. Likicki, Schneider, Frelinger, and Slomovic, *Scaffolding the Web,* 8–10.

71. Wylie Wong, "New XML E-Business Standard Introduced," CNET News.com, May 14, 2001; Mary Jo Foley, "XML Goes to the U.N., NATO: This Biz-to-Biz Infrastructure Is Not Just for Geeks Anymore," *Sm@rt Partner,* October 13, 1999.

72. Joe McGarvey, "Intranets, NT Shape Server Market," *SoftBase,* June 1, 1997.

73. Janet Kornblum, "Netscape Sets Source Code Free," *News.Com,* March 31, 1998.

74. Michael Moeller, "Fort Apache: Freeware's Spirit Outshines Commercial Products," *PC Week* 14 (June 9, 1997).

75. Glyn Moody, "The Greatest OS That (N)ever Was," *Wired,* August 1997.

76. Stephen Shankland, "Linux Growth Underscores Threat to Microsoft," CNET News.com, February 28, 2001.

77. Eamonn Sullivan, "Freedom Is Priceless, Even When It's Free," *PC Week* 13 (November 25, 1996).

78. Davidow, interview, November 11, 1995.

79. Joint Venture: Silicon Valley, *An Economy at Risk.*

80. John Markoff, "Silicon Valley Job Growth Begins to Slow," *New York Times,* January 15, 2001.

81. Marilynn S. Johnson, *The Second Gold Rush: Oakland and the East Bay in World War II* (Berkeley and Los Angeles: University of California Press, 1993).

82. Markoff, "Silicon Valley Job Growth Begins to Slow."

83. David Moberg, "Everything You Know About the New Economy Is Wrong," Salon.Com, October 26, 2000.

84. Joint Venture, *Index of Silicon Valley 1997.*

85. Mark Simon, "Down Side Of Peninsula Job Boom," *San Francisco Chronicle,* March 24, 1997.

86. Robert Hof, "Too Much of a Good Thing?" *Business Week,* August 25, 1997.

87. Joint Venture, *Index of Silicon Valley 1996.*

88. Louis Freedberg, "Race Panel Gets an Earful In San Jose; Talks Dominated by Hecklers, Complaints About Wage Gap," *San Francisco Chronicle,* February 11, 1998.

89. *Shock Absorbers in the Flexible Economy: The Rise of Contingent Labor in Silicon Valley,* report prepared by Working Partnership USA (1996).

90. Leah Beth Ward, "Tax Rules Squeezing Independent Contractors," *New York Times,* November 10, 1996.

91. *Shock Absorbers in the Flexible Economy.*

92. "Work Week," *Wall Street Journal,* March 27, 2001.

93. Hicks, interview, June 22, 1995.

94. Donald L. Barlett and James B. Steele, "Say Goodbye to High-Tech Jobs . . . Many Are Moving Offshore," *Philadelphia Inquirer* series (1996) and "Silicon, Silicon Everywhere," in "Future Perfect? A Survey of Silicon Valley," *Economist,* March 29, 1997.

95. Mark Landler, "Hi, I'm in Bangalore (but I Dare Not Tell)," *New York Times,* March 21, 2001.

96. Craig S. Smith, "City of Silk Becoming Center of Technology," *New York Times,* May 28, 2001.

97. John Markoff, "Gold Rush from Software Animates Silicon Valley," *New York*

Times, January 13, 1997; and Louis Uchitelle, "Service Sector Guru Now Having Second Thoughts," *New York Times,* May 8, 1996.

98. Davidow, interview, November 11, 1995.

99. "California's Economy: The Real Trouble," *Economist,* July 28, 2001.

100. Peter Fimrite, "Altamont Rail Plan on Track: A Central Valley-South Bay Link," *San Francisco Chronicle,* December 2, 1996.

101. Kathy Blankenship, interview by author at Smart Valley office, September 28, 1995.

102. Patricia Schnaidt, "Cruising Along the Super I-Way: Cyber-Pioneers Look to the Information Superhighway to Carry Education, Health Care, and Commerce Applications," *LAN Magazine* 9 (May 1994): S8.

103. Blankenship, interview, September 28, 1995.

104. Manuel Castells, *The Informational City: Information Technology, Economic Restructuring, and the Urban-Regional Process* (Oxford: Basil Blackwell, 1989).

105. *The Joint Venture Way,* i.13.

106. Benjamin Pimentel, "Congested Silicon Valley Slow to Board Light Rail," *San Francisco Chronicle,* February 24, 1997.

107. Maria Alicia Gaura and Marshall Wilson, "Silicon Valley Commuters Ask for More Transit: 'Unlock the Gridlock' Forum in San Jose," *San Francisco Chronicle,* March 21, 1997.

108. Peter Fimrite, "Altamont Rail Route Gets Boost; Alameda County Panel OKs Funds for Commuter Line," *San Francisco Chronicle,* December 21, 1996.

109. Catherine Bowman, "Bus Riders Union Won't Stay Seated; AC Transit Passengers Want a Say in Service," *San Francisco Chronicle,* March 31, 1997.

110. "The Changing Dream," in *Future Perfect? A Survey of Silicon Valley,* in *Economist,* March 29, 1997.

111. Greg Lucas, "Prop. 211 Loses By Wide Margin," *San Francisco Chronicle,* November 6, 1996; and Mark Simon, "Prop. 211 Foes Share Riches with Other Campaigns: $1.75 Million in Past Week Given to Anti-217 Forces," *San Francisco Chronicle,* November 1, 1996.

Chapter 5

1. See Harvey, *The Urban Experience;* and Molotch, "The City as Growth Machine.

2. Steven K. Vogel, *Freer Markets, More Rules: Regulatory Reform in Advanced Countries* (Ithaca: Cornell University Press, 1996); Jill Hills, *Deregulating Telecoms: Competition and Control in the United States, Japan, and Britain* (Westport, Conn.: Quorum Books, 1986).

3. Charles K. Meister, "Converging Trends Portend Dynamic Changes on the Banking Horizon," *Bank Marketing* 28 (July 1996): 15–21.

4. Seth Lubove, "Cyberbanking," *Forbes* 158 (October 21, 1996): 108–16.

5. "Banks Worldwide Plan to Increase Internet Services in 1997," *Bank Marketing* 29 (January 1997): 9.

6. Bill Orr, "We're Not in Kansas Anymore," *ABA Banking Journal* 88 (July 1996): 72.

7. Frances Martin, "Banking on Internet Banking's Success," *Credit Card Management* 9 (August 1996): 50; "SFNB Marks One Year on the Frontline of Internet Banking,"

ABA Banking Journal 88 (December 1996): 62; and Bruce Rule, "Internet Bank Decides the Real Money's in Consulting," *Investment Dealers Digest* 63 (January 13, 1997): 12.

8. Sandeep Junnarkar, "Fewer Bricks Mean Higher Returns at New Internet Banks," *New York Times,* February 25, 1998.

9. "More Banks Follow the Internet Route," *Banker* 146 (August 1996): 20.

10. Hicks, interview, June 22, 1995.

11. David Morgan, "Internet Begins to Demonstrate Its Value to Business," Reuters New Media, April 11, 1997.

12. Jeff Lebowitz, "The Dawning of a New Era, *Mortgage Banking* 56 (June 1996): 54–66.

13. Sam Zuckerman, "Taking Small Business Competition Nationwide," *US Banker* 106 (August 1996): 24–28.

14. Lloyd Reichenbach and Mark Hill, "Small Business Lending and the Internet," *Risk Management Association Journal,* February 1, 2001.

15. Zuckerman, "Taking Small Business Competition Nationwide."

16. Ellis Booker, "'Net to Reshape Business," *Computerworld* 29 (March 13, 1995): 14; and Orr, "'We're Not in Kansas Anymore,'" 72.

17. Joanna Smith Bers, "Banks Must Decide Whether to Be a Catalyst or Catatonic in the Face of E-Commerce," *Bank Systems and Technology* 33 (November 1996): 16.

18. Beth Davis, "Wells Fargo to Certify Net Payments," *Informationweek,* no. 610 (December 16, 1996): 30; Michael McGann, "Wells Fargo Gets SET-Compliant Certificates," *Bank Systems and Technology* 34 (March 1997): 20; Lawrence Fisher, "Hewlett in Deal for Maker of Credit-Card Devices," *New York Times,* April 24, 1997.

19. Tim Clark, "HP Leads E-Commerce Initiative," *News.Com,* December 2, 1997.

20. The early history of Bank of America comes from Gerald D. Nash, *A. P. Giannini and the Bank of America* (Norman: University of Oklahoma Press, 1992); and Joseph Nocera, *A Piece of the Action: How the Middle Class Joined the Money Class* (New York: Simon and Schuster, 1994).

21. Nash, *A. P. Giannini and the Bank of America,* 57.

22. Moira Johnston, *Roller Coaster: The Bank of America and the Future of American Banking* (New York: Ticknor and Fields, 1990), 53.

23. Nocera, *A Piece of the Action,* 139.

24. Nocera, *A Piece of the Action,* 136.

25. Johnston, in *Roller Coaster,* carefully traces the fall of Bank of America in the 1970s and 1980s.

26. Michael Lewis, *Liar's Poker: Rising Through the Wreckage on Wall Street* (New York: Penguin Books, 1989).

27. Johnston, *Roller Coaster,* 229.

28. Michael C. Perkins and Celia Nunez, "The Market's Odds Are Stacked Against You," *Newsday,* March 18, 2001.

29. Lewis, *Liar's Poker,* 104.

30. Charles K. Meister, "Converging Trends Portend Dynamic Changes on the Banking Horizon," *Bank Marketing* 28 (July 1996): 15–21.

31. Oscar Gandy Jr., "It's Discrimination, Stupid!" in *Resisting the Virtual Life: The Culture and Politics of Information,* ed. James Brook and Ian Boal (San Francisco: City Lights, 1995).

32. Cheryl J. Prince, "Last Chance to Recapture Payments," *Bank Systems and Technology* 33 (July 1996): 28–32.

33. Marc Levinson, "Get Out of Here!" *Business Week,* June 3, 1996.

34. Junnarkar, "Fewer Bricks Mean Higher Returns.

35. Meister, "Converging Trends.

36. Richard J. Ritter, "Redlining: The Justice Department Cases," *Mortgage Banking* 55 (September 1995): 16–28; Griffith L. Garwood and Dolores S. Smith, "The Community Reinvestment Act: Evolution and Current Issues," *Federal Reserve Bulletin* 79 (April 1993): 251.

37. "Wells Fargo Announces $45 Billion CRA Commitment in First Interstate Bid," *ABA Bank Compliance* 17 (January 1996): 5–6, regulatory and legislative advisory; and Ellen Rowley Monck, "How a Megabank Is Answering the CRA Challenge," *Journal of Commercial Lending* 76 (September 1993): 47–54.

38. Roger Rainbow, "Global Forces Shape the Electricity Industry," *Electricity Journal* 9 (May 1996): 14–20.

39. Joe Spiers, "Upheaval in the Electricity Business," *Fortune* 133 (June 24, 1996): 26–30.

40. Logan and Moltoch, *Urban Fortunes.*

41. Colleen Frye, "Electric Utilities Have Oasis in Sight," *Software Magazine* 16 (October 1996): 19.

42. Joseph F Schuler Jr., "Oasis: Networking on the Grid," *Public Utilities Fortnightly* 134 (November 1, 1996): 32–36; V C. Ramesh, "Information Matters: Beyond OASIS," *Electricity Journal* 10 (March 1997): 78–82; Scott M. Gawlicki, "The OASIS Horizon," *Independent Energy* 26 (November 1996): 26–29.

43. "FERC Says Use Web to Post Transmission Access Data," *Electricity Journal* 9 (January/February 1996): 5–6.

44. Sidney Mannheim Jubien, "The Regulatory Divide: Federal and State Jurisdiction in a Restructured Electricity Industry," *Electricity Journal* 9 (November 1996): 68–79.

45. Marvin Katz, "Beyond Order 636: Making the Most of Existing Pipeline Capacity," *American Gas* 77 (June 1995): 26–29.

46. Greg M. Lander, "Just in Time: EDI for Gas Nominations," *Public Utilities Fortnightly* 134 (February 1, 1996): 20–23; Sheila S. Hollis and Andrew S. Katz, "Wired or Mired? Electronic Information for the Gas Industry," *Public Utilities Fortnightly* 134 (February 1, 1996): 15–18; Brian White, "Electronic Bulletin Board Standardization Resulting from FERC EBB Working Group Actions," *Gas Energy Review* 22 (August 1994): 2–11.

47. Scott Wallace, "Power to the People: A Complex Web of Computerized Transactions Is Involved in Delivery of Electric Power," *Computerworld* 25 (June 3, 1991): 79–80; Kimberly Weisul, "Electricity Trading Mart Moves Toward Internet," *Investment Dealers Digest* 63 (January 27, 1997): 13; "FERC Says Use Web to Post Transmission Access Data," *Electricity Journal* 9 (January/February 1996): 5–6.

48. Schuler, "Oasis"; Gawlicki, "The OASIS Horizon"; Frye, "Electric Utilities Have Oasis in Sight"; Charlotte Dunlap, "Integrator Spins Web Sites for Utility Outlets," *Computer Reseller News,* no. 698 (August 26, 1996): 65–66.

49. Rick Jurgens, "Behind the Scenes of CA's Power Crisis Is a Deregulated, High-Stakes Commodities Market," *Contra Costa Times,* July 1, 2001.

50. Dan Richard and Melissa Lavinson, "Something for Everyone: The Politics of California's New Law on Electronic Restructuring," *Public Utilities Fortnightly* 134 (November 15, 1996): 37–41; Gawlicki, "The OASIS Horizon."

51. Joseph F Schuler Jr., "1996 Regulators' Forum: Consensus and Controversy," *Public Utilities Fortnightly* 134 (November 15, 1996): 14–24.

52. William H. Miller, "Electrifying Momentum," *Industry Week* 246 (February 17, 1997): 69–74.

53. Frank Clemente, "The Dark Side of Deregulation," *Public Utilities Fortnightly* 134 (May 15, 1996): 13–15.

54. Hal Plotkin, "Small Biz Gets No Charge from Electricity Deregulation," *Inc.* 18 (December 1996): 32.

55. Erik Ingram, "Deregulation—Devil's in the Details: How Competitive Market in Electricity May Shape Up," *San Francisco Chronicle*, May 8, 1997.

56. Joe Spiers, "Upheaval in the Electricity Business," *Fortune* 133 (June 24, 1996): 26–30.

57. Richard and Lavinson, "Something for Everyone."

58. Kimberly Kindy, "Power Producers Coordinated Data to Take Advantage of California Market," *Orange County Register*, March 26, 2001.

59. Jurgens, "Behind the Scenes of CA's Power Crisis."

60. John Woolfolk and Steve Johnson, "Tallying Price of California's Power Fiasco, Consumers May Pay $100 Billion over the Next Decade," *San Jose Mercury News*, July 8, 2001.

61. Clemente, "The Dark Side of Deregulation."

62. Barbara R. Barkovich, *Regulatory Interventionism in the Utility Industry: Fariness, Efficiency, and the Pursuit of Energy Conservation* (New York: Quorum Books, 1989), 157.

63. See Charles Coleman, *PGandE of California: The Centennial Story of Pacific Gas and Electric Company, 1852–1952* (New York: McGraw-Hill, 1952), which contains an early history of PG&E from the company's viewpoint.

64. Roger Rainbow, "Global Forces Shape the Electricity Industry," *Electricity Journal* 9 (May 1996): 14–20.

65. "The First Two Decades," *EPRI Journal*, January 1993.

66. Richard E. Balzhiser, "Technology—It's Only Begun to Make a Difference," *Electricity Journal* 9 (May 1996): 32–45.

67. Robert L. Hirsch, "Technology for a Competitive Industry," *Electricity Journal* 9 (May 1996): 81, 80; Cate Jones, "R&D Bracing for Economic Fallout from Deregulation," *Electrical World* 211 (January 1997): 22–23.

68. Jubien, "The Regulatory Divide."

69. Jim Drinkard, "Utility Deregulation Fierce Battle," Associated Press, April 24, 1997.

70. Barkovich, *Regulatory Interventionism in the Utility Industry*, 156.

71. Gary Locke, "Caught in the Electrical Fallout," *New York Times*, February 2, 2001.

72. Holman W. Jenkins Jr., "Maybe the Utilities Weren't So Dumb After All," *Wall Street Journal*, May 16, 2001.

73. Larry Lynch, "Large Power Users Seek Direct Access to Power Sellers," *Sacramento Bee*, March 27, 2001.

74. Jonathan Marshall, "More Failures Expected for Power Network: Increased Demand Strains Regional Transmission Lines," *San Francisco Chronicle*, August 13, 1996.

75. V. C. Ramesh, "Information Matters: Beyond OASIS," *Electricity Journal* 10 (March 1997): 78–82; John C. Hoag, "Oasis: A Mirage of Reliability," *Public Utilities Fortnightly* 134 (November 1, 1996): 38–40.

76. Jonathan Marshall and Jon Swartz, "Net Service Providers Facing Fees from Uunet," *San Francisco Chronicle*, April 25, 1997.

77. Jonathan Marshall and Jon Swartz, "Coalition Fights to Keep Net Fees Low; Phone Companies Want Higher Rates," *San Francisco Chronicle,* November 9, 1996.

78. *Surfing the "Second-Wave": Sustainable Internet Growth and Public Policy* (Pacific Telesis, 1997).

79. Sandeep Junnarkar, "Regional Phone Companies to Offer New Access Technologies for ISPs," *New York Times,* April 22, 1997.

80. Sandeep Junnarkar, "New Phone Rules Will Have Mixed Effect on Net," *New York Times,* May 8, 1997.

81. Peter Elstrom, "Telecom's New Trailblazers," *Business Week,* March 26, 1998.

82. Claude S. Fischer, *America Calling: A Social History of the Telephone to 1940* (Berkeley and Los Angeles: University of California Press, 1992).

83. Alan Stone, *Wrong Number: The Breakup of AT&T* (New York: Basic Books, 1989), 52.

84. For a broad analysis and retrospective attack on the consolidation of the Bell utility system, see Milton Mueller, "Universal Service and the Telecommunications Act: Myth Made Law," *Communications of the ACM* 40 (March 1997): 39–47.

85. Stone, *Wrong Number,* 10.

86. Along with Stone, *Wrong Number,* see Peter Temin and Louis Galambos, *The Fall of the Bell System: A Study in Prices and Politics* (Cambridge: Cambridge University Press, 1987), which gives a good economic account of the changes in the Bell system; while Steven Coll, *The Deal of the Century: The Breakup of AT&T* (New York: Atheneum, 1986) gives a good blow-by-blow history of the regulatory breakup of AT&T.

87. Larry Kahaner, *On the Line: The Men of MCI Who Took on At&T, Risked Everything, And Won!* (New York: Warner Books, 1986).

88. Temin and Galambos, *The Fall of the Bell System,* 59.

89. Stone, *Wrong Number,* 53.

90. Stone, *Wrong Number,* 53.

91. Temin and Galambos, *The Fall of the Bell System,* 149–50.

92. Temin and Galambos, *The Fall of the Bell System,* 227.

93. Torri Minton and Jon Swartz, "35 Cents Charge For Pac Bell Pay Phones," *San Francisco Chronicle,* October 18, 1997.

94. Coll, *The Deal of the Century,* 367.

95. Holly Hubbard Preston, "Minitel Reigns in Paris with Key French Connection," *Computer Reseller News,* no. 594 (September 5, 1994): 49–50; William L. Cats-Baril and Tawfik Jelassi, "The French Videotex System Minitel: A Successful Implementation of a National Information Technology Infrastructure," *MIS Quarterly* 18 (March 1994): 1–20.

96. Jack Kessler, "Electronic Networks: A View from Europe," *American Society for Information Science Bulletin* 20 (April/May 1994): 26–27.

97. Cats-Baril and Jelassi, "The French Videotex System Minitel"; and Gerard Poirot, "Minitel: Oui! Multimedia: Non!" *Communications International* 22 (July 1995): 23–24.

98. Bruno Giussani, "France Gets Along With Pre-Web Technology," *New York Times,* September 23, 1997.

99. Coll, *The Deal of the Century,* 366.

100. Pamela Fusting, "Will Universal Service Be Preserved?" *Rural Telecommunications* 15 (November/December 1996): 14–22.; Erica Schroeder, "Telecom Act Fuels Regulatory Wars," *PC Week* 13 4 (April 8, 1996): 51.

101. Amy Barrett, "But Do Aspen and Vail Really Need Phone Subsidies," *Business Week,* May 12, 1997.

102. Linda Greenhouse, "High Court to Hear Dispute on Opening Phone Markets," *New York Times,* January 27, 1998.

103. Mueller, "Universal Service and the Telecommunications Act," 39–47.

104. Melody Petersen, "Trenton Tells Bell Atlantic to Speed Up Urban Cable Connections," *New York Times,* April 22, 1997; Jonathan Marshall and Jon Swartz, "Sweeping Changes in Phone Rates Long-Distance Cuts to be Passed on to Customers," *San Francisco Chronicle,* May 8, 1997.

105. Karen Southwick, "California, Here I Come," *Upside* 6 (December 1994): 34–45.

106. Temin and Galambos, *The Fall of the Bell System,* 305.

107. California Telecommunciations Policy Forum, "Staking Out the Public Interest in the Merger Between Pacific Telesis and Southwestern Bell Corporation," White Paper, February 1997.

108. Southwick, "California, Here I Come."

109. California Telecommunciations Policy Forum, "Staking Out the Public Interest."

110. California Telecommunciations Policy Forum, "Staking Out the Public Interest."

111. Kenneth Howe, "Green Light for Takeover of Pac Bell; Refunds Cut in Deal with Texas Firm," *San Francisco Chronicle,* April 1, 1997.

112. J. P. Cassidy, "Rural, Urban Lawmakers Battle over Tauzin-Dingell Internet Bill," *Hill News,* May 9, 2001.

113. Rebecca Blumenstein, "Web Overbuilt; How the Fiber Barons Plunged the U.S. into a Telecom Glut," *Wall Street Journal,* June 18, 2001.

114. Blumenstein, "Web Overbuilt."

115. Seth Schiesel, "For Local Phone Users, Choice Isn't an Option," *New York Times,* November 21, 2000.

116. Seth Schiesel, "Sitting Pretty: How Baby Bells May Conquer Their World," *New York Times,* April 22, 2001.

117. Ronald E. Yates, "New-Look Ameritech Braves 'Change Storm,'" Chicago Tribune, April 22, 1993.

118. Chris Bucholtz, "Battle Lines Drawn for Universal Service," *Telephony* 230 (May 6, 1996): 22.

Chapter 6

1. Will Marshall, "A New Fighting Faith: Ambitious Goals for the World's Oldest Political Party." *New Democrat* (September/October 1996).

2. George A. Keyworth, "People and Society in Cyberspace," in *The Shape of Things: Exploring the Evolving Transformations in American Life,* Working Paper no. 1, Progress and Freedom Foundation, 1996.

3. Peter Huber, "Buy the Government of Your Choice," *Forbes,* December 2, 1996.

4. For a good review of the literature in this area, see Stephen Graham and Simon Marvin, *Telecommunications and the City: Electronic Spaces, Urban Places* (London: Routledge, 1996).

5. *Taxation of Interstate Mail Order Sales: 1994 Revenue Estimates,* U.S. Advisory Commission on Intergovernmental Relations, Washington, D.C., 1994, SR-18.

6. Wally Dean, interview by author at CityNet office, April 5, 1995.

7. Tim W. Ferguson, "Web Grab? If It Moves, Tax It: State And Local Governments

Smell Revenue in the Internet," *Forbes,* March 9, 1998; Margaret Kane, "Online Spending to Hit $65 Billion," CNET News.com, May 2, 2001.

8. "A River Runs Through It," *Economist,* May 10, 1997.

9. Masotto, interview, June 13, 1995.

10. Daniel O'Connell, "U.S. Supreme Court Reviews State and Local Taxation Issues," *CPA Journal* 62 (March 1992): 16–21.

11. James Srodes, "Murdering Mail Order," *Financial World* 161 (March 31, 1992): 64–67.

12. U.S. General Accounting Office, *Sales Taxes: Electronic Commerce Growth Presents Challenges; Revenue Losses are Uncertain,* report to Congressional requesters, June 2000.

13. David Cay Johnston, "Online Sales Collide with Off-Line Tax Questions," *New York Times,* November 10, 1997.

14. U.S. General Accounting Office, *Sales Taxes.*

15. Richard Stevenson, "Governors Stress Tax Cuts and Austerity," *New York Times,* May 8, 1997.

16. Maria Alicia Gaura, "Santa Clara County Coffers Should Jingle with $8 Million Surplus; Lower Welfare Costs, Higher Property Values—$8 Million Surplus," *San Francisco Chronicle,* February 17, 1998.

17. Figures in this paragraph are from Municipal Analysis Services, Incorporated, *Governments of California: 1993 Annual Financial and Employee Analysis* (Austin, Tex.: 1993). Note that San Francisco is unique in that there is no difference between city and county borders, unlike with Los Angeles and San Diego, where the county and city borders are separate.

18. Richard W. Genetelli, David B. Zigman, and Cesar E. Bencosme, "Recent U.S. Supreme Court Decisions on State and Local Tax Issues," *CPA Journal* 62 (November 1992): 38–44.

19. Greg Gattuso, "Tax Fairness Act Unfair: DMA," *Direct Marketing* 57 (June 1994): 6.

20. David Cay Johnston, "Sales Tax Proposal Angers Mail-Order Customers," *New York Times,* November 7, 1997.

21. "Bill to Prohibit Internet Taxation Moving Forward," Reuters New Media, May 12, 1997.

22. "Lawmakers Criticize Internet Bill," Associated Press, March 10, 1998.

23. Ferguson, "Web Grab?"

24. Gattuso, "Tax Fairness Act Unfair."

25. Brett Glass, "The Real Cost of Mail-Order PCs," *InfoWorld* 13 (December 2, 1991): 45–46.

26. Cyndee Miller, "Catalogs Alive, Thriving," *Marketing News* 28 (February 28, 1994): 1, 6.

27. Srodes, "Murdering Mail Order."

28. John R. Gwaltney, "Fallacies of Sales-Tax Loophole for Mail-Order Firms," *Small Business Reports* 15 (April 1990): 26–29.

29. Srodes, "Murdering Mail Order."

30. Kaye K. Caldwell, "Solving State and Local Use Tax Collection Problems: A Necessary First Step Before Dealing with Use Tax Problems of Electronic Commerce," discussion draft by Software Industry Coalition, 1996.

31. Glass, "The Real Cost of Mail-Order PCs," 45–46.

32. Citizens for Tax Justice, *A Far Cry from Fair: CTJ's Guide to State Tax Reform,* a

joint project with the Institute for Taxation and Economic Policy, Washington, D.C., April 1991.

33. Michael Mazerov and Iris J. Lay, *A Federal "Moratorium" on Internet Commerce Taxes Would Erode State and Local Revenues and Shift Burdens to Lower-Income Households*, report by the Center on Budget and Policy Priorities, May 11, 1998.

34. Anders Schneiderman, "The Hidden Handout" (Ph.D. diss., University of California, Berkeley, 1995).

35. Ned Eichler, *The Merchant Builders* (Cambridge: MIT Press, 1982), 219, 259.

36. Robert Kuttner, *Revolt of the Haves: Tax Rebellions and Hard Times* (New York: Simon and Schuster, 1980), 51.

37. Mike Davis, *City of Quartz: Excavating the Future in Los Angeles* (New York: Vintage Books, 1990), 130–31.

38. Davis, *City of Quartz*, 166.

39. Logan and Molotch, *Urban Fortunes*, 187.

40. George Will, "'Slow Growth' Is the Liberalism of the Privileged," *New York Times*, August 30, 1987.

41. Martin Mayer, *The Builders: Houses, People, Neighborhoods, Governments, Money* (New York: Norton, 1978).

42. See Schneiderman, "The Hidden Handout," chap. 6.

43. Lenny Goldberg, *Taxation with Representation: A Citizen's Guide to Reforming Proposition 13* (Sacramento: California Tax Reform Association and New California Alliance, 1991).

44. Clarence Y. H. Lo, *Small Property Versus Big Government: Social Origins of the Property Tax Revolt* (Berkeley and Los Angeles: University of California Press, 1990), 154.

45. Kuttner, *Revolt of the Haves*, 193.

46. Goldberg, *Taxation with Representation*, 42–43.

47. Goldberg, *Taxation with Representation*, 50, 78.

48. "Tax Facts: The System Sets Cities Up to Be Squeezed by Superstores," editorial, *San Jose Mercury News*, July 24, 1995.

49. Greg Leroy, "No More Candy Store: States Move to End Corporate Welfare as We Know It," *Dollars and Sense*, no. 199 (May–June 1995): 10.

50. Melvin L Burstein and Arthur J. Rolnick, "Congress Should End the Economic War Among the States," *Federal Reserve Bank of Minneapolis: The Region* 9 (March 1995): 3–20.

51. LeRoy, "No More Candy Store."

52. Curt Harler, "Why Governments Want Your Network Center," *Communications News* 29 (December 1992): 30.

53. Lowenthal, interview, July 21, 1995.

54. Landler, "Hi, I'm in Bangalore."

55. "Telecommunications Infrastructure," *East Asian Executive Reports* 18 (September 15, 1996): 14–16.

56. "In Search of State-of-the-Art Technologies: Korea's Drive to the Information Superhighway," *East Asian Executive Reports* 18 (September 15, 1996): 8, 18.

57. See President William J. Clinton and Vice President Albert Gore Jr., *A Framework for Global Electronic Commerce* (Washington, D.C. July 1, 1997).

58. *Information Technology: An Important Tool for a More Effective Government*, Legislative Analyst's Office, State of California, June 16, 1994; *California Integrated Information Network: A Strategic Plan for Calnet and All State Telecommunications Networks*, Depart-

ment of Information Technology, Department of General Services, State of California, December 1996; *1996 Annual Report: Department of Information Technology,* State of California, 1996.

59. "California Wants Bids to Privatize State Telephone System," *Newsbytes News Network,* October 9, 1997.

60. *California State University's Technology Infrastructure Initiative,* business plan by GTE, Microsoft, Fujitsu, and Hughes, August 25, 1997.

61. Barbara Gengler, "California's Big Picture: Serious About the Potential of the Internet, Pacific Bell Has Developed an Experimental Backbone," *LAN Computing* 5 (April 1994).

62. Syd Leung, interview, November 30, 1995.

63. Deborah Lutterbeck, "Muni Telecoms," *Infrastructure Finance* 6 (April 1997): 14–17.

64. Jim Duffy, "IP Net Helps Texas Dole Out Benefits," *Network World* 10 (November 29, 1993): 1.

65. Lutterbeck, "Muni Telecoms"; "Computer Networking: The Wiring of Iowa," *Economist,* July 15, 1995, 19.

66. Michael McCabe, "Palo Alto Moves Fast to Center of Internet; Fiber-Optic Ring Nearly Completed," *San Francisco Chronicle,* June 26, 1997; John Markoff, "Will Commerce Flourish Where Rivers of Wire Converge?" *New York Times,* December 8, 1997.

67. Lutterbeck, "Muni Telecoms," 14–17.

68. "Bill Could Divide Telco Customers into Haves And Have-Nots," *InfoWorld* 18 (September 16, 1996): 72.

69. Brian Moura, "The Telecommunications Revolution Comes to City Hall," *Western City* (League of California Cities), April 1996; Wayne C. Lusvardi and Charles B. Warren, "What Price an Easement? Setting Market Value in Fiber Optic Corridors," *Public Utilities Fortnightly,* July 1, 2001; "Indirect Tax; Poway Seeks Internet Franchise Fees," *San Diego Union-Tribune,* June 1, 2001.

70. Donald W. McClellan Jr., "The FCC's $13 Billion Tax Hike" paper by The Progress and Freedom Foundation, 1997.

71. Delaine Eastin, state superintendent of Public Instruction, State of California, "Discounts for Telecommunications," letter, November 20, 1996.

72. Nanette Asimov, "California Schools Rate D-Minus in Report; Exhaustive Study Blames Prop. 13 for the Damage," *San Francisco Chronicle,* January 16, 1997.

73. Pamela Mendels, "Study Faults Net Training for Teachers," *New York Times,* July 17, 1997.

74. Pamela Mendels, "Not Everyone Is Eager to Jump on the Net-Day Bandwagon," *New York Times,* October 19, 1996.

75. John Gartner, "Net Access May Increase Inequalities," *TechWeb,* May 11, 1998.

76. Jerry Borrell, "America's Shame: How We've Abandoned Our Children's Future," *Macworld,* September 1992.

77. Charles Piller, "Separate Realities," *Macworld,* September 1992.

78. Larry Slonaker and Howard Bryant, "Wealthy Schools Draw Abundant Help in Net-Wiring Project, but Have-Nots Come Up Short in Volunteers and Funding," *San Jose Mercury News,* March 7, 1996.

79. Robert B. Gunnison, "Ex-Workers Say Giveaway Totals Were Inflated; Wilson-Backed Program to Equip State's Schools," *San Francisco Chronicle,* May 20, 1997.

80. Gunnison, "Ex-Workers Say Giveaway Totals Were Inflated"; Robert B. Gunni-

son, "Computer Giveaway Favors Silicon Valley Schools; Foundation Ignores Most Other Areas," *San Francisco Chronicle,* March 1, 1997.

81. Peter Sinton, "Big Haul Of PCs For Local Schools," *San Francisco Chronicle,* November 14, 1996.

82. Michael Hanley, "Small Towns, Big Plans," *Telephony* 232 (June 2, 1997): 246–56.

83. Joel A. Tarr, "Sewerage and the Development of the Networked City in the United States, 1850–1930," in *Technology and the Rise of the Networked City in Europe and America,* ed. Joel A. Tarr, and Gabriel Dupuy (Philadelphia: Temple University Press, 1988).

84. Andre Guillerme, "The Genesis of Water Supply Distribution, and Sewerage Systems in France, 1800–1850," in *Technology and the Rise of the Networked City in Europe and America,* ed. Joel A. Tarr and Gabriel Dupuy (Philadelphia: Temple University Press, 1988).

85. Anthony Sutcliffe, "Street Transport in the Second Half of the Nineteenth Century: Mechanization Delayed?" in *Technology and the Rise of the Networked City in Europe and America,* ed. Joel A. Tarr and Gabriel Dupuy (Philadelphia: Temple University Press, 1988).

86. Bill Schechter, "Metropolitan Governance: A Bibliographic Essay," *National Civic Review* 85 (Spring–Summer, 1996): 63.

87. Terry Bryzinsky, interview by author at ABAG office, March 7, 1995.

88. Warren Slocum, interview by author at county assessor's office, March 30, 1995.

89. Kurt Handelman, interview by author at Sacramento EDD office, March 29, 1995.

90. Dean, interview, April 5, 1995.

91. Lutterbeck, "Muni Telecoms."

92. Jim Duffy, "IP Net Helps Texas Dole Out Benefits," *Network World* 10 (November 29, 1993).

93. Lisa Corbin, "Contractors Suffer Blows in Budget Battle," *Government Executive* ("Top 200 Federal Contractors" supplement), August 1996, 41–48.

94. Slocum, interview, March 30, 1995.

95. Harvey, *The Condition of Postmodernity.*

96. Bryzinsky, interview, March 7, 1995.

97. Julia King, "Electronic Forms Are New Wave in Pollution Control," *Computerworld* 29 (July 10, 1995): 42.

98. See the Silicon Valley Toxics Coalition Web site, <http://www.svtc.org/svtc/>.

99. Greg Karras, interview by author by telephone, December 1, 1997.

100. Cary Paul Peck, interview by author by telephone, April 10, 1995.

101. Peck, interview, April 10, 1995.

102. Bryzinsky, interview, March 7, 1995.

103. Peck, interview, April 10, 1995.

104. "Labor Plans Big Sweep Up," *San Francisco Chronicle,* May 31, 1996.

105. Handelman, interview, March 29, 1995.

106. A too visible labor market would without question also create greater scrutiny of racism in hiring practices that is often obscured through the secondary hiring markets of contingent employment. Sociologist William Julius Wilson has emphasized that the decline of large factory employment and the flight of core jobs from the inner city has left many inner-city residents, particularly African Americans, cut off from knowledge or access to job openings. See Wilson, *When Work Disappears.* Despite the attacks on the "work ethic" of inner-city blacks, Wilson colleague Martha Van Haitsma's research has highlighted how low-wage employers often use extended social networks of employees in

their informal hiring practices, further isolating unemployed blacks with few job networks. One result of the underground and word-of-mouth nature of most low-wage hiring is dramatic differences in job access by immigrants and African Americans in the inner city. Immigrant families are much more likely to have multiple adults in the household along with extended family networks, thereby creating greater opportunities to hear about job openings often not advertised publicly. Employers can then use these Latino networks for hiring, reinforcing the broader employer prejudice against hiring African Americans (Martha Van Haitsma, "Attitudes, Social Context, and Labor Force Attachment: Blacks and Immigrant Mexicans in Chicago Poverty Areas" [paper prepared for "Urban Poverty and Family Life" conference, Chicago, 1994]).

107. "Calif. Public Records Not So Public," Associated Press, August 18, 1997.

108. Michael Duffner, interview by author by telephone, May 2, 1996.

109. Bryzinsky, interview, September 13, 1995.

110. Rhonda Mitschele, "Share and Share Alike: Creating a Cost-Effective GIS," *American City and County* (March 1996).

111. Sharen Kindel, "Geographic Information Systems," *Financial World*, January 19, 1993; "The Delight of Digital Maps," *Economist*, March 21, 1992.

112. Nina Bernstein, "Lives on File: Privacy Devalued in Information Economy," *New York Times*, June 12, 1997.

113. Victor Pottoroff, interview by author at CSAC office, August 29, 1995.

114. Slocum, interview, March 30, 1995.

115. Julia Angwin and Jon Swartz, "P-Trak Defends Database; Other Companies Offer Lengthier Personal Files," *San Francisco Chronicle*, September 20, 1996.

116. Nina Bernstein, "Lives on File: Privacy Devalued in Information Economy," *New York Times*, June 12, 1997.

117. Charles Cook, *The Cook Report on Internet-NREN; NII: The Dark Side in Washington State*, available [online], <http://pobox.com/cook/washington.html>.

118. Cook, *The Cook Report on Internet-NREN*.

119. Slocum, interview, March 30, 1995.

120. Don Wimberly, BASIC manager, interview by author at BASIC office, July 27, 1995.

121. Steven A. Holmes, "Budget Cuts May Curtail Census in Year 2000," *New York Times*, August 23, 1995.

122. Maurine A. Haver, "Presidential Address: Economic Statistics: A Call for Action!" *Business Economics*, January 1, 1996.

123. Duffner, interview, May 2, 1996.

124. Bill Kovach, "When Public Business Goes Private," *New York Times*, December 4, 1996.

125. For a full review of the West/JURIS controversy, see the Taxpayers Assets Project Web site, <http://www.essential.org/tap/>.

126. David Milgram, interview by author at Lockheed office, July 6, 1995.

Chapter 7

1. General Agreement on a New Economy (GANE), *For Full Employment, Equity and Environmental Sustainability: Summary of Working Draft #1* (Washington, D.C.), available [online], <http://www.igc.org/econwg/gane/>.

2. Carl Davidson and Jerry Harris, "The Third Wave and the Republicans: Is Newt Gingrich a Closet New Leftist?" *cy.Rev: A Journal of Cybernetic Revolution, Sustainable Socialism and Radical Democracy*, no. 2 (March 1995).

3. Immanuel Maurice Wallerstein, *The Modern World-System* (New York: Academic Press, 1974).

4. Alexis de Tocqueville, *Democracy in America*, ed. J. P. Mayer, trans. George Lawrence (Garden City, N.Y.: Anchor Books, 1969), 517–18.

5. Leslie Miller, "Activism Goes On-line: Plugging in to Electronic Organizing," *USA Today*, April 25, 1995; and Rose George, "Drop the Prop," *Nation*, January 9/16, 1995. As a co-director of the Center for Community Economic Research, I have to acknowledge that my involvement in this and a few other of the cyberorganizing actions described here was more than as a disinterested researcher.

6. See <http://www.organizenow.net>.

7. Rich Cowan, interview by author by telephone, June 13, 1995.

8. James Johnson, Sacramento-based organizer for CTWO, interview by author, March 29, 1995.

9. Emilee Whitehurst, interview by author at ETS office, May 26, 1995.

10. Pat Bourne, interview by author at SFUNET office, April 8, 1995.

11. For more information, see <http://www.techrocks.org/>.

12. Interviews by author at union office with Jim Dupont, president, and Stephanie Ruby, organizer, Local 2850, HERE, 1996.

13. "Labor Plans Big Sweep in Valley: Janitors Just First Wave of Organizing 'Renaissance.'" *San Francisco Chronicle*, May 31, 1996.

14. Katie Buller, interview by author by telephone, August 1995.

15. Alexander Cockburn and Ken Silverstein, "War and Peso (Mexico)," *New Statesman and Society*, February 24, 1995.

16. Gary Chapman, "Mexico: Window on Technology and the Poor," *Los Angeles Times*, October 28, 1996.

17. Matt Richtel, "The Left Side of the Web Seeds Global Grass Roots," *New York Times*, June 14, 1997.

18. Steven Levy, *Hackers: Heroes of the Computer Revolution* (Garden City, N.Y.: Anchor Press, 1984).

19. Susanne Sallin, *The Association for Progressive Communications: A Cooperative Effort to Meet the Information Needs of Non-Governmental Organization*, case study prepared for the Harvard-CIESIN Project on Global Environmental Change Information Policy, February 14, 1994.

20. Craig Warkentin, *Reshaping World Politics: NGOS, the Internet, and Global Civil Society* (Rowman and Littlefield, 2001), 143–55.

21. Michael Stein, interview by author at Stein's home, March 8, 1995.

22. Stein, interview, March 8, 1995.

23. Warkenstin, *Reshaping World Politics*, 64.

24. Mark Ritchie, interview by author by telephone, June 8, 1995.

25. Ritchie, interview, June 8, 1995.

26. Peter Beinart, "TRB from Washington: The Next NAFTA," *New Republic*, December 15, 1997.

27. "The Sinking of the MAI," *Economist*, Mar 14, 1998.

28. "Citizens' Groups: The Non-Governmental Order: Will Ngos Democratise, or

Merely Disrupt, Global Governance?" *Economist,* December 11, 1999. NGO is the abbreviation for *nongovernmental organization.*

29. For an example of the call to action, see <http://flag.blackened.net/global/aacon vergence.htm>.

30. For a listing of all sites, see <www.indymedia.org>.

31. Peter Phillips, "The New Progressive Movement and Corporate Media," in *Project Censored,* available [online], <http://www.projectcensored.org/frontpagenews/7_16_oopeter.html>, July 17, 2000.

32. Janet Wells, "Graduate School for Protesters: D.C. is Next for the Ruckus Society and Its Well-Trained Troublemakers," *San Francisco Chronicle,* April 8, 2000.

33. Judd Slivka, "Ruckus Society Getting Set to Disrupt WTO Conference; Group Emphasizes Need for Environmental Laws," *Seattle Post-Intelligencer,* September 15, 1999.

34. "Globalizing from Below: An Interview with Anuradha Mittal," *Said It,* September 2000.

35. Jim Gardner, "Bay Area Activist Turns WTO into Medea's Medium," *San Francisco Business Times,* December 10, 1999.

36. Medea Benjamin, "The Price of Power," *San Francisco Bay Guardian,* July 26, 2000.

37. Simon Rodberg, "The CIO Without the CIA," *American Prospect,* July 2, 2001.

38. Robert Taylor, "Employment: Rights Are 'Vital to Trade,'" *Financial Times,* November 19, 1999; "A New Strategy for Trade and Development," in *Statement on the Agenda for the Third Ministerial Conference of the World Trade Organisation (WTO),* International Confederation of Free Trade Unions, Seattle, Washington, November 30–December 3, 1999.

39. Aaron Bernstein, "Labor Inches Toward a Milestone at the WTO Talks," *Business Week,* December 2, 1999.

40. Jeffy Crosby, "'The Kids Are All Right'. . . and Other Thoughts from IUE Visitors to Seattle," *LaborNotes,* January 2000.

41. Tyler Marshall and Norman Kempster, "Albright Envisions New Global Approach to Advance Democracy," *Los Angeles Times,* January 17, 1999.

42. For a pro-"fix" view from the movement, see Susan George, "Fixing or Nixing the WTO," *Le Monde Diplomatique,* January 2000. For a "nix it" strategic view, see Philippine economist Walden Bellow of Focus on the South (http://www.focusweb.org/publications/2000/publications_2000_index.htm).

43. Thomas L. Friedman, "Globalization Foes Are No Ally of Poor," *New York Times,* April 26, 2001.

44. For a series of essays in this approach, see David G. Becker and Richard Sklar, eds. *Postimperialism and World Politics* (London: Praeger, 1999); see also Scott Bowman, *The Modern Corporation and American Political Thought* (University Park: Pennsylvania State University Press, 1996).

45. Michael Hardt and Antonio Negri, *Empire* (Cambridge: Harvard University Press: 2000).

46. Hardt and Negri, *Empire,* 45.

47. For a discussion of various approaches to understanding NGOs in global politics, see Kenneth Stiles, ed., *Global Institutions and Local Empowerment: Competing Theoretical Perspectives* (New York: Macmillan, 2000).

48. Tocqueville, *Democracy in America,* 555–58.

Bibliography

Working Papers, Reports, Theses, and Dissertations

AFL-CIO Working America: The Current Economic Situation. Available [Online], <http://www.aflcio.org/cse/mod5/index.htm>.

Bar, Francois. "Configuring the Telecommunications Infrastructure for the Computer Age: The Economics of Network Control." Ph.D diss., University of California, Berkeley, 1990.

Bell Atlantic. Filing in response to WorldCom-MCI merger, Federal Communications Commission, Washington, D.C., January 5, 1998.

Boyd, Donald J. "State Fiscal Issues and Risks at the Start of a New Century." *Multistate Tax Commission Review* 2001 (April 2001).

Brady, Raymond J. "Intraregional Economic Linkages of the High Tech Industry in the San Francisco Bay Area." Report prepared for the conference "Silicon Valley: The Future?" University of California, Santa Cruz. Working Paper no. 6, Association of Bay Area Governments, Center for Analysis and Information Services, Oakland, Calif., 1986.

Caldwell, Kaye K. "Solving State and Local Use Tax Collection Problems: A Necessary First Step Before Dealing with Use Tax Problems of Electronic Commerce." Discussion draft for Software Industry Coalition, 1996.

California Integrated Information Network: A Strategic Plan for Calnet and All State Telecommunications Networks. Department of Information Technology, Department of General Services, State of California, December 1996.

California State University's Technology Infrastructure Initiative. Business plan by GTE, Microsoft, Fujitsu, and Hughes. August 25, 1997.

California Telecommunciations Policy Forum. "Staking Out the Public Interest in the Merger Between Pacific Telesis and Southwestern Bell Corporation." White Paper. February 1997.

Citizens for Tax Justice. *A Far Cry from Fair: CTJ's Guide to State Tax Reform.* A joint project with the Institute for Taxation and Economic Policy, Washington, D.C., April 1991.

Clinton, Bill. *Technology: The Engine of Economic Growth. A National Technology Policy for America.* Campaign document, September 18, 1992.

Clinton, President William J., and Vice President Albert Gore Jr. *A Framework for Global Electronic Commerce.* Washington, D.C., July 1, 1997.

Cook, Charles. *The Cook Report on Internet-NREN; NII: The Dark Side in Washington State.* Available [Online], <http://pobox.com/cook/washington.html>.

The Cook Report—NSFnet "Privatization" and the Public Interest: Can Misguided Policy Be Corrected? 1992. Available [Online], <http://cookreport.com/p.index.html# part1>.

Cross-Industry Working Team (XIWT) of the Corporation for National Research Initiatives. "Electronic Commerce in the NII." White Paper. October 1995. Available [Online], <www.xiwt.org>.

Eastin, Delaine, State Superintendent of Public Instruction, State of California. "Discounts for Telecommunications." Public letter, November 20, 1996.

Gates, Jeff. "Statistics on Poverty and Inequality." Global Policy Forum (1999). Available [Online], <http://www.igc.apc.org/globalpolicy/socecon/inequal/gates99.htm>.

General Agreement on a New Economy (GANE). *For Full Employment, Equity, and Environmental Sustainability: Summary of Working Draft #1.* Washington, D.C. Available [Online], <http://www.igc.org/econwg/gane/g11sum.html>.

Goldberg, Lenny. *Taxation with Representation: A Citizen's Guide to Reforming Proposition 13.* Sacramento: California Tax Reform Association and New California Alliance, 1991.

Government Policies and the Diffusion of Microelectronics. Paris: Organization of Economic Co-Operation and Development (OECD), 1989.

Hardy, Henry Edward. "The History of the Net." Master's thesis, Grand Valley State University, September 28, 1993.

Information Technology: An Important Tool for a More Effective Government. Legislative Analyst's Office, State of California, June 16, 1994.

Joint Venture: Silicon Valley. *An Economy at Risk: The Phase I Diagnostic Report.* Report prepared by the Center for Economic Competitiveness, SRI International. 1992.

Joint Venture. *Index of Silicon Valley: Measuring Progress Towards a Twenty-first Century Community.* Prepared by Collaborative Economics for Joint Venture: Silicon Valley. 1995.

———. *Index of Silicon Valley 1996: Measuring Progress Towards a Twenty-first Century Community.* Prepared by Collaborative Economics for Joint Venture: Silicon Valley. 1996.

———. *Index of Silicon Valley 1997: Measuring Progress Towards a Twenty-first Century Community.* Prepared by Collaborative Economics for Joint Venture: Silicon Valley. 1997.

The Joint Venture Way: Lessons for Regional Rejuvenation. Report produced by Joint Venture: Silicon Valley Network, San Jose, 1995.

Keefe, Jeff. *Monopoly.com: Will the WorldCom-MCI Merger Tangle the Web?* Economic Policy Institute, 1998.

Keyworth, George A. "People and Society in Cyberspace." In *The Shape of Things: Exploring the Evolving Transformations in American Life.* Working Paper no. 1, Progress and Freedom Foundation, 1996.

McClellan, Donald W., Jr. "The FCC's $13 Billion Tax Hike." Progress and Freedom Foundation, 1997.

Mazerov, Michael, and Iris J. Lay. *A Federal "Moratorium" on Internet Commerce Taxes Would Erode State and Local Revenues and Shift Burdens to Lower-Income Households.* Report by the Center on Budget and Policy Priorities, May 11, 1998.

Municipal Analysis Services. *Governments of California: 1993.* Annual financial and employee analysis. Austin, Tex., 1993.

Newman, Nathan. "Net Loss: Government, Technology, and the Political Economy of Community in the Age of the Internet." Ph.D. diss., University of California, Berkeley, 1998.

1996 Annual Report. Department of Information Technology, State of California, 1996.

Sallin, Susanne. *The Association for Progressive Communications: A Cooperative Effort to Meet the Information Needs of Non-Governmental Organization.* Case study prepared for the Harvard-CIESIN Project on Global Environmental Change Information Policy, February 14, 1994.

Saxenian, AnnaLee. "Contrasting Patterns of Business Organization in Silicon Valley." Working Paper no. 535, Institute of Urban and Regional Development, University of California, Berkeley, April 1991.

Schiller, Dan. *Bad Deal of the Century: The Worrisome Implications of the WorldCom-MCI Merger.* Economic Policy Institute, 1998.

Schneiderman, Anders. "The Hidden Handout." Ph.D. diss., University of California, Berkeley, 1995.

Shock Absorbers in the Flexible Economy: The Rise of Contingent Labor in Silicon Valley. Report prepared by Working Partnership USA. 1995.

State Information Technology: An Update. Legislative Analyst's Office. State of California. January 23, 1996.

Surfing the "Second-Wave": Sustainable Internet Growth and Public Policy. Pacific Telesis, 1997.

Taxation of Interstate Mail Order Sales: 1994 Revenue Estimates. U.S. Advisory Commission on Intergovernmental Relations. Washington, D.C., 1994, SR-18.

National Academy of Sciences. *Evolving the High Performance Computing and Communications Initiative to Support the Nation's Information Infrastructure.* Washington, D.C.: National Academy Press, 1995.

NII 2000 Steering Committee of the Computer Science and Telecommunications Board, Commission on Physical Sciences, Mathematics, and Applications and the National Research Council. *The Unpredictable Certainty: Information Infrastructure Through 2000.* Washington, D.C.: National Academy Press, 1996.

U.S. General Accounting Office. *Sales Taxes: Electronic Commerce Growth Presents Challenges; Revenue Losses Are Uncertain.* Report to congressional requesters, June 2000.

Van Haitsma, Martha. "Attitudes, Social Context, and Labor Force Attachment: Blacks and Immigrant Mexicans in Chicago Poverty Areas." Paper prepared for "Urban Poverty and Family Life" conference, Chicago, 1994.

Books

Barber, Benjamin R. *Jihad vs. McWorld: How the Planet Is Both Falling Apart and Coming Together—and What This Means for Democracy.* New York: Times Books, 1995.

Barkovich, Barbara R. *Regulatory Interventionism in the Utility Industry: Fariness, Efficiency, and the Pursuit of Energy Conservation.* New York: Quorum Books, 1989.

Becker, David G., and Richard Sklar, eds. *Postimperialism and World Politics.* London: Praeger, 1999.

Berners-Lee, Tim. *Weaving the Web: The Original Design and Ultimate Destiny of the World Wide Web by Its Inventor.* San Francisco: HarperSanFrancisco, 1999.

Best, Michael. *The New Competition: Institutions of Industrial Restructuring.* Cambridge: Harvard University Press, 1990.

Birmingham, Stephen. *California Rich: The Lives, the Times, the Scandals, and the Fortunes of the Men and Women Who Made and Kept California's Wealth.* New York: Simon and Schuster, 1980.

Blackford, Mansel. *The Politics of Business in California, 1890–1920.* Columbus: Ohio State University Press, 1977.

Blumenthal, Sidney. *The Rise of the Counter-Establishment: From Conservative Ideology to Political Power.* New York: Times Books, 1986.

Bowman, Scott R. *The Modern Corporation and American Political Thought: Law, Power, and Ideology.* University Park: Pennsylvania State University Press, 1996.

Brown, John Seely, and Paul Duguid. *The Social Life of Information.* Boston: Harvard Business School Press, 2000.

Castells, Manuel. *The Informational City: Information Technology, Economic Restructuring, and the Urban-Regional Process.* Cambridge, Mass.: Basil Blackwell, 1989.

———. *The Rise of the Network Society.* Cambridge, Mass.: Basil Blackwell, 1996.

Chandler, Alfred, Jr. *The Visible Hand: The Managerial Revolution in American Business.* Cambridge: Harvard University Press, Belknap Press, 1977.

Coleman, Charles. *PGandE of California: The Centennial Story of Pacific Gas and Electric Company, 1852–1952.* New York: McGraw-Hill, 1952.

Coll, Steven. *The Deal of the Century: The Breakup of AT&T.* New York: Atheneum, 1986.

Crew, Michael A., ed. *Competition and the Regulation of Utilities.* Boston: Kluwer Academic Publishers, 1991.

Davis, Mike. *City of Quartz: Excavating the Future in Los Angeles.* New York: Vintage Books, 1990.

Eichler, Ned. *The Merchant Builders.* Cambridge: MIT Press, 1982.

Evans, Peter. *Embedded Autonomy: States and Industrial Transformation.* Princeton: Princeton University Press, 1995.

Ferguson, Thomas, and Joel Rogers. *Right Turn: The Decline of the Democrats and the Future of American Politics.* New York: Hill and Wang, 1986.

Fischer, Claude S. *America Calling: A Social History of the Telephone to 1940.* Berkeley and Los Angeles: University of California Press, 1992.

Flamm, Kenneth. *Creating the Computer: Government, Industry, and High Technology.* Washington, D.C.: Brookings Institute, 1988.

Florida, Richard, and Martin Kenney. *The Breakthrough Illusion: Corporate America's Failure to Move from Innovation to Mass Production.* New York: Basic Books, 1990.

Frank, Robert H., and Philip J. Cook. *The Winner-Take-All Society: Why the Few at the Top Get So Much More Than the Rest of Us.* New York: Penguin Books, 1996.

Friedman, Thomas L. *The Lexus and the Olive Tree.* New York: Farrar, Straus and Giroux, 1999.

Garfinkel, Simson. *Architects of the Information Society: Thirty-five Years of the Laboratory for Computer Science at MIT.* Cambridge: MIT Press, 1999.

Gibson, David, and Everett Rogers. *R&D Collaboration on Trial: The Microelectronics and Computer Technology Corporation.* Boston: Harvard Business School Press, 1994.

Gordon, David. *Fat and Mean: The Corporate Squeeze of Working Americans and the Myth of Managerial "Downsizing."* New York: Free Press, 1996.

Graham, Stephen, and Simon Marvin. *Telecommunications and the City: Electronic Spaces, Urban Places.* London: Routledge, 1996.

Habermas, Jürgen. *Legitimation Crisis.* Boston: Beacon Press, 1975.

Hafner, Katie, and Matthew Lyon. *Where Wizards Stay Up Late: The Origins of the Internet.* New York: Simon and Schuster, 1996.

Hall, Mark, and John Barry. *Sunburst: The Ascent of Sun Microsystems.* Chicago: Contemporary Books, 1990.

Hanson, Dirk. *The New Alchemists: Silicon Valley and the Microelectronics Revolution.* Boston: Little, Brown, 1982.

Hardt, Michael, and Antonio Negri. *Empire.* Cambridge: Harvard University Press, 2000.

Harrison, Bennett. *Lean and Mean: The Changing Landscape of Corporate Power in the Age of Flexibility.* New York: Basic Books, 1994.

Harvey, David. *The Condition of Postmodernity.* Cambridge, Mass.: Blackwell, 1989.

———. *The Urban Experience.* Baltimore: Johns Hopkins University Press, 1989.

Hauben, Michael, and Ronda Hauben. *Netizens: On the History and Impact of Usenet and the Internet.* [Netbook]. Available [Online], <http://www.columbia.edu/~hauben/netbook/>.

Hills, Jill. *Deregulating Telecoms: Competition and Control in the United States, Japan, and Britain.* Westport, Conn: Quorum Books, 1986.

Hiltzik, Michael. *Dealers of Lightning: Xerox PARC and the Dawn of the Computer Age.* New York: HarperBusiness, 1999.

Johnson, Marilynn S. *The Second Gold Rush: Oakland and the East Bay in World War II.* Berkeley and Los Angeles: University of California Press, 1993.

Johnston, Moira. *Roller Coaster: The Bank of America and the Future of American Banking.* New York: Ticknor and Fields, 1990.

Kahaner, Larry. *On the Line: The Men of MCI Who Took On AT&T, Risked Everything, and Won!* New York: Warner Books, 1986.

Kotkin, Joel, and Paul Grabowicz. *California Inc.—The People Who Own California Control Your Future—Whether You Like It or Not!* New York: Rawson, Wade, 1982.

Kuttner, Robert. *Revolt of the Haves: Tax Rebellions and Hard Times.* New York: Simon and Schuster, 1980.

Lehne, Richard. *Industry and Politics—United States in Comparative Perspective.* Englewood Cliffs, N.J.: Prentice Hall, 1993.

Levy, Steven. *Hackers: Heroes of the Computer Revolution.* Garden City, N.Y.: Anchor Press, 1984.

Lewis, Michael. *Liar's Poker: Rising Through the Wreckage on Wall Street.* New York: Penguin Books, 1989.

Likicki, Martin, James Schneider, David Frelinger, and Anna Slomovic. *Scaffolding the Web: Standards and Standards Policy for the Digital Economy.* Science and Technology Policy Institute Rand, 2000.

Lipietz, Alan. *Mirages and Miracles: The Crises of Global Fordism.* Translated by David Macey. London: Verso, 1987.

Lo, Clarence Y. H. *Small Property Versus Big Government: Social Origins of the Property Tax Revolt.* Berkeley and Los Angeles: University of California Press, 1990.

Logan, John R., and Harvey Molotch. *Urban Fortunes: The Political Economy of Place.* Berkeley and Los Angeles: University of California Press, 1987.

Mahon, Thomas. *Charged Bodies: People, Power, and Paradox in Silicon Valley.* New York: New American Library Books, 1985.

Malone, Michael. *The Big Score: The Billion-Dollar Story of Silicon Valley.* Garden City, N.Y.: Doubleday, 1985.

Manes, Stephen, and Paul Andrews. *Gates: How Microsoft's Mogul Reinvented an Industry—and Made Himself the Richest Man in America.* New York: Simon and Schuster, 1993.

Markusen, Ann, Peter Hall, and Amy Glasmeier. *High Tech America: The What, How, Where, and Why of the Sunrise Industries.* Boston: Allen and Unwin, 1996.

Markusen, Ann, Peter Hall, Scott Campbell, and Sabina Deitrick. *The Rise of the Gunbelt: The Military Remapping of Industrial America.* New York: Oxford University Press, 1991.

Massey, Doreen, Paul Quintas, and David Wield. *High-Tech Fantasies: Science Parks in Society, Science, and Space.* London: Routledge, 1992.

Mayer, Martin. *The Builders: Houses, People, Neighborhoods, Governments, Money.* New York: Norton, 1978.

Moschovitis, Christos, Hilary Poole, Tami Schuyler, and Theresa Senft. *The History of the Internet: A Chronology, 1843 to the Present.* Santa Barbara, Calif.: ABC-CLIO, 1999.

Nash, Gerald D. *A. P. Giannini and the Bank of America.* Norman: University of Oklahoma Press, 1992.

Negroponte, Nicholas. *Being Digital.* New York: Vintage Books, 1995.

Nocera, Joseph. *A Piece of the Action: How the Middle Class Joined the Money Class.* New York: Simon and Schuster, 1994.

Norris, Frank. *The Octopus.* New York: Penguin Books, 1901.

Osborne, David. *Laboratories of Democracy.* Boston: Harvard Business School Press, 1988.

Osborne, David, and Ted Gaebler. *Reinventing Government: How the Entrepreneurial Spirit Is Transforming the Public Sector.* New York: Addison-Wesley, 1992.

Piore, Michael, and Charles Sabel. *The Second Industrial Divide: Possibilities for Prosperity.* New York: Basic Books, 1984.

Polanyi, Karl. *The Great Transformation: The Political and Economic Origins of Our Time.* Boston: Beacon Press, 1944.

Putnam, Robert D., with Robert Leonardi and Raffaella Y. Nanetti. *Making Democracy Work: Civic Traditions in Modern Italy.* Princeton: Princeton University Press, 1993.

Reich, Robert. *The Work of Nations: Preparing Ourselves for Twenty-First Century Capitalism.* New York: Alfred Knopf, 1991.

Rheingold, Howard. *Tools for Thought: The People and Ideas Behind the Next Computer Revolution.* New York: Simon and Schuster, 1985.

Rifkin, Jeremy. *The End of Work: The Decline of the Global Labor Force and the Dawn of the Post-Market Era.* New York: G. P. Putnam's Sons, 1995.

Rogers, Everett, and Judith Larsen. *Silicon Valley Fever: Growth of High-Technology Culture.* New York: Basic Books, 1986.

Sassen, Saskia. *Cities in a World Economy.* Thousand Oaks, Calif.: Pine Forge Press, 1994.

Saxenian, Annalee. *Regional Advantage: Culture and Competition in Silicon Valley and Route 128.* Cambridge: Harvard University Press, 1994.

Slomovic, Anna. *An Analysis of Military and Commercial Microelectronics: Had DoD's*

R&D Funding Had the Desired Effect. A Rand Graduate School Dissertation. Santa Monica, Calif.: Rand, 1991.

Stiles, Kenneth, ed. *Global Institutions and Local Empowerment: Competing Theoretical Perspectives.* New York: Macmillan, 2000.

Stone, Alan. *Wrong Number: The Breakup of AT&T.* New York: Basic Books, 1989.

Tarr, Joel A., and Gabriel Dupuy, eds. *Technology and the Rise of the Networked City in Europe and America.* Philadelphia: Temple University Press, 1988.

Temin, Peter, and Louis Galambos. *The Fall of the Bell System: A Study in Prices and Politics.* Cambridge: Cambridge University Press, 1987.

Tocqueville, Alexis de. *Democracy in America.* Edited by J. P. Mayer. Translated by George Lawrence. Garden City, N.Y.: Anchor Books, 1969.

Toffler, Alvin. *The Third Wave.* New York: William Morrow, 1980.

Vogel, Steven K. *Freer Markets, More Rules: Regulatory Reform in Advanced Countries.* Ithaca: Cornell University Press, 1996.

Wallerstein, Immanuel Maurice. *The Modern World-System.* New York: Academic Press, 1974.

Warkentin, Craig. *Reshaping World Politics: NGOS, the Internet, and Global Civil Society.* Rowman and Littlefield, 2001.

Williamson, Oliver. *The Economic Institutions of Capitalism: Firms, Markets, Relational Contracting.* New York: Free Press, 1986.

Wilson, William Julius. *When Work Disappears: The World of the New Urban Poor.* New York: Alfred A. Knopf, 1996.

Articles

Abate, Tom. "Internet Infighting." *Upside Today,* September 30, 1995.

Aboba, Bernard, with Vinton G. Cerf. "How the Internet Came to Be." In Bernard Aboba, *The Online User's Encyclopedia.* Reading, Mass.: Addison-Wesley, 1993.

"AFL-CIO, Church Groups, Urge FCC Reject Worldcom-MCI Merger." *Communications Daily,* August 13, 1998.

Alper, Alan. "State DOA Bids Poky Paper Adieu." *Computerworld* (*Electronic Commerce Journal* supplement), April 29, 1996, 12–14.

Alsop, Stewart. "Please Save Me from My Editors." *Fortune* 134 (October 28, 1996): 203–4.

"AMP Unit Targets Online Commerce." *Electronic Engineering Times* 919 (September 16, 1996): 24.

Anderson, Jerry. "Cashing In on GIS." *American City and County* 108 (April 1993).

Angwin, Julia. "Online Catalogs: Technology Has Surpassed How Shoppers Shop." *San Francisco Chronicle,* October 8, 1996.

Angwin, Julia, and Jon Swartz. "New Devices Threaten PCs: Message at Comdex: Computers Must Evolve—Fast." *San Francisco Chronicle,* November 19, 1996.

———. "P-Trak Defends Database; Other Companies Offer Lengthier Personal Files." *San Francisco Chronicle,* September 20, 1996.

Anthes, Gary H. "City Deploys $2M ATM Net." *Computerworld* 30 (September 16, 1996): 85.

———. "Commercial Users Move onto Internet." *Computerworld* 25 (November 25, 1991): 50.

———. "Feds Ax High-Tech Point Man: DARPA Move Seen as Slap at Industry Funding." *Computerworld* 24 (April 30, 1990): 1.

———. "NSF to Solicit Backbone Operators." *Computerworld* 25 (December 9, 1991): 49, 56.

Armstrong, Lisa. "A New Tune for the Internet Pipers?" *Communications International* 21 (December 1994): 8.

Arthur, W. Brian. "Increasing Returns and the Two Worlds of Business." *Harvard Business Review,* July–August 1996.

Asimov, Nanette. "California Schools Rate D-Minus in Report. Exhaustive Study Blames Prop. 13 for the Damage." *San Francisco Chronicle,* January 16, 1997.

"Assembly OKs Bill to Post Campaign Reports on Net." *San Francisco Chronicle,* September 9, 1997.

Aversa, Jeannine. "Feds Clear Way for Takeover of MCI." Associated Press, September 15, 1998.

Avery, Simon. "Linux Takes Hollywood, Beating Microsoft, SGI Animation Systems." *Wall Street Journal,* May 17, 2001.

Baker, Stan. "Industry Wonders: What's Next?" *Electronic Engineering Times* 588 (April 30, 1990): 84.

Baker, Steven. "The Evolving Internet Backbone: History of the Internet Computer Network." *UNIX Review* 11 (September 1993): 15.

Balderston, Jim. "GTE, Uunet Merge Internet Services." *InfoWorld* 18 (July 15, 1996): 12.

Balzhiser, Richard E. "Technology—It's Only Begun to Make a Difference." *Electricity Journal* 9 (May 1996): 32–45.

Bank, David. "Public Schools Get Wired via High-Tech Generosity." *San Jose Mercury News,* May 13, 1995.

———. "Smart Valley to Get $8 Million; Goal of U.S. Grant Is to Help Turn Internet into Electronic Marketplace." *San Jose Mercury News,* November 24, 1993, 8D.

"Banks Worldwide Plan to Increase Internet Services in 1997." *Bank Marketing* 29 (January 1997): 9.

Barbrook, Richard, and Andy Cameron. "The California Ideology." (1995). Reprinted in *Crypto Anarchy, Cyberstates, and Pirate Utopias,* edited by Peter Ludlow. Cambridge: MIT Press, 2001.

Barlett, Donald L., and James B. Steele. "Say Goodbye to High-Tech Jobs . . . Many Are Moving Offshore." Series. *Philadelphia Inquirer,* 1996.

Barrett, Amy. "But Do Aspen and Vail Really Need Phone Subsidies." *Business Week,* May 12, 1997.

"BBN to Acquire Barrnet from Stanford." *Cambridge Telecom Report,* June 27, 1994.

"BBN Planet Launches Internet Services Nationwide." *Cambridge Work-Group Computing Report,* March 20, 1995.

Beinart, Peter. "TRB from Washington: The Next NAFTA." *New Republic,* December 15, 1997.

Belluck, Pam, and David Barboza. "After Summer's Power Failures, Concerns About Large Utilities." *New York Times,* September 13, 1999.

Bernier, Paula. "The Science of High-Speed Computing." *Telephony* 228 (May 1, 1995): 7, 15.

Bernstein, Nina. "Lives on File: Privacy Devalued in Information Economy." *New York Times,* June 12, 1997.

Bers, Joanna Smith. "Banks Must Decide Whether to Be a Catalyst or Catatonic in the Face of E-Commerce." *Bank Systems and Technology* 33 (November 1996): 16.

"Bigger, Faster, Better Ears." *Economist,* December 7, 1996, 59–60.

"Bill Could Divide Telco Customers into Haves and Have-Nots." *InfoWorld* 18 (September 16, 1996): 72.

"Bill to Prohibit Internet Taxation Moving Forward." *Reuters New Media,* May 12, 1997.

Birkhead, Evan. "Happy 25th to the Arpanet." *Internetwork* 5 (September 1994): 48.

Blake, Pat. "The New Matchmakers." *Telephony* 232 (March 3, 1997): 42–48.

Blumenstein, Rebecca. "Web Overbuilt; How the Fiber Barons Plunged the U.S. into a Telecom Glut." *Wall Street Journal,* June 18, 2001.

Booker, Ellis. "'Net to Reshape Business." *Computerworld* 29 (March 13, 1995): 14.

———. "Spyglass to Commercialize Future Mosaic Versions." *Computerworld* 28 (August 29, 1994): 16.

"Boring, Boring, Boring: Business-to-Business E-Commerce Is a Revolution in a Ball Valve." In "In Search of the Perfect Market: A Survey of Electronic Commerce." series. *Economist,* May 10, 1997.

Borland, John. "Net Blackout Marks Web's Achilles Heel." CNET News.com, June 6, 2001.

Borrell, Jerry. "America's Shame: How We've Abandoned Our Children's Future." *MacWorld,* September 1992.

Borsook, Paulina, and Joanne Cummings. "The Twenty-Five Most Powerful People in Networking." *Network World* 13 (January 1, 1996): 31–53.

Bosco, Pearl. "Data Goes Dynamic, and Complex." *Bank Systems and Technology* 34 (February 1997): 39.

Boslet, Mark. "Faced with Energy Shortfalls and an Aging Power Grid, Internet Economy Companies Search for Solutions." *TheStandard.com,* September 12, 2000.

Bowman, Catherine. "Bus Riders Union Won't Stay Seated. AC Transit Passengers Want a Say in Service." *San Francisco Chronicle,* March 31, 1997.

Bowman, Scott. "Mergers, Markets, and the Transformation of International Law." In *Postimperialism and World Politics.* London: Praeger, 1999.

Bradner, Scott. "Future Structures on the Internet." *Network World* 11 (July 25, 1994): 16.

Brandel, Mary. "On-line Catalogs Are Booting Up." *Computerworld* (*Electronic Commerce Journal* supplement), April 29, 1996, 5.

Brewin, Bob. "DARPA's Fields Folly: What's Next for U.S. High-Tech? (Reassignment of Defense Advanced Research Projects Agency Director Craig I. Fields Draws Sharp Criticism)." *Federal Computer Week* 4 (April 30, 1990): 6.

Bright, Julian. "The Smart City: Communications Utopia or Future Reality?" *Telecommunications* (international edition), 29 (September 1995): 175–81.

Broder, John M. "AT&T Takes Lead on a Plan to Cut Long-Distance Costs." *New York Times,* May 4, 1997.

Brothers, Art. "Minitel Revisited: An Update on the 'French Connection.'" *America's Network* 100 (April 15, 1996): 62.

Brown, Jim. "More Public/Private Sharing in Mass." *Network World* 12 (May 15, 1995): 45.

Bucholtz, Chris. "Battle Lines Drawn for Universal Service." *Telephony* 230 (May 6, 1996): 22.

Buckley, Sean. "Back Talk." *Telecommunications,* April 1, 2001.

Bulkeley, William M. "Linux, Maverick of Computing, Gets Respectable." *Wall Street Journal,* April 9, 2001.

"Bulletin." *R & D*, February 1, 2000.

Burger, Dale. "SMART Toronto Promoting City as a Technology Hub (Non-profit Organization Promoting Toronto Support for Information Superhighway)." *Computing Canada* 21 (February 15, 1995): 32.

Burrows, Peter. "Craig Fields's Not-So-Excellent Adventure." *Business Week* (industrial/technology edition) 3362 (March 14, 1994): 32.

Burstein, Melvin L., and Arthur J. Rolnick. "Congress Should End the Economic War Among the States." *Federal Reserve Bank of Minneapolis: The Region* 9 (March 1995): 3–20.

Bushaus, Dawn. "Project Provides a Breeding Ground for Developing Multicasting, Quality-of-Service Solutions." *InformationWeek*, March 20, 2000.

Buswell, R. J., R. P. Easterbrook, and C. S. Morphet. "Geography, Regions, and Research and Development Activity: The Case of the United Kingdom." In *The Regional Impact of Technological Change*, edited by A. T. Thwaites and R. P. Oakey. New York: St. Martin's Press, 1985.

Button, Kate. "Hub of the Solar System. Sun Microsystems' Scott McNealy's Attitude Towards Cloning." *Computer Weekly*, December 3, 1992, 39.

Bylinsky, Gene. "DARPA: A Big Pot of Unrestricted Money." *Fortune* 123 (Spring–Summer, 1991): 65.

"Calif. Public Records Not So Public." Associated Press, August 18, 1997.

"California Wants Bids to Privatize State Telephone System." *Newsbytes News Network*, October 9, 1997.

"California's Economy: The Real Trouble." *Economist*, July 28, 2001.

Carey, John, and Jim Bartimo. "If You Control . . . Computers, You Control the World." *Business Week* (industrial/technology edition) 3170 (July 23, 1990): 31.

Carroll, J. A. "Wired to the World." *CA Magazine* 126 (August 1993): 28–31.

Cassidy, J. P. "Rural, Urban Lawmakers Battle over Tauzin-Dingell Internet Bill." *Hill News*, May 9, 2001.

Cats-Baril, William L., and Tawfik Jelassi. "The French Videotex System Minitel: A Successful Implementation of a National Information Technology Infrastructure." *MIS Quarterly* 18 (March 1994): 1–20.

Cerf, Vinton G. "Computer Networking: Global Infrastructure for the Twenty-First Century." Available [Online], <http://cra.org/research.impact>.

Chakravarty, Subrata N. "The Quiet Corner." *Forbes*, February 23, 1998, 76.

"The Changing Dream." In "Future Perfect? A Survey of Silicon Valley." *Economist*, March 29, 1997.

Chapman, Gary. "Mexico: Window on Technology and the Poor." *Los Angeles Times*, October 28, 1996.

Charles, Petit. "Praise for California at Science Summit—but Leaders Worry About Low Test Scores." *San Francisco Chronicle*, May 29, 1996, A15.

Chernicoff, David P. "Networking History in a Nutshell." *PC Week* 11 (September 12, 1994): N12.

"Chip Makers, U.S. Announce Venture." Associated Press, September 11, 1997.

Chris, Tracy. "Maligning Free Software Is a Growing Web Tradition." *New York Times*, August 13, 1997.

Clark, Tim. "HP Leads E-Commerce Initiative." CNET News.com, December 2, 1997.

———. "VeriFone Furthers Online Payments." CNET News.com, December 15, 1997.

Clarke, Peter. "Drumbeat Sounds for CPU Equivalent of Open-Source Software Movement—Free Cores Marching into Market." *Electronic Engineering Times,* February 5, 2001.

———. "European Spacemen Launch Free Sparc-Like Core." *Electronic Engineering Times,* March 6, 2000.

Clausing, Jeri. "Senate Panel Supports Ban on Internet Taxes." *New York Times,* November 5, 1997.

Clemente, Frank. "The Dark Side of Deregulation." *Public Utilities Fortnightly* 134 (May 15, 1996): 13–15.

Cocheo, Steve. "Breakaway Strategies." *ABA Banking Journal* 88 (January 1996): 32–37.

Cockburn, Alexander, and Ken Silverstein. "War and Peso (Mexico)." *New Statesman and Society,* February 24, 1995.

"A Collision on the Iowa I-Way." *Business Week,* May 19, 1997.

"CommerceNet Moves East." *Telecommunications* (American edition) 29 (February 1995): 12.

"Communities and Workers Beware!! Did You Know:" Silicon Valley Toxics Coalition. Available [Online], <http://www.svtc.org/svtc/>.

"Compaq's Power Play: How the Compaq-Digital Deal Will Reshape the Entire World of Computers." *Business Week,* January 29, 1998.

"Computer Networking: The Wiring of Iowa." *Economist,* July 15, 1995, 19.

Cooney, Michael, Adam Gaffin, and Ellen Messmer. "Internet Surge Strains Already Shaky Structure." *Network World* 12 (April 3, 1995): 1, 67.

Corbin, Lisa. "Contractors Suffer Blows in Budget Battle." *Government Executive* ("Top 200 Federal Contractors" supplement), August 1996, 41–48.

Courant, Paul N., Edward M. Gramlich, and Susanna Loeb. "Michigan's Recent School Finance Reforms: A Preliminary Report." *American Economic Review* 85 (May 1995): 372–77.

Courneen, Michael. "Raising Support for GIS Development." *American City and County* 108 (July 1993).

Coursey, David. "'New' 3Com Is Branching Out and Turning Green." *InfoWorld* 14 (July 20, 1992): 106.

Coy, Peter, and Gary McWilliams. "Electricity: The Power Shift Ahead." *Business Week,* December 2, 1996.

Cross, Phillip S. "California Backs Loans for WEPEX Software." *Public Utilities Fortnightly* 134 (October 15, 1996): 52.

Cushman, John H., Jr. "E.P.A. Is Pressing Plan to Publicize Pollution Data." *New York Times,* August 12, 1997.

"DARPA Official Hits Policy." *Chilton's Electronic News* 37 (July 29, 1991): 4.

Davidson, Carl, and Jerry Harris. "The Third Wave and the Republicans: Is Newt Gingrich a Closet New Leftist?" *cy.Rev: A Journal of Cybernetic Revolution, Sustainable Socialism, and Radical Democracy* 2 (March 1995).

Davis, Beth. "Wells Fargo to Certify Net Payments." *Informationweek* 610 (December 16, 1996): 30.

Deitz, Dan. "Agile Manufacturing on the Internet." *Mechanical Engineering* 117 (March 1995): 26.

"The Delight of Digital Maps." *Economist,* March 21, 1992.

Denison, D. C. "Woburn, Mass.–Based Internet Services Firm Genuity to Cut Jobs." *Boston Globe,* May 3, 2001.

Deutschman, Alan. "The Next Big Info Tech Battle." *Fortune* 128 (November 29, 1993): 38–50; European ed., 24–31.

DeVoe, Deborah. "nCube to Unveil Multimedia Servers." *InfoWorld* 18 (February 19, 1996): 38.

Drinkard, Jim. "Utility Deregulation Fierce Battle." Associated Press, April 24, 1997.

Duffy, Jim. "IP Net Helps Texas Dole Out Benefits." *Network World* 10 (November 29, 1993): 1.

Dunlap, Charlotte. "Internet Prepares for Commercialization." *Computer Reseller News* 603 (November 7, 1994): 55, 59.

———. "Integrator Spins Web Sites for Utility Outlets." *Computer Reseller News* 698 (August 26, 1996): 65–66.

Edmondson, Brad, "Your Local Future (Local Area Demographic Data Firms)." *American Demographics,* April 1, 1996.

Edwards, Morris. "The Network Computer Comes of Age." *Communications News* 33 (July 1996): 44–45.

Edwards, Owen. "Douglas Engelbart—ASAP Legends." *Forbes,* October 10, 1994, S130.

"Effort to Forge Government/Business Bond Has Merit." *Network World* 10 (March 1, 1993): 26.

Einstein, David. "New Chief at Bay Networks—Veteran Intel Exec to Lead Faltering Networking Firm." *San Francisco Chronicle,* October 31, 1996, D3.

Elliot, Paul. "No Slump for These Sales—Technical Workstation Market." *Government Computer News* 11 (February 17, 1992):S4.

Elstrom, Peter. "Telecom's New Trailblazers." *Business Week,* March 26, 1998.

Evans, Peter. "Government Action, Social Capital, and Development: Reviewing the Evidence on Synergy." *World Development* 24 (June 1996): 1119–1132.

———. "Introduction: Development Strategies Across the Public-Private Divide." *World Development* 24(June 1996): 1033–37.

Feder, Barnaby J. "Compressed Data: Report Challenges E-Commerce Lore." *New York Times,* March 5, 2001.

Feller, Gordon. "Users and Vendors to Split Cost of Multimedia Apps Development." *Network World* 12 (May 15, 1995): 45.

"FERC Issues Advance NOPR over Hebert's Objections to Begin Phase II of Oasis Implementation." *Foster Electric Report,* July 19, 2000.

"FERC Says Use Web to Post Transmission Access Data." *Electricity Journal* 9 (January/February 1996): 5–6.

Ferguson, Tim W. "Web Grab? If It Moves, Tax It. State and Local Governments Smell Revenue in the Internet." *Forbes,* March 9, 1998.

Fimrite, Peter. "Altamont Rail Plan on Track: A Central Valley-South Bay Link." *San Francisco Chronicle,* December 2, 1996.

———. "Altamont Rail Route Gets Boost. Alameda County Panel OKs Funds for Commuter Line." *San Francisco Chronicle,* December 21, 1996.

"The First Two Decades." *EPRI Journal,* January 1993.

Fisher, Lawrence. "Hewlett in Deal for Maker of Credit-Card Devices." *New York Times,* April 24, 1997.

———. "Routing Makes Cisco Systems a Powerhouse of Computing." *New York Times,* November 11, 1996.

Flynn, Laurie J. "New Economy: Open-Source Movement Advances." *New York Times,* June 4, 2001.

Foley, Mary Jo. "Microsoft Releases BizTalk E-Commerce Software." CNET News.com, December 12, 2000.

———. "XML Goes to the U.N., NATO: This Biz-to-Biz Infrastructure Is Not Just for Geeks Anymore." Sm@rt Partner, October 13, 1999.

Freedberg, Louis. "Race Panel Gets an Earful in San Jose; Talks Dominated by Hecklers, Complaints About Wage Gap." San Francisco Chronicle, February 11, 1998.

Frye, Colleen. "Electric Utilities Have Oasis in Sight." Software Magazine 16 (October 1996): 19.

Fusting, Pamela. "Will Universal Service Be Preserved?" Rural Telecommunications 15 (November/December 1996): 14–22.

Gaffin, Adam. "Net Pioneers See No End to Their Grand Experiment." Network World 11 (August 22, 1994): 1, 61.

Gage, Deborah. "Sun Launches New Servers, Support Boost." Computer Reseller News no. 681 (April 29, 1996): 67, 70.

Gandy, Anthony. "Community Link-Up." Banker 147 (February 1997): 74–75.

Gandy, Oscar, Jr. "It's Discrimination, Stupid!" In Resisting the Virtual Life: The Culture and Politics of Information, edited by James Brook and Ian Boal. San Francisco: City Lights, 1995.

Gannes, Stuart. "Can GTE's Sprint Keep Pace in Long Distance?" Fortune 111 (April 29, 1985): 98–104.

Garten, Jeffrey. "Why the Global Economy Is Here to Stay." Business Week, March 12, 1998.

Garwood, Griffith L. "Should New CRA Have 'Tolerance for Ambiguity'?" ABA Banking Journal 89 (February 1997): 26–32.

Garwood, Griffith L., and Dolores S. Smith. "The Community Reinvestment Act: Evolution and Current Issues." Federal Reserve Bulletin 79 (April, 1993): 251.

Gattuso, Greg. "Tax Fairness Act Unfair: DMA." Direct Marketing 57 (June 1994): 6.

Gaura, Maria Alicia. "Santa Clara County Coffers Should Jingle with $8 Million Surplus; Lower Welfare Costs, Higher Property Values—$8 Million Surplus." San Francisco Chronicle, February 17, 1998.

———. "2 Tax Measures Attack Silicon Valley Gridlock." San Francisco Chronicle, October 14, 1996.

Gaura, Maria Alicia, and Marshall Wilson. "Silicon Valley Commuters Ask for More Transit: 'Unlock the Gridlock' Forum in San Jose." San Francisco Chronicle, March 21, 1997.

Gawlicki, Scott M. "The OASIS Horizon." Independent Energy 26 (November 1996): 26–29.

Genetelli, Richard W., David B. Zigman, and Cesar E. Bencosme. "Recent U.S. Supreme Court Decisions on State and Local Tax Issues." CPA Journal 62 (November 1992): 38–44.

Gengler, Barbara. "California's Big Picture: Serious About the Potential of the Internet, Pacific Bell Has Developed an Experimental Backbone." LAN Computing 5 (April, 1994).

George, Rose. "Drop the Prop." Nation, January 9/16, 1995.

Gerth, Jeff, and Joseph Kahn. "Critics Say U.S. Energy Agency Is Weak in Oversight of Utilities." New York Times, March 23, 2001.

Giussani, Bruno. "France Gets Along with Pre-Web Technology." New York Times, September 23, 1997.

Glass, Brett. "The Real Cost of Mail-Order PCs." *InfoWorld* 13 (December 2, 1991): 45–46.

Glyn Moody. "The Greatest OS That (N)ever Was." *Wired,* August 1997.

Goodin, Dan. "The Nonprofit Millionaires." *Industry Standard,* February 26 2001.

Gordon, R., and L. Kimball. "The Impact of Industrial Structure on Global High Technology Location." In *The Spatial Impact of Technological Change,* edited by John Brotchie, Peter Hall, and Peter Newton. London: Croom Helm, 1987.

Gottlieb, Joseph. "Education: The Right Formula." *Network World* 5 (May 30, 1988): 28–30.

Gramsci, Antonio. "Americanism and Fordism." In *Selections from the Prison Notebooks of Antonio Gramsci.* New York: International Publishers, 1971.

Greco, Monica. "AMTEX Bears First Fruit." *Apparel Industry Magazine* 56 (April 1995): 32–38.

Green-Armytage, Jonathan. "Second Look at the Future." *Computer Weekly,* June 16, 1994, 54.

Greenhouse, Linda. "High Court to Hear Dispute on Opening Phone Markets." *New York Times,* January 27, 1998.

Greenhouse, Steven. "Contract Accord Averts Strike by Doormen." *New York Times,* April 20, 2000.

Gunnison, Robert B. "Computer Giveaway Favors Silicon Valley Schools; Foundation Ignores Most Other Areas." *San Francisco Chronicle,* March 1, 1997.

———. "Ex-Workers Say Giveaway Totals Were Inflated; Wilson-Backed Program to Equip State's Schools." *San Francisco Chronicle,* May 20, 1997.

Gurley, J. William, and Michael H. Martin. "The Price Isn't Right on the Internet." *Fortune* 135 (January 13, 1997): 152–54.

Gwaltney, John R. "Fallacies of Sales-Tax Loophole for Mail-Order Firms." *Small Business Reports* 15 (April 1990): 26–29.

Haavind, Robert. "Lean and Limber Will Describe the Company of the Future." *Electronic Business* 16 (April 30, 1990): 58–60.

Hall, Mark. "CIO Goes Beyond the Cutting Edge." *ComputerWorld Canada,* March 23, 2001.

Hall, Peter. "The Geography of High Technology: An Anglo-American Comparison." In *The Spatial Impact of Technological Change,* edited by John Brotchie, Peter Hall, and Peter Newton. London: Croom Helm, 1987.

———. "The Geography of the Post-Industrial Economy." In *The Spatial Impact of Technological Change,* edited by John Brotchie, Peter Hall, and Peter Newton. London: Croom Helm, 1987.

Hanley, Michael. "Small Towns, Big Plans." *Telephony* 232 (June 2, 1997): 246–56.

Harler, Curt. "Why Governments Want Your Network Center." *Communications News* 29 (December 1992): 30.

Hatlestad, Luc. "Vendors Shop Products for Easy 'Net Commerce at IEC." *InfoWorld* 18 (September 2, 1996): 9.

Hayes, John R. "Blackout.: Disputes Among Telecommunications Companies Jeopardize Customer Service." *Forbes,* November 4, 1996, 346.

Hellerstein, Judith. "The NTIA Needs to Rethink Its Role in the New Telecommunications Environment." *Telecommunications* (Americas edition) 30 (August 1996): 22.

Herhold, Scott. "Tandem Helped Invent a Unique Valley Culture." *San Jose Mercury News,* June 24, 1997.

Hettinga, Bob. "Walking 'Down' the Hierarchy; Reliability, Differentiation, and the Emergence of Internet Syndicalism." Red Rock Eater News Service (RRE), February 29, 1996.

Hirsch, Robert L. "Technology for a Competitive Industry." *Electricity Journal* 9 (May 1996): 81, 80.

Hirschman, Carolyn. "Equality Assurance." *Telephony* 232 (March 17, 1997): 114.

———. "Jockeying for Position." *Telephony* 231 (December 9, 1996): 110–11.

Hoag, John C. "Oasis: A Mirage of Reliability." *Public Utilities Fortnightly* 134 (November 1, 1996): 38–40.

Hodgson, Bob. "Technopolis: Challenges and Issues: A Tale of (at Least) Three Cities." In *The Technopolis Phenomenon: Smart Cities, Fast Systems, Global Networks*, edited by David V. Gibson, George Kozmetsky, and Raymond W. Smilor. Lanham, Md.: Rowman and Littlefield, 1992.

Hof, Robert. "Too Much of a Good Thing?" *Business Week*, August 25, 1997.

Holding, Reynolds. "Bill to Allow Resale of Public Records: First Amendment Lawyers Say Free Access Is Threatened." *San Francisco Chronicle*, January 22, 1996.

Hollis, Sheila S., and Andrew S. Katz. "Wired or Mired? Electronic Information for the Gas Industry." *Public Utilities Fortnightly* 134 (February 1, 1996): 15–18.

Holmes, Steven A. "Budget Cuts May Curtail Census in Year 2000." *New York Times*, August 23, 1995.

Holson, Laura M. "New BankAmerica Will Have a National Reach." *New York Times*, April 14, 1998.

Holzinger, Albert G. "Netscape Founder Points, and It Clicks." *Nation's Business* 84 (January 1996): 32.

Horwitt, Elisabeth. "It Ain't No Superhighway." *CommunicationsWeek* 547 (March 13, 1995): S23–S24.

Hotchkiss, D. Anne. "Brickless Banking." *Bank Marketing* 29 (January 1997): 36–37.

"The House That Larry Built." *Economist*, February 19, 1994, 73.

Howe, Kenneth. "Green Light for Takeover of Pac Bell; Refunds Cut in Deal with Texas Firm." *San Francisco Chronicle*, April 1, 1997.

———. "Northern California Cities Vary Greatly in Business Taxes." *San Francisco Chronicle*, March 13, 1997.

———. "PacTel Is No More—It's SBC; Company Plans to Be One-Stop Phone Shop." *San Francisco Chronicle*, April 2, 1997.

Huber, Peter. "Buy the Government of Your Choice." *Forbes*, December. 2, 1996.

Hulett, Stanley W. "Universal Service: Theory or Reality?" *Public Utilities Fortnightly* 134 (October 15, 1996): 18–19.

Hunter, Roselle. "Big Boost for Small Business: Wells Fargo $45 Billion Pledge Should Spur Urban Economic Development." *Black Enterprise* 27 (November 1996): 18.

Hutheesing, Nikhil. "HP's Giant ATM. Why Did Hewlett-Packard Spend $1.3 Billion for a Credit Card Operation? For a Weapon to Take On IBM in Electronic Commerce." *Forbes*, February 9, 1998.

Hyman, Leonard. "A Timid Proposal: Follow the Internet, Not Poolco." *Electricity Journal* 8 (January/February 1995): 21–27.

Hymer, Stephen. "The Multinational Corporation and the Law of Uneven Development." In *Economics and the World Order*, edited by J. Bhagwati. New York: Free Press, 1972.

"In Search of the Perfect Market." *Economist*, May 10, 1997.

"In Search of State-of-the-Art Technologies: Korea's Drive to the Information Superhighway." *East Asian Executive Reports* 18 (September 15, 1996).

"Indirect Tax; Poway Seeks Internet Franchise Fees." *San Diego Union-Tribune*, June 1, 2001.

Ingram, Erik. "Deregulation—Devil's in the Details: How Competitive Market in Electricity May Shape Up." *San Francisco Chronicle*, May 8, 1997.

———. "Marin Group Files Suit to Block Expansion of Lucasfilm Empire: Plan's Opponents Ask for Overturn of Vote Approving New Facility." *San Francisco Chronicle*, December 4, 1996.

Ironmonger, D. "The Impact of Technology on the Household Economy." In *The Spatial Impact of Technological Change*, edited by John Brotchie, Peter Hall, and Peter Newton. London: Croom Helm, 1987.

Jackson, William. "Next-Generation vBNS Gets Three-Year Extension." *Government Computer News*, June 5, 2000.

Janah, Monua, and Mary E. Thyfault. "Bandwidth Quest." *Informationweek* 625 (April 7, 1997): 38–48.

Jenkins, Holman W., Jr. "Maybe the Utilities Weren't So Dumb After All." *Wall Street Journal*, May 16, 2001.

Johnston, David Cay. "Deal to Close Mail-Order Tax Loophole Said to Be Imminent." *New York Times*, November 6, 1997.

———. "Online Sales Collide with Off-Line Tax Questions." *New York Times*, November 10, 1997.

———. "Sales Tax Proposal Angers Mail-Order Customers." *New York Times*, November 7, 1997.

Jones, Cate. "R&D Bracing for Economic Fallout from Deregulation." *Electrical World* 211 (January 1997): 22–23.

Jubien, Sidney Mannheim. "The Regulatory Divide: Federal and State Jurisdiction in a Restructured Electricity Industry." *Electricity Journal* 9 (November 1996): 68–79.

Junnarkar, Sandeep. "Fewer Bricks Mean Higher Returns at New Internet Banks." *New York Times*, February 25, 1998.

———. "New Phone Rules Will Have Mixed Effect on Net." *New York Times*, May 8, 1997.

———. "Regional Phone Companies to Offer New Access Technologies for ISPs." *New York Times*, April 22, 1997.

Jurgens, Rick. "Behind the Scenes of CA's Power Crisis Is a Deregulated, High-Stakes Commodities Market." *Contra Costa Times*, July 1, 2001.

Kady, Martin. "Energy Gives Qwest Breakthrough Deal." *Denver Business Journal*, January 14, 2000.

Kahn, Robert E. "The Role of Government in the Evolution of the Internet." *Communications of the ACM* 37 (August 1994): 15–19.

Kane, Margaret. "Online Spending to Hit $65 Billion." CNET News.com, May 2, 2001.

Katz, Marvin. "Beyond Order 636: Making the Most of Existing Pipeline Capacity." *American Gas* 77 (June 1995): 26–29.

Kemp, William J. "The Western Systems Power Pool: A Bulk Power Free Market Experiment." *Public Utilities Fortnightly* 119 (April 30, 1987): 23–27.

Kerr, Susan. "3Com Corp. (the Datamation 100 Company Profile)." *Datamation* 38 (June 15, 1992): 141.

Kessler, Jack. "Electronic Networks: A View from Europe." *American Society for Information Science Bulletin* 20 (April/May 1994): 26–27.

Kindel, Sharen. "Geographic Information Systems." *Financial World*, January 19, 1993).

Kindy, Kimberly. "Power Producers Coordinated Data to Take Advantage of California Market." *Orange County Register*, March 26, 2001.

———. "Studies Show Pitfalls in Power Market." *Orange County Register*, March 25, 2001.

King, Julia. "Electronic Forms Are New Wave in Pollution Control." *Computerworld* 29 (July 10, 1995): 42.

Kirsner, Scott. "WorldCom/MCI: Good Giant or Bad?" *Wired News*, October 2, 1997.

"Knit Your Own Superhighway." *Economist*, October 16, 1993.

Kong, Deborah. "Palo Alto Generous in Budgeting; but City Plans 12% Increase in Electric Rates." *San Jose Mercury News*, June 24, 1997.

Kornblum, Janet. "Domain Players Prepare for Competition." *News.com*, January 27, 1998.

———. "Netscape Sets Source Code Free." *News.Com*, March 31, 1998.

Kovach, Bill. "When Public Business Goes Private." *New York Times*, December 4, 1996.

Krantz, Michael. "Doing Business on the Net Is Hard Because the Underlying Software Is So Dumb. XML Will Fix That." *Time*, November 10, 1997.

Krugman, Paul. "Complex Landscapes in Economic Geography." *American Economic Review* 84 (May 1994).

Kupfer, Andrew. "Son of Internet GTE Is Buying the Firm That Built the Original. It's Also Betting $4 Billion to Build Its Own Data Network." *Fortune*, June 23, 1997.

"Labor Plans Big Sweep in Valley: Janitors Just First Wave of Organizing 'Renaissance.'" *San Francisco Chronicle*, May 31, 1996.

Lander, Greg M. "Just in Time: EDI for Gas Nominations." *Public Utilities Fortnightly* 134 (February 1, 1996): 20–23.

Landler, Mark. "Hi, I'm in Bangalore (but I Dare Not Tell)." *New York Times*, March 21, 2001.

Lannon, Larry. "Short Shots." *Telephony* 232 (February 10, 1997): 72.

Lash, Alex. "JavaOS Headed to Consumer Gear." CNET *News.Com*, March 25, 1998.

———. "Making Money with Free Software." *News.Com*, February 2, 1998.

"Lawmakers Criticize Internet Bill." Associated Press, March 10, 1998.

Lawson, Stephen. "Internet Gridlock Escalates." *InfoWorld* 18 (July 15, 1996): 1, 24.

Lebowitz, Jeff. "The Dawning of a New Era." *Mortgage Banking* 56 (June 1996): 54–66.

Lee, Jennifer. "A New Silicon Valley in Utah?" *New York Times*, June 13, 2001.

Leonard, Andrew. "Gobbling Up the Net." *Salon*, June 12, 1997.

Leong, Kathy Chin. "Siliconnected Valley (Silicon Valley Companies Form Smart Valley Inc. Non-Profit to Build Giant Information Network)." *InformationWeek* 438 (August 16, 1993): 22.

Leroy, Greg. "No More Candy Store: States Move to End Corporate Welfare as We Know It." *Dollars and Sense* 199 (May–June, 1995): 10.

Levine, Shira. "Fiber to the Mansion?" *Telephony* 232 (April 14, 1997): 38.

Levinson, Marc. "Get Out of Here!" *Business Week*, June 3, 1996.

Levy, Steven. "How the Propeller Heads Stole the Electronic Future." *New York Times Magazine*, September 24, 1995, 59.

Lewis, Peter. "Alliance Formed Around Sun's Java Network." *New York Times*, December 11, 1996.

Lindstrom, Ann H. "Carriers Lose Debate over Iowa Educational Network." *Telephony* 220 (April 29, 1991): 8.

Lock, Robert K., Jr. "Transients: The Next Market to Tap." *Telephony* 232 (March 3, 1997): 76–78.

Locke, Gary. "Caught in the Electrical Fallout." *New York Times,* February 2, 2001.

Lohr, Steve. "Spyglass, a Pioneer, Learns Hard Lessons About Microsoft." *New York Times,* March 2, 1998.

Loizos, Constance. "Ecologies of Scale." *Red Herring,* February 1, 1998.

Lubove, Seth. "Cyberbanking." *Forbes,* October 21, 1996, 108–16.

Lucas, Greg. "Prop. 211 Loses by Wide Margin." *San Francisco Chronicle,* November 6, 1996.

———. "State Data Chief Quits After Audit; He's Accused of Taking Golf Junkets." *San Francisco Chronicle,* August 22, 1997.

Lusvardi, Wayne C., and Charles B. Warren. "What Price an Easement? Setting Market Value in Fiber Optic Corridors." *Public Utilities Fortnightly,* July 1, 2001.

Lutterbeck, Deborah. "Muni Telecoms." *Infrastructure Finance* 6 (April 1997): 14–17.

Lynch, Larry. "Large Power Users Seek Direct Access to Power Sellers." *Sacramento Bee,* March 27, 2001.

McCabe, Michael. "Palo Alto Moves Fast to Center of Internet; Fiber-Optic Ring Nearly Completed." *San Francisco Chronicle,* June 26, 1997.

McGarvey, Joe. "Intranets, NT Shape Server Market." *SoftBase,* June 1, 1997.

McGhee, Tom. "Corporate Opposites Attract." *Albuquerque Journal,* August 30, 1999.

Mack, Toni. "Empty Pipes the Delicious Dilemma for Phil Anschutz and Joe Nacchio: Fill Qwest Communications' Capacious New Network or Sell It Off." *Forbes,* November 30, 1998.

McKay, John. "Comparative Perspectives on Transit in Europe and the United States, 1850–1914." In *Technology and the Rise of the Networked City in Europe and America,* edited by Joel A. Tarr and Gabriel Dupuy. Philadelphia: Temple University Press, 1988.

Maize, Kennedy. "Calif. Competitors See Threat from Utility Monopolies." *Electricity Journal* 10 (March 1997): 8.

Malecki, E. J. "Public Sector Research and Development and Regional Economic Performance in the United States." In *The Regional Impact of Technological Change,* edited by A. T. Thwaites and R. P. Oakey. New York: St. Martin's Press, 1985.

Mandel, Michael. "Taking Its Place in the Pantheon." *Business Week,* August 25, 1997.

Manes, Stephen. "The Info Footpath: How the Internet Got Its Start in the Early Days of the Computer Age." *New York Times,* September 8, 1996.

Markoff, John. "Gold Rush from Software Animates Silicon Valley." *New York Times,* January 13, 1997.

———. "Ignore the Label, It's Flextronics: Inside Outsourcing's New Cachet in Silicon Valley." *New York Times,* February 15, 2001.

———. "Rivals Cooperate on Chip Equipment." *New York Times,* December 12, 2000.

———. "Several Big Deals Near for Sun's Java Language." *New York Times,* March 24, 1998.

———. "Silicon Valley Job Growth Begins to Slow." *New York Times,* January 15, 2001.

Marquis, Christopher. "U.S. Approves Formation of Supply Web Site for Automakers." *New York Times,* September 12, 2000.

Marshall, Jonathan. "Economist's Auction Theory Goes to Market." *San Francisco Chronicle,* April 21, 1997.

———. "More Failures Expected for Power Network: Increased Demand Strains Regional Transmission Lines." *San Francisco Chronicle,* August 13, 1996.

———. "PG&E Is First to Use the Net to Sell Surplus." *San Francisco Chronicle,* November 12, 1996.

———. "Rate Cuts Could Hurt PacTel Deal; Quigley Says Demands Threaten Merger." *San Francisco Chronicle,* October 2, 1996.

———. "Silicon Valley Boom Means Trade-Off for Residents." *San Francisco Chronicle,* January 13, 1997.

Marshall, Jonathan, and Jon Swartz. "Coalition Fights to Keep Net Fees Low; Phone Companies Want Higher Rates." *San Francisco Chronicle,* November 9, 1996.

Marshall, Jonathan, and Jon Swartz. "Sweeping Changes in Phone Rates Long-Distance Cuts to Be Passed on to Customers." *San Francisco Chronicle,* May 8, 1997.

Marshall, Will. "A New Fighting Faith: Ambitious Goals for the World's Oldest Political Party." *New Democrat,* September/October 1996.

Martin, Frances. "Banking on Internet Banking's Success." *Credit Card Management* 9 (August 1996): 50.

Martinez, Carlos. "Structuring Strong Document Access." *Managing Office Technology* 40 (October 1995): 32.

"MCC Digs Out of the 'Celestial Sandbox of Computer Science.'" *Electronic Business* 18 (May 18, 1992): 67–68.

Meister, Charles K. "Converging Trends Portend Dynamic Changes on the Banking Horizon." *Bank Marketing* 28 (July 1996): 15–21.

Mendels, Pamela. "Not Everyone Is Eager to Jump on the Net-Day Bandwagon." *New York Times,* October 19, 1996.

Messmer, Ellen. "America Online Buys Internet Service Provider." *Network World* 11 (December 5, 1994): 4.

———. "Automakers Drive Private IP Network." *Network World* 13 (September 30, 1996): 33.

———. "Ford Testing a 'Private Internet' to Link Suppliers." *Network World* 12 (March 13, 1995): 14.

———. "Industry Heavyweights Team Up to Design Information Superhighway." *Network World* 10 (December 20, 1993): 20.

———. "NSF Changes Course on Its Internet Plan." *Network World* 9 (December 21, 1992): 1, 4.

———. "Spyglass Captures Mosaic Licensing." *Network World* 11 (August 29, 1994): 4.

Metcalfe, Bob. "Old Fogies to Duke It Out for Credit at Internet's 25th Anniversary." *InfoWorld* 16 (August 29, 1994): 50.

———. "Thanks, NCSA, for Graduating a Few of Your Mosaic Cyberstars." *InfoWorld* 16 (June 6, 1994): 50.

Michalski, Jerry. "O Pioneers!" *RELease 1.0* 94 (January 31, 1994): 5.

"Microsoft Steps Up Calif. Presence." Associated Press, March 26, 2001.

Miller, Cyndee. "Catalogs Alive, Thriving." *Marketing News* 28 (February 28, 1994): 1, 6.

Miller, Leslie. "Activism Goes On-line: Plugging in to Electronic Organizing." *USA Today,* April 25, 1995.

Miller, William H. "Electrifying Momentum." *Industry Week* 246 (February 17, 1997): 69–74.

Minton, Torri, and Jon Swartz. "35 Cents Charge for Pac Bell Pay Phones." *San Francisco Chronicle*, October 18, 1997.

Mitra, Steve. "Community Development Banks: Urban Renewal That Works." *Business and Society Review* 86 (Summer 1993): 38–42.

Mitschele, Rhonda. "Share and Share Alike: Creating a Cost-Effective GIS." *American City and County*, March 1996.

Moberg, David. "Everything You Know About the New Economy Is Wrong." *Salon.Com*, October 26, 2000.

Moeller, Michael. "Fort Apache: Freeware's Spirit Outshines Commercial Products." *PC Week* 14 (June 9, 1997).

Molotch, Harvey. "The City as Growth Machine: Toward a Political Economy of Place." *American Journal of Sociology* 82 (1976): 309–32.

Moltzen, Edward F. "IBM Puts 20% of R&D into Networking." *Computer Reseller News* 705 (October 14, 1996): 290.

Monck, Ellen Rowley. "How a Megabank Is Answering the CRA Challenge." *Journal of Commercial Lending* 76 (September 1993): 47–54.

"More Banks Follow the Internet Route." *Banker* 146 (August 1996): 20.

Morisi, Teresa L. "Commercial Banking Transformed by Computer Technology." *Monthly Labor Review* 119 (August 1996): 30–36.

Morton, Oliver. "Private Spy." *Wired*, August 1997.

Moss, M. "Telecommunications and International Financial Centers." In *The Spatial Impact of Technological Change*, edited by John Brotchie, Peter Hall, and Peter Newton. London: Croom Helm, 1987.

Moukheiber, Zina. "Windows NT—Never!" *Forbes*, September 23, 1996, 42–43.

Moura, Brian. "San Carlos Discovers the Internet." *Public Management* 78 (January, 1996): 31.

———. "The Telecommunications Revolution Comes to City Hall." *Western City* (League of California Cities), April 1996.

Mueller, Milton. "Universal Service and the Telecommunications Act: Myth Made Law." *Communications of the ACM* 40 (March 1997): 39–47.

Muhammad, Tariq K., and Ronald Roach. "Doing Business the Paperless Way." *Black Enterprise* 27 (November 1996): 96–102.

Munk, Nina. "Voluntary Simplicity." *Forbes*, October 21, 1996, 284.

Murphy, Jamie, and Brian Massey. "No Shortage of Bottlenecks on Information Superhighway." *New York Times*, July 29–30, 1996.

Murphy, Jamie, and Charlie Hofhacker. "Explosive Growth Clogs Internet Backbone." *New York Times*, July 29–30, 1996.

Nash, Kim S. "NCube Piggybacks on Oracle Plans." *Computerworld* 28 (January 31, 1994): 33.

Nee, Eric. "Jim Clark." *Upside* 7 (October 1995): 28–48.

"Netting New York: http://manhattan." *Economist*, May 25, 1996.

"A New Player in Set-Top Boxes." *Business Week*, March 12, 1998.

Nieves, Evelyn. "As the New Economy Starts to Wheeze, San Francisco's Party Mood Fizzles." *New York Times*, March 26, 2001.

Nocera, Joseph. "Cooking with Cisco." *Fortune* 132 (December 25, 1995): 114–22.

Noer, Michael. "Stillborn." *Forbes*, March 19, 1998.

Noll, A. Michael. "Cybernetwork Technology: Issues and Uncertainties." *Communications of the ACM* 39 (December 1996): 27–31.

Nolle, Tom. "Deregulation Takes a Turn for the Better." *INTERNETWORK* 7 (December, 1996): 12.

"Northern Telecom Joins ANS to Encourage Development of National Info Infrastructure." *Information Today* 10 (December 1993): 64.

"Nsfnet Backbone Decommissioned Nsfnet: Program Takes Next Steps in Advancing Networking." NSF PR 95–37 (May 15, 1995).

O'Connell, Daniel. "U.S. Supreme Court Reviews State and Local Taxation Issues." *CPA Journal* 62 (March 1992): 16–21.

Orr, Bill. "We're Not in Kansas Anymore." *ABA Banking Journal* 88 (July 1996): 72.

"Pac Bell Unit's President Resigns." *San Francisco Chronicle,* May 30, 1997.

"Pacific Gas & Electric to Unload 4 Plants, 3,059 MW." *Electricity Journal* 9 (December 1996): 4–5.

Page, Rob. "The Virtual Commute." *LAN Magazine* 9 (October, 1994): S37.

Pappalardo, Denise. "WorldCom Adds to Its 'Net Riches." *Network World* 14 (September 15, 1997): 6.

Passell, Peter. "Taxing Internet Sales—Governors Versus Tax Freedom Act." *New York Times,* March 16, 1998.

Patch, Kimberly. "Spyglass Takes on Mosaic Licensing: Will Focus on Support and Security." *PC Week* 11 (August 29, 1994): 123.

Patterson, Tom. "E-Commerce Electrifies Utility Industry." *Communications News* 35 (March 1998): 72–73.

Pelline, Jeff. "Safer Internet Shopping Offered." *San Francisco Chronicle,* June 23, 1995.

Perera, Rick. "Dot-Com Crash Doesn't Dent Online Spending." *CNN.com,* May 25, 2001.

Perkins, Michael C., and Celia Nunez. "The Market's Odds Are Stacked Against You." *Newsday,* March 18, 2001.

Petersen, Melody. "Trenton Tells Bell Atlantic to Speed Up Urban Cable Connections." *New York Times,* April 22, 1997.

Petit, Charles. "Spending More to Use Less Energy; U.S. Urged to Double Research." *San Francisco Chronicle,* November 29, 1997.

"PG&E to Purchase 18 Plants." Associated Press, August 6, 1997.

Pimentel, Benjamin. "Congested Silicon Valley Slow to Board Light Rail." *San Francisco Chronicle,* February 24, 1997.

Pitta, Julie. "Long Distance Relationship." *Forbes,* March 16, 1992, 136–37.

Plotkin, Hal. "Small Biz Gets No Charge from Electricity Deregulation." *Inc.* 18 (December 1996): 32.

Poirot, Gerard. "Minitel: Oui! Multimedia: Non!" *Communications International* 22 (July 1995): 23–24.

Pontin, Jason, and Tom Quinlan. "Java Is Getting Stronger: Sun Previews Devices to Bolster Java Applets." *InfoWorld* 18 (February 5, 1996): 1, 22.

Port, Otis, and Peter Burrows. "R&D, with a Reality Check." *Business Week* 3355 (January 24, 1994): 62–65.

Press, Larry. "Before the Altair: The History of Personal Computing." *Communications of the ACM* 36 (September 1993): 27–33.

———. "Seeding Networks: The Federal Role." *Communications of the ACM* 39 (October 1996): 11–18.

Presti, Ken. "Chambers Foresees a Growing Reseller Role." *Computer Reseller News* ("Benchmarks" supplement), September 16, 1996, 50.

Preston, Holly Hubbard. "Minitel Reigns in Paris with Key French Connection." *Computer Reseller News* 594 (September 5, 1994): 49–50.

Prince, Cheryl J. "Last Chance to Recapture Payments." *Bank Systems and Technology* 33 (July 1996): 28–32.

Radford, Bruce W. "Must Run, Must Take." *Public Utilities Fortnightly* 134 (September 15, 1996): 4–5.

Rafter, Michelle V. "A Transcontinental Undertaking Company Follows Railroads in Building a National Network." *Chicago Tribune,* December 1, 1997.

Rainbow, Roger. "Global Forces Shape the Electricity Industry." *Electricity Journal* 9 (May 1996): 14–20.

Ramesh, V. C. "Information Matters: Beyond OASIS." *Electricity Journal* 10 (March 1997): 78–82.

Randall, Neil. "Leaning Toward Linux." *PC Magazine* 16 (July 1997).

Raynovich, R. Scott. "*LAN Times* Talks to Psinet's William Schrader About Internet Connectivity and the Competition." *LAN Times* 12 (October 23, 1995): 8.

Reichenbach, Lloyd, and Mark Hill. "Small Business Lending and the Internet." *Risk Management Association Journal,* February 1, 2001.

"Revised Internet Tax Bill Gets OK." Associated Press, March 19, 1998.

Reynolds, Larry. "Speeding Toward the Information Superhighway." *Management Review* 82 (July 1993): 61–63.

"Rich Man, Poor Man." *Forbes* (*ASAP* supplement), October 9, 1995, 100–101.

Richard, Gordon. "Collaborative Linkages, Transnational Networks and New Structures of Innovation in Silicon Valley's High-Technology Industry." Report no. 4, "Industrial Suppliers/Services in Silicon Valley." Silicon Valley Research Group. University of California at Santa Cruz. January 1993.

Richard, Dan, and Melissa Lavinson. "Something for Everyone: The Politics of California's New Law on Electronic Restructuring." *Public Utilities Fortnightly* 134 (November 15, 1996): 37–41.

Richtel, Matt. "The Left Side of the Web Seeds Global Grass Roots." *New York Times,* June 14, 1997.

Riechmann, Deb. "Teachers Warned on Packaged Lessons Businesses, Other Special Interests Tap into School Market." *San Jose Mercury News,* January 16, 1996.

Ritter, Richard J. "Redlining: The Justice Department Cases." *Mortgage Banking* 55 (September 1995): 16–28.

"A River Runs Through It." *Economist,* May 10, 1997.

Robertson, Jack. "Transfer of DARPA Head Ignites Furor." *Electronic News* 36 (April 30, 1990): 1.

Robinson, Brian. "Bush, DoD Attacked for Firing Fields." *Electronic Engineering Times* 588 (April 30, 1990): 1 (2 pages).

Rogers, Adam. "Getting Faster by the Second." *Newsweek,* December 9, 1996, 86–89.

Romero, Simon. "Once-Bright Future of Optical Fiber Dims." *New York Times,* June 18, 2001.

Rosenberg, Scott. "WorldCom Buys Up the Internet. The Net Becomes WorldCom's Fiefdom." *Salon,* October 9, 1997.

Rosenblum, Harvey. "Electronic Money: Hype and Reality." *Banking Strategies* 72 (November/December 1996): 6–16.

Rule, Bruce. "Internet Bank Decides the Real Money's in Consulting." *Investment Dealers Digest* 63 (January 13, 1997): 12.

Russo, Kimberly. "Deregulation: Making Sense of the De-Tariffing Wars." *Telecommunications* 30 (December, 1996): 56.

Sabel, Charles. "Bootstrapping Reform: Rebuilding Firms, the Welfare State, and Unions." *Politics and Society* 23 (March 1995): 5–48.

———. "Learning by Monitoring: The Institutions of Economic Development." In *The Handbook of Economic Sociology.* Princeton: Princeton University Press, 1994.

Saffo, Paul. "Racing Change on a Merry-Go-Round: MIT Management in the Nineties Program Reports Industry Overall Is Not More Productive Because of Computing Technology." *Personal Computing* 14 (May 25, 1990): 67.

Sandburg, Brenda. "The Standards Bearers: When It Comes to the Internet, Patents Are Made to Trade—Even for Mighty Microsoft." *Recorder,* March 10, 1999.

Satran, Dick. "In(ternet) Crowd Is Biggest On-line Advertiser." *San Jose Mercury News,* December 11, 1995.

Saunders, Renee. "Territorial Wars." *Telephony* 232 (March 17, 1997): 16.

Schechter, Bill. "Metropolitan Governance: A Bibliographic Essay." *National Civic Review* 85 (Spring–Summer, 1996): 63 (4 pages).

Schiesel, Seth. "FCC Urges That Internet Phone Service Be Fee-Based." *New York Times,* April 11, 1998.

———. "For Local Phone Users, Choice Isn't an Option." *New York Times,* November 21, 2000.

———. "MCI Accepts $36.5 Billion Worldcom Offer." *New York Times,* November 11, 1997.

———. "Sitting Pretty: How Baby Bells May Conquer Their World." *New York Times,* April 22, 2001.

Schlender, Brent. "Sun's Java: The Threat to Microsoft Is Real." *Fortune* 134 (November 11, 1996): 165–70; European ed., 87–91.

———. "Whose Internet Is It, Anyway?" *Fortune* 132 (December 11, 1995): 120–42.

Schnaidt, Patricia. "Cruising Along the Super I-Way: Cyber-Pioneers Look to the Information Superhighway to Carry Education, Health Care, and Commerce Applications." *LAN Magazine* 9 (May, 1994): S8.

Schrader, William, and Mitchell Kapor. "The Significance and Impact of the Commercial Internet." In "Global InteNet: Global Internetworking Strategies and Applications," special section of *Telecommunications* 26 (February, 1992): S17 (2 pages).

Schroeder, Erica. "Telecom Act Fuels Regulatory Wars." *PC Week* 13 (April 8, 1996): 51 (2 pages).

Schuler, Joseph F., Jr. "1996 Regulators' Forum: Consensus and Controversy." *Public Utilities Fortnightly* 134 (November 15, 1996): 14–24.

———. "Oasis: Networking on the Grid." *Public Utilities Fortnightly* 134 (November 1, 1996): 32–36.

"SFNB Marks One Year on the Frontline of Internet Banking." *ABA Banking Journal* 88 (December 1996): 62.

"Shakeout in Fiber Optics." *New York Times,* July 29, 2001.

Shankland, Stephen. "Linux Growth Underscores Threat to Microsoft." CNET News.com, February 28, 2001.

Shea, Dan. "MCI Focuses on the Internet." *Telephony* 227 (November 28, 1994): 6.

Shim, Richard. "Dell Throws Its Weight Behind DVD + RW." CNET News.com, June 28, 2001).

Sibley, Kathleen. "Linux: More Than an OS, It's Like a Religion." *Computing Canada* 23 (July 21, 1997).

"Silicon, Silicon Everywhere." In "Future Perfect? A Survey of Silicon Valley." *Economist,* March 29, 1997.

"Silicon Valley Telecommuting Program Gets Underway as Part of 'Smart Valley' Initiative. *Telecommuting Review: The Gordon Report* 11 (April 1994): 1.

Silverthorne, Sean. "Science Friction.—Need for R&D Labs to Focus More on Creating Actual Products." *PC Week* 11 (April 11, 1994):A1.

Simon, Mark. "Down Side of Peninsula Job Boom." *San Francisco Chronicle,* March 24, 1997).

———. "Prop. 211 Foes Share Riches with Other Campaigns: $1.75 Million in Past Week Given to Anti-217 Forces." *San Francisco Chronicle,* November 1, 1996.

"The Sinking of the MAI." *Economist,* March 14, 1998.

Sinton, Peter. "Big Haul of PCs for Local Schools." *San Francisco Chronicle,* November 14, 1996.

———. "Businesses Wired on Intranets." *San Francisco Chronicle,* September 26, 1996.

Smith, Craig S. "City of Silk Becoming Center of Technology." *New York Times,* May 28, 2001.

Smith, Tom. "Internetworking Hardware." *Computer Reseller News* 627 (April 24, 1995): 80.

Somogyi, Stephan. "Sages or Stooges?" *Upside* 9 (June 1997): 62–68.

Southwick, Karen. "California, Here I Come." *Upside* 6 (December 1994): 34–45.

Sowa, Frank X. "Who's Keeping the U.S. Domain Name System? (Network Solutions' Mismanagement of Internet Domain Names)." *Boardwatch Magazine* 10 (June 1996): 74.

Spiers, Joe. "Upheaval in the Electricity Business." *Fortune* 133 (June 24, 1996): 26–30.

Sraeel, Holly. "VeriFone's VeriSmart Strategy Could Be Just What Banks Need to Jump-Start the U.S. Market." *Bank Systems and Technology* 33 (October 1996): 44.

Srodes, James. "Murdering Mail Order." *Financial World* 161 (March 31, 1992): 64–67.

"State Spotlights Computer Woes." *San Jose Mercury News,* June 24, 1997.

Stedman, Craig. "Sun Scales up via Cray." *Computerworld* 30 (June 3, 1996): 39–41.

Steinert-Threlkeld, Tom. "The Internet Shouldn't Be a Breeding Ground for Monopolies—Mosaic Communications' Netscape Giveaway Could Be Prelude to Market Dominance." *InterActive Week* 1 (November 7, 1994): 44.

Sterling, Bruce. "Short History of the Internet." *Magazine of Fantasy and Science Fiction,* February 1993.

Stevens, Tim. "Multiplication by Addition." *Industry Week* 245 (July 1, 1996): 20–24.

———. "NCSA: National Center for Supercomputing Applications." *Industry Week* 243 (December 19, 1994): 56–58.

Stevenson, Richard. "Governors Stress Tax Cuts and Austerity." *New York Times,* May 8, 1997.

Sullivan, Eamonn. "Freedom Is Priceless, Even When It's Free." *PC Week* 13 (November 25, 1996).

"Sun Spots Its Chance." *Network World* 13 (August 12, 1996): I33.

"Sun Workstations: State and Local Governments Turning to Sun; State and Local Customers Now Span 46 States." *EDGE: Work-Group Computing Report* 3 (February 17, 1992): 39.

"Survey of the Internet: The Accidental Superhighway." *Economist,* July 1, 1995, 4.

Sutcliffe, Anthony. "Street Transport in the Second Half of the Nineteenth Century: Mechanization Delayed?" In *Technology and the Rise of the Networked City in Europe and America,* edited by Joel A. Tarr and Gabriel Dupuy. Philadelphia: Temple University Press, 1988.

Taft, Darryl K. "Opportunity Lies in Government Downsizing." *Government Computer News* 11 (February 17, 1992): S14.

———. "Rivals to Govt. Workstation Crown Strain to Find Chinks in Sun Armor." *Government Computer News* 11 (February 17, 1992): 43.

"Tax Facts: The System Sets Cities Up to Be Squeezed by Superstores." Editorial. *San Jose Mercury News,* July 24, 1995.

Taylor, M. J. "Organizational Growth, Spatial Interaction, and Locational Decision-Making." *Regional Studies* 9 (1975): 313–23.

Tedeschi, Bob. "The Net's Real Business Happens .Com to .Com." *New York Times,* April 19, 1999.

"Telecommunications Infrastructure." *East Asian Executive Reports* 18 (September 15, 1996): 14–16.

"Telecoms: FCC Warns Future Mergers After Worldcom/MCI." *Network Briefing,* September 16, 1998.

Temin, Thomas R., and Shawn P. McCarthy. "Sun Chief Sells Unix and Minces No Words." *Government Computer News* 12 (December 6, 1993): 20.

Thomas, Margan D. "Regional Economic Development and the Role of Innovation and Technological Change." In *The Regional Impact of Technological Change,* edited by A. T. Thwaites and R. P. Oakey. New York: St. Martin's Press, 1985.

Thoren-Peden, Deborah, and Julius L. Loeser. "Is Your Bank Going down the Interstate Highway?" *ABA Bank Compliance* 17 (September/October 1996): 3–13.

Tosi, Umberto. "Adele Goldberg." *Forbes,* October 7, 1996, S136.

Uchitelle, Louis. "Companies Spending More on Research and Development." *New York Times,* November 7, 1997.

———. "Corporate Outlays for Basic Research Cut Back Significantly." *New York Times,* October 8, 1996.

Uchitelle, Louis. "Service Sector Guru Now Having Second Thoughts." *New York Times* May 8, 1996.

Uimonen, Terho. "Netscape's Chairman Clark Attacks Microsoft's ActiveX." *Network World* 13 (September 9, 1996): 14.

"University of Illinois and Netscape Communications Reach Agreement." *Information Today* 12 (March 1995): 39.

Vasilash, Gary S. "SRI International—Where Innovation Is a Way of Life." *Production* 99 (December 1987): 54–61.

Vasilash, Gary S., and Robin Yale Bergstrom. "The Five Hottest Manufacturers in Silicon Valley: Customer Obsession at Solectron." *Production* 107 (May 1995): 56–58.

Wagner, Mitch. "Electric Exchange to Buy, Sell over 'Net." *Computerworld* 30 (September 2, 1996): 28.

Wallace, Bob. "High-Speed MCI 'New Internet' Mapped Out." *InfoWorld* 17 (May 1, 1995): 43.

Wallace, David J. "Data Mining and the New Marketing." *Mortgage Banking* 57 (February 1997): 58–63.

Wallace, Scott. "Power to the People: A Complex Web of Computerized Transactions Is Involved in Delivery of Electric Power." *Computerworld* 25 (June 3, 1991): 79–80.

Ward, Leah Beth. "Tax Rules Squeezing Independent Contractors." *New York Times,* November 10, 1996.

Warren, Jim. "We, the People, in the Information Age: Early Times in Silicon Valley." *Dr. Dobb's Journal* 16 (January, 1991): 96D.

Watson, Tom. "Click Here for a Slab of Peanut Pie." *Restaurant Business* 94 (March 20, 1995): 15–18.

Weinberg, Neil. "Backbone Bullies Beneath the Internet's Happy Communal Culture, a Cadre of Giant Carriers Is Mercilessly Squeezing Every Last Dime It Can out of Smaller Players. Users Are Picking Up the Tab." *Forbes,* June 12, 2000.

Weisul, Kimberly. "Electricity Trading Mart Moves Toward Internet." *Investment Dealers Digest* 63 (January 27, 1997): 13.

"Wells Fargo Announces $45 Billion CRA Commitment in First Interstate Bid." *ABA Bank Compliance* 17 (January 1996): 5–6. Regulatory and legislative advisory.

Wexler, Joanie. "State of Envy." *Network World* 11 (November 21, 1994): 64–66.

"What That TCI-Bell Atlantic Merger Means for You." *Fortune* 128 (November 15, 1993): 82–90

White, Brian. "Electronic Bulletin Board Standardization Resulting from FERC EBB Working Group Actions." *Gas Energy Review* 22 (August 1994): 2–11.

Will, George. "'Slow Growth' Is the Liberalism of the Privileged." *New York Times,* August 30, 1987.

Wilson, Marshall. "Boom Means Scarce, Expensive Housing." *San Francisco Chronicle,* March 19, 1997.

Winkler, Connie. "Wells Fargo Stakes Out New Frontiers." *Computerworld* (*Financial Services Journal* supplement), November 1996, F14–F18.

Wolfe, Alexander. "Cray Enters Race for Teraflops Computer." *Electronic Engineering Times* 922 (October 7, 1996): 1, 16.

Wong, Wylie. "New XML E-Business Standard Introduced." CNET News.com, May 14, 2001.

Woodall, Pam. "The Beginning of a Great Adventure: Globalisation and Information Technology Were Made for Each Other." *Economist,* September 23, 2000.

Woolfolk, John, and Steve Johnson. "Tallying Price of California's Power Fiasco, Consumers May Pay $100 Billion over the Next Decade." *San Jose Mercury News,* July 8, 2001.

"Work Week." *Wall Street Journal,* March 27, 2001.

Yates, Ronald E. "New-Look Ameritech Braves 'Change Storm.'" *Chicago Tribune,* April 22, 1993.

Young, Monica. "Finocchio on 3Com." *Computer Reseller News* 673 (March 4, 1996): 73–74.

Yung, Katherine. "American Joins Trend: Home PCs; Job Perk Called Way to Link with Workers." *Dallas Morning News,* March 2, 2000.

Zakon, Robert Hobbes. "Hobbes' Internet Timeline v2.5." 1993–96. Available [Online], <http://info.isoc.org/guest/zakon/Internet/History/HIT.html>.

Zuckerman, Sam. "Taking Small Business Competition Nationwide." *US Banker* 106 (August 1996): 24–28.

Interviews and Other Communications

Bill Davidow, Smart Valley founder, venture capitalist, November 11, 1995.

Cary Paul Peck, vendor relations for the Los Angeles Metropolitan Transit Authority, speaking at ABAG meeting, April 10, 1995.

Daniel Wall, California State Association of Counties, deputy director Revenue and Taxation and Federal Affairs, August 29, 1995.

David Milgram, Lockheed principal investigator for BADGER, July 6, 1995.

Don Wimberly, BASIC manager, July 27, 1995

Emilee Whitehurst, coordinator of Empty the Shelters Oakland, May 26, 1995.

Eran Gross interview, BAMTA project coordinator on health care and aerospace, November 15, 1995.

Eric Benhamou, CEO 3COM and board of directors, Smart Valley.

Greg Karras, senior scientist at Communities for a Better Environment, December 1, 1997.

James Johnson, organizer with STAND (Sacramento Communities Taking Action for Neighborhood Dignity), March 29, 1995.

Jim Dupont, president Local 2850 HERE, and Stephanie Ruby, organizer Local 2850, multiple interviews, 1996.

Karen Greenwood, BAMTA project coordinator at Smart Valley, September 28, 1995.

Kathy Blankenship, Smart Valley director of Communications And External Affairs, September 28, 1995.

Katie Buller, PUBLABOR moderator, August 1995.

Kurt Handelman, Employment Development Department, March 29, 1995.

Leslie Saal, Smart Valley staff, speaking at ABAG meeting, November 13, 1995.

Mack Hicks, Bank of America vice president for Electronic Delivery Services, June 22, 1995.

Mark Masotto, CommerceNet Working Catalogs program manager, June 13, 1995.

Mark Ritchie, Institute for Agriculture and Trade Policy executive director, June 8, 1995.

Michael Duffner, ABAG and San Francisco Internet manager, May 2, 1996.

Michael McRay, CommerceNet program manager at Smart Valley, June 5, 1995.

Michael Stein, IGC Program Manager, March 8, 1995.

Pat Bourne, director SFUnet, April 8, 1995.

Randy Whiting, CommerceNet chair of Sponsored Projects, July 21, 1995.

Rich Cowan, Center for Campus Organizing, interview, June 13, 1995.

Richard Eposito, executive director of the Sacramento Cable Television Commission, August 29, 1995.

Sean Garcia, Sunnvale director of Information Technology, speaking at ABAG meeting, November 13, 1995.

Syd Leung, Pacific Bell, CalREN project manager, November 30, 1995.

Terry Bryzinsky, head of Information Services, Association of Bay Area Governments, March 7, 1995 and September 13, 1995.

Victor Pottoroff, deputy directory of the California State Association of Counties (CSAC), August 29, 1995.

Wally Dean, mayor of Cupertino, April 5, 1995.

Warren Slocum, San Mateo county assessor, March 30, 1995.

Index